Artificial Intelligence-Enhanced Software and Systems Engineering

Volume 4

The book series AI-SSE publishes new developments and advances on all aspects of Artificial Intelligence-enhanced Software and Systems Engineering—quickly and with a high quality. The series provides a concise coverage of the particular topics from both the vantage point of a newcomer and that of a highly specialized researcher in these scientific disciplines, which results in a significant cross-fertilization and research dissemination. To maximize dissemination of research results and knowledge in these disciplines, the series will publish edited books, monographs, handbooks, textbooks and conference proceedings. Of particular value to both the contributors and the readership are the short publication timeframe and the world-wide distribution, which enable both wide and rapid dissemination of research output.

Stamatios Theocharis · George A. Tsihrintzis

Semantic Knowledge Modelling via Open Linked Ontologies

Ontologies in E-Governance

Springer

Stamatios Theocharis
Department of Informatics
University of Piraeus
Piraeus, Greece

George A. Tsihrintzis
Department of Informatics
University of Piraeus
Piraeus, Greece

ISSN 2731-6025 ISSN 2731-6033 (electronic)
Artificial Intelligence-Enhanced Software and Systems Engineering
ISBN 978-3-031-20584-2 ISBN 978-3-031-20585-9 (eBook)
https://doi.org/10.1007/978-3-031-20585-9

This Springer imprint is published by the registered company Springer Nature Switzerland AG
The registered company address is: Gewerbestrasse 11, 6330 Cham, Switzerland

Series Editor's Foreword

During the past few years, the entire world has gone through several crises in finance, the pandemic and, recently, energy problems that have led society to confront serious consequences on economy and social cohesion. To overcome the crises, public sectors around the Globe were required to implement extensive administrative reforms and to increase their efficiency via the exploitation and incorporation in their various bodies of Information and Communication Technologies. This was particularly the case in Greece, which had to reduce public expenses to a very large extent as well as to implement remodelling and innovative approaches to its public sector.

As the first steps of modernisation of computer infrastructure and information systems have already been taken, evolving technological advances in *Artificial Intelligence-empowered Software* present significant potential to lead *e-Government* towards more collective efforts, exchange of experiences on best practices both at national and international levels and dissemination of secluded administrative knowledge.

In the monograph at hand, Dr. Stamatios Theocharis and Prof. George A. Tsihrintzis develop and present novel *semantic web-based* and *linked open data-based* approaches for the modelling and management of the huge volume of administrative data and the procedures followed by public sector bodies and for the production and management of relevant administrative knowledge. The monograph benefits from the double extended expertise of Dr. Theocharis in the public sector and in e-Government as well as Professor Tsihrintzis's deep knowledge and outstanding contributions in artificial intelligence and applications. As such, the monograph describes the authors' innovative research by analysing both the public sector expertise and the technical novelties that were developed and evaluated. The result is an excellent manuscript of explainable artificial intelligence-enhanced software methods in e-Government, which are based on the real data of the Greek case and can serve as a model for the solution of similar problems in other countries.

Overall, the book consists of eight well-written chapters, each of which includes relevant bibliographic references for deeper probing. Appendices complement this work with sections of configuration files of the applications developed and used. Thus, Professors, researchers, scientists, engineers and students in e-Government, artificial intelligence, and other computer science-related disciplines are expected to benefit greatly from it, along with non-specialist readers from other disciplines who are interested in getting versed in the recent developments in e-Government.

As an Editor of the *Artificial Intelligence-Enhanced Software and Systems Engineering* series of Springer, I am happily presenting this monograph to research communities around the World. I congratulate the authors for their superb work and I encourage them to extend this work further and update the research communities with their further research results.

October 2022

Maria Virvou
Department of Informatics
University of Piraeus
Piraeus, Greece

Preface

The subject of this book is the application of the solutions offered by the Semantic Web in the production, modelling and management of knowledge in the specialised field of Electronic Government (e-Government) as well as in the management of the huge amount of data and procedures followed by public sector bodies. The authors' work includes the development of specific applications and methodologies in the field of linked open data with an emphasis on government-administrative data.

The most specific research area of the work carried out for the writing of this book includes the Semantic Web, linked open data and the tools that support them with the aim of developing applications for Modelling and Knowledge Management in the field of e-Government.

In more detail in this work, the authors, firstly, present the necessary theoretical background of the fields of e-Government, the Web and the Semantic Web through relevant bibliographic references. Next, they present examples of innovative applications in the field of e-Government through the special tools and applications provided by the Semantic Web. This way, an attempt is made to highlight the potential of the Semantic Web and its tools in the field of knowledge production and management, as well as the opportunities through them for the further development and improvement of e-Government services.

Especially in the field of knowledge management and in the context of the research carried out, specific tools of the Semantic Web were used and presented, while specific methodologies, procedures and algorithms are proposed regarding the modelling of knowledge through the development of ontologies and their publication as linked open data in specialised open data repositories.

The contribution of the work carried out in the context of writing this book in the field of Semantic Web, linked open data and Knowledge Modelling is summarised in the following areas:

- Deeper study of Web and Semantic Web theory, tools and applications.
- Comparative study of the Web, the existing Web 2.0 and Web 3.0 in specific areas of interest.

- Deeper study of knowledge modelling with an emphasis on the use of semantic concepts, technologies and tools such as ontologies and linked data.
- Knowledge modelling through the development of specific ontologies in the field of e-Government.
- Development of a methodology for publishing linked open data and ontologies.
- Highlighting the capabilities of Semantic Web tools and the opportunities through them for e-Government.

Piraeus, Greece
September 2022

Stamatios Theocharis
George A. Tsihrintzis

Contents

Chapter 1
Introduction

Abstract In this book, software applications are developed as solutions offered by the Semantic Web in the production, modelling and management of knowledge in the specialised field of e-Government as well as in the management of the huge amount of data and procedures followed by public sector bodies. Particular emphasis is placed on the field of linked open data and government-administrative data and the software tools that support them. The book consists of eight chapters. At the end of each chapter, extensive relevant bibliographic references are included for deeper probing. Moreover, appendices are attached at the book end that include sections of configuration files of the applications developed. Professors, researchers, scientists, engineers and students in artificial intelligence, e-Government and other computer science-related disciplines are expected to benefit greatly from this book, along with non-specialist readers from other disciplines who are interested in getting versed in the recent developments in e-Government.

Keywords e-Government · Semantic Web · Semantic tools · Linked open data · Open ontologies

1.1 Subject of the Book

In a particularly critical period for Greece and the entire world, characterised by an extended economic crisis with direct effects on social cohesion, the public sector is charged with the obligation to implement administrative reforms. Parts of these reforms are expected to be achieved through the further implementation of e-Government as a means to increase the efficiency, effectiveness and transparency of the public sector. During the first phase of the e-Government project in Greece, Information and Communication Technologies (ICT) were introduced in almost all public sector bodies. Specifically, computers were installed and equipped with software applications to support key management processes. At the same time, the necessary network was created and websites, portals and various information systems were developed. The next step aims at using information systems and their connection to continuing to provide services to citizens and businesses. The application requires not

S. Theocharis and G. A. Tsihrintzis, *Semantic Knowledge Modelling via Open Linked Ontologies*, Artificial Intelligence-Enhanced Software and Systems Engineering 4,
https://doi.org/10.1007/978-3-031-20585-9_1

only the implementation of infrastructure and communications technology and the interconnection of information systems, but also ensuring interoperability of information systems and processes. Institutional and organisational changes in the public sector that will enable public bodies to meet the needs of the new digital era, are also considered necessary.

The further promotion of e-Government includes collective effort, exchange of experiences on best practices both at national and European level in order to disseminate secluded experience and knowledge. The environment in which public administration operates is characterised by production and distribution of vast size information resulting from the processing of an equally large dataset. The whole situation is exacerbated when employees leaving the service do not leave behind them the valuable knowledge gained during the years of service. Even in the case where organisational knowledge is "somewhere" inside the organisation-body, the recovery is an extremely difficult and time-consuming task. The result is obvious: waste of time, low service quality, reduced efficiency, and poor image to the outside.

Key success factor of the overall system is a fully functional, efficient and effective back-office. For the improvement or development of existing systems, emphasis should be given, apart from institutional reforms, to innovation, cutting-edge technologies and knowledge management systems. The digitization of administrative documents and data in combination with the development of new electronic public services to citizens and businesses has highlighted as a key issue the need to manage a huge amount of data already stored in "traditional" relational databases. We believe that the goal of increasing the efficiency, effectiveness and transparency of public sector services is not fully served by existing solutions. The challenge in the near future is the creation of mechanisms that will discover, draw, exploit and distribute the knowledge of each individual employee in order to increase productivity and efficiency of services.

The present work is an attempt to identify issues related to the concept of knowledge, consolidation and knowledge management systems, as initial steps of the procedure of developing a management mechanism in the public sector. Also in the present work we deal with the development of specific applications in the specialised field of modelling administrative data through ontologies and the production of linked open administrative data through tools, technologies and protocols of the Semantic Web. Finally, since every transformation requires, among other things, the participation and cooperation of the human factor, we believe that education and communication at the same time, through the exchange of information and knowledge between public sector executives, requires innovative solutions. In this area, in this work we present as a case study the possibilities provided through the development of a wiki through free software.

1.2 Purpose of the Book and Contribution to Science

The purpose of this book is to highlight the solutions offered by the Semantic Web in the management of the huge volume of data and the procedures followed by the public sector bodies as well as in the production and management of knowledge in this field, by implementing specific applications and methodologies in the field of linked open data with emphasis on government—administrative data.

This purpose is served:

A. through the presentation of the relevant tools, methodologies and the relevant bibliographic and technological reports, which mainly include the following:

- Presentation of the field of Governance of both e-Government and the emerging Open Government.
- Presentation of the protocols, tools and languages used in the various levels of development of the Semantic Web with a comparative presentation of the corresponding ones of the existing Web 2.
- Presentation of the steps required for the production of Open Data.
- Presentation of methodology for the production and publication of linked open data.
- Development and presentation of methodology and related code for the production of RDF triples from open data of JSON format.
- Development and presentation of methodology and related code for the production of RDF triples from data stored in relational databases.

B. through the application of the above tools and methodologies in specific areas of interest which include:

- The development, evaluation and publication as Online Open Data via a publicly accessible URL of our version of the e-Government ontology in open and editable form through free—open software.
- Development and presentation of the operation of a publicly accessible SparqL Endpoint which is hosted on a dedicated server. Through this point it is possible to pose semantic queries about the aforementioned ontology and retrieve relevant information.
- The development and evaluation of ontology for the so-called official statistics of the Greek Statistical System.
- The development and evaluation of ontology in the field of Internal Audit in the light of the Greek Public Administration.

Furthermore, the contribution in the field of e-Government and Open Government as an extension of it, lies in the development and implementation of solutions of the Semantic Web and linked open data in the specific field of modelling of administrative data and management of administrative knowledge. We believe that the proposed applications can be the trigger for the transformation of Public Sector systems in response to the question related to the prospects and the evolution of e-Government. We believe that in order to achieve the best result, both in terms of resource economy and effort for the digital transformation of the state, it will be achieved by adopting the new perspective through the prism of the Semantic Web and the linked data. This is because we believe that the enormous size, complexity, diversity and variability of the data managed by the public sector are not adequately addressed by existing systems, while they are more effectively addressed using the tools offered by the Semantic Web.

1.3 Structure of the Book

This work is developed in eight chapters at the end of which the relevant bibliographic references are included. Also attached at the end are appendices that include sections of configuration files of the applications used.

This First Chapter concerns the Introduction to the Book, which presents, in addition to its subject, the documentation for the purpose and contribution of this work in science in the field of e-Government with solutions of the Semantic Web and the linked data. References are also made to publications made either in international conferences or in international scientific journals.

The Second Chapter presents the theoretical model of e-Government and mentions the critical points for its further development and refinement. There are also issues related to the back-office systems of Public Administration with emphasis on Electronic Administration.

The Third Chapter first presents the existing Web, its evolution over time, as well as its evolution towards the Web3. Next, we present the basic tools, technologies and protocols used for the final comparison between Web2 and Web3. Finally, we present examples of applications of the Semantic Web while the advantages, the challenges as well as the benefits from the application of the solutions of the Semantic Web are of particular interest. At the heart of this chapter and this work are the opportunities for e-Government from the implementation of the solutions offered by the Semantic Web.

The Fourth Chapter presents issues related to knowledge management with emphasis on knowledge in the field of e-Government. First, the theoretical background on knowledge and knowledge management is presented, followed by specific tools and technologies of the Semantic Web that can be used in these areas. The following is the modelling and export of knowledge in the field of e-Government through the development of ontologies such as the ontology for the description of e-Government, official statistics and internal control. For each ontology we present the analysis, design and development of ontologies and then we present examples of their use through queries as well as their evaluation.

The Fifth Chapter presents issues related to the semantic representation and modelling of open data. Initially, a theoretical background is presented for the open data and subsequently the expected benefits from the opening of the data, as well as more specific issues from the Greek open administrative data are presented. At the heart of this chapter is the modelling of open data through the development of a relevant ontology. A more specific application for the use of open data in the field of e-Government is the production of RDF data from a specific dataset of open administrative data. Finally, a relevant methodology is presented as well as results from the publication of open data using special software as a case study.

The subject of the Sixth Chapter is the linked open data with emphasis on the data that are modelled and published as open ontologies. First the theoretical background of this field is covered and then references are made to the basic rules and technologies used for the production and publication of linked data. The methodology and implementation of the production and publication of RDF data from administrative data stored in relational databases as well as from open administrative data repositories is presented. The chapter closes with a presentation of the process and results of publishing an open ontology through specific management tools.

The Seventh Chapter examines the case of e-distance learning and evaluation of public sector staff. Of particular interest is the development of a wiki using open source software as well as its prospects as a means of communication and training of public sector staff.

As a conclusion of this work, the Eighth Chapter includes the conclusions from the study and development of the aforementioned applications with tools and technologies of the Semantic Web in the specific field of knowledge management in the e-Government environment.

Further Reading

1. Theocharis, S. A., & Tsihrintzis, G. A. (2012). Semantic web technologies in e-Government. In *Proceedings of International Conference ICDGS 2012—International Conference on e-Democracy, e-Government and e-Society, World Academy of Science, Engineering and Technology. International Journal of Computer and Information Engineering, 6*(4).
2. Theocharis, S. A., & Tsihrintzis, G. A. (2012). E-administration: Back-office systems upgrading. In *Proceedings of International Conference on Information Society (i-Society 2012)* (pp. 205–208). IEEE. Electronic ISBN: 978-1-908320-05-6.
3. Theocharis, S. A., & Tsihrintzis, G. A. (2013). E-government: Back-office systems development—Critical issues for the Greek case. *International Journal of Digital Society (IJDS), 4*(1), 738–744. https://doi.org/10.20533/ijds.2040.2570.2013.0090
4. Theocharis, S. A., & Tsihrintzis, G. A. (2012). Knowledge acquisition for the further development of e-Government. In *Proceedings of 10th Joint Conference on Knowledge-Based Software Engineering (JCKBSE 2012)* (pp. 286–297). IEEE. https://doi.org/10.3233/978-1-61499-094-9-286
5. Theocharis, S. A., & Tsihrintzis, G. A. (2013). Personalization as a means to improve e-services. In *Proceedings of the International Conference on Computer, Information and Telecommunication Systems—CITS2013* (pp. 1–5). IEEE. https://doi.org/10.1109/CITS.2013.6705726
6. Theocharis, S. A., & Tsihrintzis, G. A. (2013). Open data for e-Government—The Greek case. In *Proceedings of the Fourth International Conference on Information, Intelligence, Systems and Applications-IISA2013*. IEEE. https://doi.org/10.1109/IISA.2013.6623722

7. Theocharis, S. A., & Tsihrintzis, G. A. (2013). Education and assessment of civil employees in e-Government: The case of a Moodle based platform. In *Proceedings of International Conference on Computer, Information and Education Technology—2013, World Academy of Science, Engineering and Technology. International Journal of Humanities and Social Sciences, 7*(11).
8. Kalogeraki, E.-M., Theocharis, S., Apostolou, D., Tsihrintzis, G., & Panayiotopoulos, T. (2014). Semantic concepts in BPMN 2.0. In *Proceedings of the Fifth International Conference on Information, Intelligence, Systems and Applications-IISA2014*. IEEE. https://doi.org/10.1109/IISA.2014.6878791
9. Theocharis, S. A., & Tsihrintzis, G. A. (2014). Ontology development to support the open public data—The Greek case. In *Proceedings of the Fifth International Conference on Information, Intelligence, Systems and Applications-IISA2014*. IEEE. https://doi.org/10.1109/IISA.2014.6878820
10. Theocharis, S. A., & Tsihrintzis, G. A. (2014). Knowledge management systems in public sector: Critical issues. In *Proceedings of the International Conference on Knowledge and Software Engineering—ICKSE 2014, LNSE* (Vol. 4, no.1, pp. 59–65). ISSN: 2301-3559.
11. Theocharis, S. A., & Tsihrintzis, G. A. (2015). Open data triplification—The case of the Greek open public data. In *Proceedings of the Sixth International Conference on Information, Intelligence, Systems and Applications-IISA2015*. IEEE. https://doi.org/10.1109/IISA.2015.7388063
12. Theocharis, S. A., & Tsihrintzis, G. A. (2016). Semantic tools; their use for knowledge management in the public sector. In *Intelligent computing systems—Emerging application areas* (pp. 5–37). Springer. https://doi.org/10.1007/978-3-662-49179-9_2
13. Kalogeraki, E.-M., Theocharis, S., Apostolou, D., Tsihrintzis, G., & Panayiotopoulos, T. (2016). A semantic approach for representing and querying business processes. In *Intelligent computing systems—Emerging application areas* (pp. 87–114). Springer. https://doi.org/10.1007/978-3-662-49179-9_4
14. Theocharis, S. A., & Tsihrintzis, G. A. (2016). RDF serialization from JSON data—The case of JSON data in Diavgeia.gov.gr. In *Proceedings of the Seventh International Conference on Information, Intelligence, Systems and Applications-IISA2016*. IEEE. https://doi.org/10.1109/IISA.2016.7785386
15. Theocharis, S. A., & Tsihrintzis, G. A. (2017). Semantic modeling of internal audit field. In *Proceedings of the International Conference on Knowledge and Software Engineering—ICKSE 2017. International Journal on Software Engineering—IJSE, 3*(1), 1–7. https://doi.org/10.18178/ijke.2017.3.1.078. ISSN: 2382-6185.
16. Theocharis, S. A. (2019). Semantic modeling of official statistics—The case of the Greek statistics. *Poster in New Techniques and Technologies for Statistics—NTTS 2019, European conference, Eurostat.* https://ec.europa.eu/eurostat/cros/NTTS2019_en

Chapter 2
e-Government: The Concept, the Environment and Critical Issues for the Back-Office Systems

Abstract e-Government and the Open Government initiative are the main pillars for improving the efficiency, effectiveness and quality of public administration services to citizens and businesses as well as for the transition of public administration to the digital state. This chapter presents the broader environment of public administration in which e-Government is developed, the models and levels of development of e-Government, the critical points for the successful implementation of electronic and Open Government as well as the characteristics and critical points for the development of a system to support the back-office of a paperless public administration.

Keywords e-Government · E-Government architecture · Back-office systems

2.1 Introduction

The 1990s were marked by the growth of the Internet and the increase in the power and capabilities of personal computers that had already appeared two decades earlier. During this period, the introduction and widespread use of Information and Communication Technology (ICT) first in businesses and then in households has resulted in the public sector following the example of the market, and gradually introducing the use of ICT in everyday life of public administration. At the same time, new concepts of governance such as e-Government, e-democracy, e-administration, etc. have emerged worldwide, in some cases as a continuation—an extension of the existing administrative system, while in others as something completely new. After the first two decades of the transformation of public administration and the maturation of new concepts of governance and the consolidation of the range of possibilities provided by ICT to serve the public, new opportunities, challenges and critical points emerged that require special attention from the public administration. It has become clear that the transition to e-Government does not simply require the installation and use of computers and related ICT infrastructure, but also substantial administrative and organisational reforms. The use of new technologies and methods must necessarily be linked to the adaptation of procedures in the back-office, staff training and

S. Theocharis and G. A. Tsihrintzis, *Semantic Knowledge Modelling via Open Linked Ontologies*, Artificial Intelligence-Enhanced Software and Systems Engineering 4,
https://doi.org/10.1007/978-3-031-20585-9_2

the development of a new administrative culture. The integration/completion of these changes is linked to the degree of maturity of the national public administrations in combination with the administrative and organisational models followed.

Nowadays, citizens and businesses, as equal pillars of the tripolic social model, demand more transparency and accountability of governments and public administration, through the opportunities now provided by new ICTs. A typical example is the global Open Government Partnership (OGP) initiative and its members, who aspire to meet the public's demand for more transparency and accountability to civil society. The members of this initiative share the administrative data they hold, based on specific common open standards and procedures and share them as open, offering the opportunity to produce new information and knowledge and add value to the primary data. At the same time, special points of attention are highlighted, such as ensuring the protection of personal data, the protection of data from malicious attacks or unauthorised access and processing of data.

One of the emerging challenges for both the e-Government and open government is related to the management and processing of a really huge amount of data (open or not) and information managed and/or shared by public sector entities throughout the range of its activities. This situation makes it imperative to apply new innovative methods and applications for data and information management in order to produce knowledge and increase their added value.

The recording and analysis of the components and parameters of the management system of the modern electronic state as well as its environment is a necessary initial step in order to further investigate in depth the issues related to the management and representation of knowledge in the field of e-Government and for this reason they are analysed in the following paragraphs of this chapter.

2.2 The Environment of Public Administration

In recent decades, there has been a sharp shift in both the status quo and the form of organisations that carry out public activities, resulting in the increasingly difficult demarcation of the state and the public sector and, consequently, their responsibilities and obligations. This is because in an effort to increase the efficiency of management, political leaders create or transform existing organisational structures, in market terms, such as public sector public limited companies or private legal entities. These new structures interact with the existing structures of the public sector and act in accordance with the political decisions of the leadership, and produce public products in parallel with the other organisational structures of the state.

Nowadays, the significant changes that have taken place at the socio-economic and technological level have shaped a new environment of action for public administration which is characterised by:

- Exacerbating the constant problems that the public administration has always faced, as the needs of the society are constantly expanding and at the same time evolving, demanding a renewal of public policies and the search for administrative innovations.
- New, more difficult and complex problems that are added to the old ones, which are related to the heterogeneity and the peculiarities that characterise the various social groups and that require further specialised provision of public products.
- Problems stemming from the globalisation of the economy and the movement of capital and persons.
- The transformation of economic activity so far and the emergence of new economic activities that lead society to an economy of knowledge and information.
- The increase of the demands of the citizens from the public sector in terms of their efficiency and effectiveness, having as a measure of comparison the high level of electronic services provided by the companies.
- Citizens' dissatisfaction with politics as a result of incomplete accountability and transparency resulting in the declining public participation.
- Constantly declining government spending as a result of the economic crisis.

In response to these new conditions, innovative administrative and economic methods have been adopted by the advanced administrative and economic states in Europe and the United States related to the use of electronic means in transactions between citizens and public administration. These include e-Government, e-Governance, e-administration, e-democracy, etc. These concepts have often been used incorrectly in the past as synonymous or different from their true meaning, while from the first years of their establishment, international organisations such as the OECD and the European Commission, as well as the scientific community, have been involved. This way, various definitions were given and relevant integration models were developed. The degree of integration of the reforms was linked to specific indicators, in order to establish in the national administrations that e-Government is gradually developing and is not exclusively identified with the introduction and use of ICT infrastructure but requires simultaneous implementation of reforms and adjustments to existing internal processes.

The terms "Government" and "Governance" are often confused with each other, although they constitute two distinct concepts. In contrast to the traditional concept of "Government"—considered as Public Administration, governance has been established in recent years as a broader concept or system characterised by two additional elements. On the one hand, the state must develop into a smaller, more citizen-friendly formation, which through transparent procedures consults and co-decides with civil society and the market. From another point of view, the state must clearly be established as the most basic but not exclusively decisive part of a tripolic social system, in which three distinct spaces coexist and interact: state, citizens and civil

Fig. 2.1 The tripolic social model

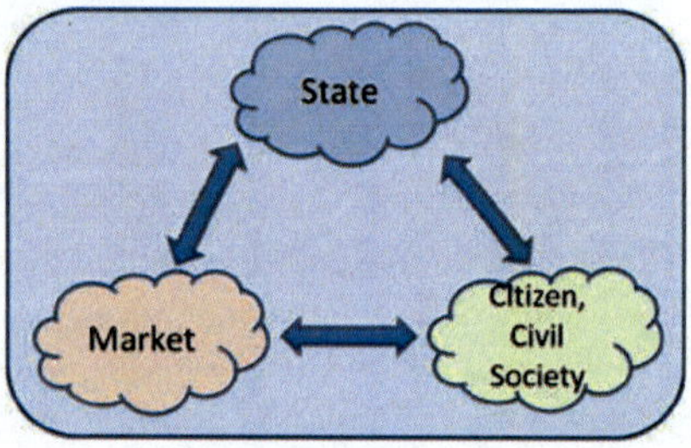

society, market (Fig. 2.1). The new relationship of the State with Civil Society is the hallmark of Governance by the Government—Public Administration.

Contrary to the traditional notion of Government in a representative democracy, represented graphically by a pyramid based on the electorate/people and at the top the elected government representing exclusively the legitimate expression of organised society, in the eyes of the government without to challenge the primacy of the legitimately elected government, the other two poles/spaces of modern societies are added to the chain of procedures of public decision-making. In this context, crucial issues for the further analysis of e-Government are the interactions between the components of the overall system and those between the individual components of public administration (Figs. 2.2 and 2.3).

Given that what is needed nowadays is the balancing of the forces of the three poles of the social system mainly to weaken the "colonisation" tendencies of civil society either by the state itself (phenomenon of statism) or by the private sector—market, the issue of the emergence, operation and strengthening of new institutions that interact with these three pillars is crucial. Such institutions may be:

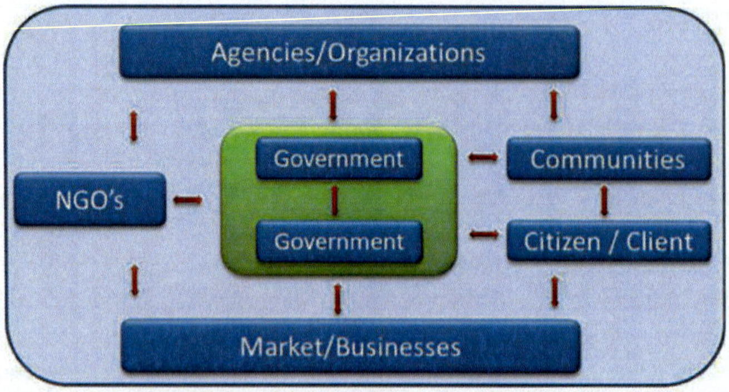

Fig. 2.2 Special components of the overall e-Government system

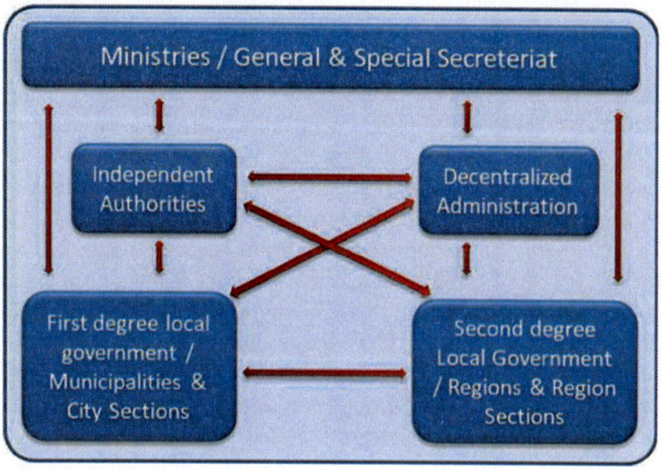

Fig. 2.3 Interactions within the Greek Public Administration

- Official organisations independent of the state e.g. Independent Administrative Authorities such as the National Broadcasting Council, the National Telecommunications and Post Commission, the Citizen's Advocate, the Personal Data Protection Authority, etc., in the logic of the transfer of public power.
- Semi-official, without public authority but with partial regulatory competence (bodies of social dialogue between employers and employees), or consulting (consulting bodies with NGOs and Civil Society).
- Informal (citizen forums, groups organised through social networks).

New processes that have already emerged as innovative include:

- Public consultation.
- Electronic procurement.
- Online auctions.
- Open availability of public data/open data repositories.
- Electronic voting.
- Local referendums.
- Pre-election elections with an open ballot to nominate candidates to the parties.

With the emergence of these new institutions, the critical difference between Governance and the classic logic of Government stands out. As an application of the above, the electronic expression of governance, e-Governance, relied on the development of concepts such as e-Democracy and e-consultation. These institutions, on the one hand, act in addition to the traditional forms of representative democracy and, on the other hand, provide the opportunity for citizens to participate in the entire chain of the legislative process. For example, in drafting legislation, it is now common for public consultation to take place through specific transparent electronic procedures provided by government electronic portals.

The introduction of ICTs in public administration produces different results in the two distinct concepts of e-Governance and e-Government. In the first case, the state radically changes its operation, especially in the field of state-citizen relations, while in the second case, there is an improvement in the performance of administrative operations and transactions with improvements in the functioning of the state.

The concept of e-Government in practice is identified with the Electronic Public Administration or simply the electronic provision of public services. In a number of developed countries since the early 1990s and through the beginnings of "New Public Administration", a central trend has emerged which is the use of new information and communication technologies by the state and public administration in transactions with citizens and businesses. The central idea was to transform the state into a smaller, cheaper, more strategic and customer-centric state within the framework of market operation. The rapidly growing economy in the technologically advanced and globalised environment combined with the ever-increasing needs and expectations of the public for better management of time and public resources imposed this development. From the very first steps of the new institutions, it became clear that the introduction of ICT in the operation of the Public Administration alone cannot bring about any radical change. On the contrary, significant changes are needed in the internal mode of operation in the so-called "back-office" and in the interface with the public, by increasing efficiency, transparency and by reducing operating costs.

The definitions and theories that have been formulated in recent years concerning e-Government are many and have many similarities and differences. Indicatively, we mention the following.

Theoretically, two are the dominant perceptions about e-Government. One considers e-Government as "the application of e-commerce tools and techniques to the functioning of governance". This perception focuses on practical efficiency and cost reduction, such as those that can come from electronic services such as e.g. the electronic filing of the tax return or the electronic procurement. According to another view, e-Government has the potential to improve the so-called "democratic participation" and to overcome political distancing. This perception focuses on those initiatives that will establish another level of citizen-administration interaction.

From a more practical point of view, e-Government can be seen in two main versions. One concerns the implementation of technological solutions in the operation of public administration, in the sense of the use of information and communication technologies to offer better public services to citizens and businesses, but also to strengthen the "civil society" that is, the participation of citizens in public. The other concerns the implementation of business methods regarding the organization and structure of public administration in order to improve and utilise any kind of resource and exchange information between all stakeholders, such as businesses, public bodies and citizens, through the use of technological solutions. Thus a government—in the sense of public administration, with business orientations—should meet the following principles:

- Be catalytic and direct and not just produce services.
- Be property of the community of citizens and empower instead of serving them.

- Be competitive by introducing competition in the provision of services.
- Be guided through its mission by properly transforming its various bodies.
- Be focused on the results through supporting them rather than supporting the inputs.
- Have the citizen in the spotlight—that is, to be customer-centric, in the sense of satisfying the citizen or the companies by providing appropriate services.
- Be businesslike in the sense that they should pursue profit and reduce costs.
- Be focused on prevention, not treatment.
- Be decentralised with the aim of participation and teamwork.
- Be market-oriented, in the sense of promoting change through the market.

Among the definitions for e-Government we mention the following:

A. e-Government has been designated by the European Commission [1] as the use of information and communication technologies in public administration, combined with organisational change and new skills, to improve the provision of public services and democratic processes and to strengthen public policy support. According to the Commission, e-Government is a catalyst for the establishment of a better and more efficient administration. It improves the formulation and implementation of policies pursued by the State and helps the public sector to meet the conflicting requirements to provide more and better services with fewer resources. This term was created with the general practices, according to which the adjective "electronic" is placed in order to emphasise the electronic way of production and distribution of services such as e.g. e-commerce, e-business, e-learning, etc.
"Online" public administration (or "e-Government") is one of the priorities set at European level by the e-Europe 2005 action plan and has been updated since then until the announcement of the European e-Government Action Plan 2016–2020 which states that the digital transformation of government is a key element to the success of the Single Market; helping to remove existing digital barriers, reduce administrative burdens and improve the quality of interactions with government [1]. It is an essential lever for the provision of more efficient and better quality public services, for the reduction of the waiting time of the users, the improvement of the transparency and the responsibility of the services. According to a similar definition [2], e-Government aims to use information and communication technologies to improve the quality and accessibility of public services. e-Government is able to reduce costs for both businesses and governments and facilitate trade between administrations and administrators. In addition, it helps to make the public sector more open and transparent as well as governments more understanding and accountable to their citizens.
B. According to US Government Announcements (2002), e-Government is the government's use of Internet applications and other technologies, in conjunction with procedures that incorporate these technologies to enhance access to government information and public service, other services and government entities, or the improvement in government functions in terms of efficiency, quality of services and their transformation [3].

C. According to the OECD, e-Government is linked to the use of information and telecommunications technologies, especially the Internet, as a tool for better governance [4].
D. e-Government—according to Osborne and Gaebler [5] is the use of Information and Communication Technologies (ICT) in the public sector in order to improve:

   - Internal public sector procedures.
   - Communication and services provided to citizens.
   - Relations among public sector organisations with other public or private sector organisations with non-profit enterprises and with organisations of other states.

E. e-Government according to Burn and Robins [6] is the modernization of Public Administration with the introduction of cutting-edge technologies and the digital connection of government information infrastructures to save resources and provide quality services to citizens and businesses and which can be distinguished into the following four basic models:

   - Government-to-Government: concerns all those activities both in the back-office and among different public bodies, which will improve and upgrade the services provided.
   - Government-to-Citizen: refers to the government's relationship with citizens. It deals mainly with issues of legislation, democracy and the provision of information or services.
   - Government-to-Business: refers to government-to-business relations. It focuses mainly on collaborations and joint activities both nationally and internationally.
   - Government-to-Employees: concerns electronic Services to Employees (e.g. Corporate Portals). In particular, government-to-government (G2G) e-services are a relatively new model. The short history shows that there are theories that focus only on the technical aspect of intergovernmental cooperation, while others, on the contrary, try to give a broader perspective, including organisational issues. Communication and exchange of information between public sector bodies can be considered as a minor G2G service. The completion of such services requires organisational and structural reforms aimed at transforming the classic bureaucratic body into a procedure-oriented structure.

The experience accumulated from the application of various methods and techniques in e-commerce is an important advantage for e-Government. This is because basic methods can be used accordingly. For example, the electronic identification of customers and the provision of electronic services on databases have already been adopted by public bodies. However, there are important differences that delay the completion of the project for the transition to the e-state. It has now become clear that it is not enough to just develop databases, publish service directories on websites or portals. The alignment of procedures and the complete interconnection of public

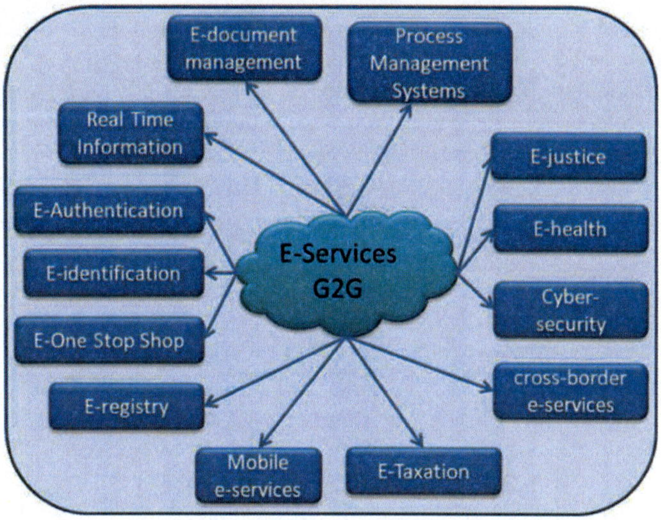

Fig. 2.4 The distribution of electronic services from government to government

systems require a reorganisation of the public sector, so that the back-offices of the different public sector organisations can work together.

Reorganising public services requires an integrated government effort and will be one of the most challenging issues for e-Government. Figure 2.4 shows some general G2G e-services. The introduction of G2G e-Government systems aims to improve the quality of these basic services and to strengthen the three basic principles of separation of powers (freedom, equality and legitimacy). Systems such as electronic document management portals will enable the creation of more dynamic and democratic services.

2.3 Developing e-Government—Models and Levels of Development

Nowadays, all the governments of the developed countries have recognized e-Government as a necessary action to improve the services provided and the further development of the economy and have already taken important steps to complete it. Transactions with public sector services have already been diversified from the past. Public services provide the public with innovative electronic services and real-time information. However, there is heterogeneity of the services provided and the level of their completion as a consequence of non-compliance with common development models and management philosophy. This phenomenon occurs both within states as well as in their interaction. The existence and use of specific common models will help in the following directions:

- Achievement of collaborations between the different digital service systems since they will have been developed based on common models.
- Dissemination of best practices among countries already developing digital services in order to achieve maximum service development.
- Provision of significant know-how to countries that either do not follow or are following the development of digital services at a slower pace.

Below, for the sake of completeness of this work, we present elements of the state of the art regarding the development models of digital services in the context of e-Government [7].

2.3.1 The Three Rings Model

The three-ring model was proposed by Koh and Prybutok [8] and it is a simple and understandable model that groups the typical functions of the Internet. This model categorises Internet applications into three main user categories: (a) Information use (b) Transactions (c) Procedures (Fig. 2.5).

Particularly:

- *Information Use.* It is the simplest and most original application of the technological solutions offered by ICT. Following the successful implementation by companies, public bodies initially used their websites (and still do) to inform the public, to educate remotely, or simply to establish a point of contact with the public. The first government applications concerned the information of the public about the responsibilities of public bodies or about the services provided to the citizens.
- *Transactions.* The provision of electronic services by companies, such as e-shops, e-banking, etc. led to similar developments in the public sector. This development

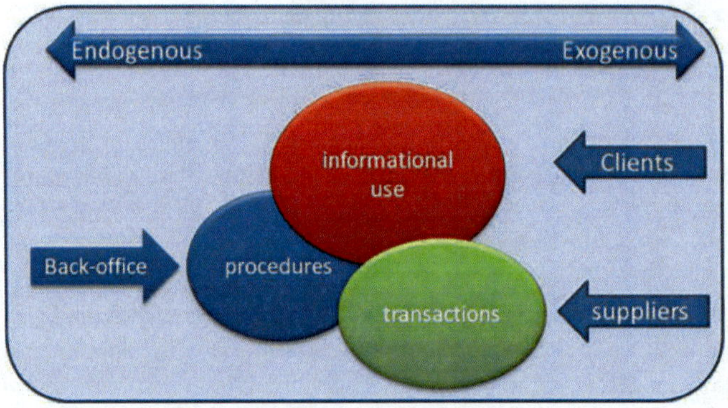

Fig. 2.5 The three uses of Internet applications

created problems that did not exist in the early stages of providing electronic services. We refer to the problem of the security of electronic transactions and the protection of personal data. For example, through the Internet, a citizen of a country is not only able to monitor and be informed about his tax return or his accounts to the public, but he can even order payment. On the other hand, the various public bodies keep and manage sensitive personal data of the citizens, which are always the target of malicious individuals. Security is now the crucial issue for the further development of e-commerce and the development of e-commerce and e-economy.

- *Procedures.* The provision of electronic services by public bodies is a challenge for the public administration. The provision of innovative services requires, on the one hand, the review or reform of the internal procedures that were designed for the classical bureaucratic system and, on the other hand, the use of specialised applications for this purpose. Various software applications are used to display and manage workflows, project management, customer/audience service (CRM) processes for both front-office and back-office. This includes all those applications that are used to evaluate processes and make decisions.

This model enables those designing and managing e-Government projects to have a comprehensive and multifaceted view of the ever-changing and growing range of web applications, resulting in a better focus on critical issues. Knowing which issues are critical and focusing on them will reduce costs and allow e-Government to provide services and information in a more efficient and effective manner.

2.3.2 The Model of Focus and Centrality

The Model of Focus and Centrality [9] includes a two-dimensional frame of reference in which the effects of the Internet on public administration are evident. In the horizontal axis of this theoretical framework, the dimension of e-Government is given, in contrast (right) with e-Government. In the vertical axis, the relationship where the focus of policies is on the citizen of the country (politico-central point of view) is given in contrast (downwards) with the one, where the focus is on the organisation (organo-central point of view). This creates four quadrants (Fig. 2.6), each of which we examine and study below.

- *1st Quadrant.* There are two main issues in this area. First, is the improvement of the provision of value-added services by the government to the citizens and second, the increase of the efficiency of a government website. In cases where websites have been created that are of informative nature only, the issues that arise are related to their content and the policy pursued by the government. The main factors that determine the success of e-Government are related to the design of websites, the access to government portals, any links to other government portals

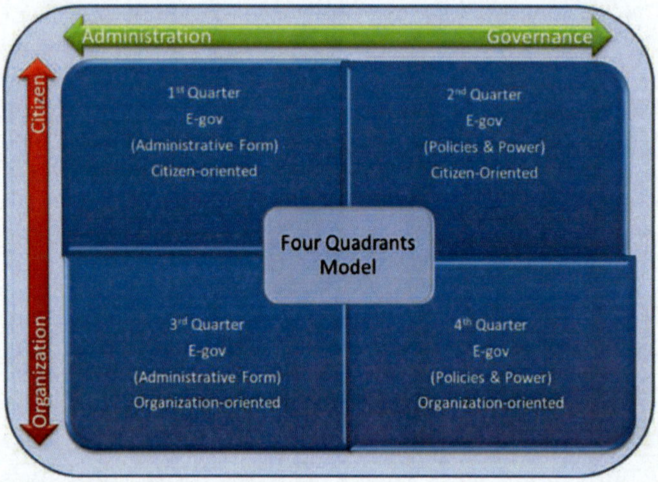

Fig. 2.6 The model of the four quadrants

and the ability to use search engines that will allow the citizen to easily and quickly find the administrative area he is interested in.

- *2nd Quadrant.* One of the most important aspects of government-centred governance is the political role that the Internet can play. Undoubtedly innovative Internet communication tools e.g. e-mail, virtual chat rooms, mailing lists etc. have made expressing and widely disseminating views on government policies, clearly easier than ever. The Internet is presented as the means by which new technologies offer new forms of communication and consulting locally, nationally and internationally. Thus the role of governance seems more complex in conjunction with ever-evolving social and political structures.

A second important aspect of citizen-centred governance is related to the development of so-called "online communities". Various sections of the population with the same or similar interests create virtual, electronic communities and either formally or informally act as an advisory mechanism or as providers of information from officially established bodies. However, the emergence of these communities has made societies face the problem of the "digital gap". It is obvious that not all people have the same access to technology, the skills to use it, or the education and training required. Thus it is very possible to create marginal groups that will be completely ignored, and even more so than in the past. The role of governments is linked to promoting appropriate policies to assist such groups in balancing trends.

A third aspect is related to the role of the Internet in the intensity of competition to distract the attention and interest of the public. Therefore, in order to be of interest and due importance to the people, governments must take the conditions of this type of economy into serious consideration.

Governments should therefore focus on meeting citizens' access and information needs in such a way as to mitigate differences between social groups, while at the same time redefining their relationships with communities of interest and concentrating as much as possible the attention of the general public.

- *3rd Quadrant.* The area of the third quadrant examines the impact that the Internet may have on the form of a public body or even the government itself. It is a fact that government reshuffles aimed at increasing the efficiency and effectiveness of government when not accompanied by structural or other changes have generally had little success. On the contrary, the implementation of e-Government appears as an attempt to promote a climate of change at the governmental level.

The practical implications of a network environment of government structural changes are enormous, but they require a shift in power and responsibilities, a threat to those who have prospered under the traditional regime. Networking and interconnection technologies favour a form of approach and transparency based on the provision of services through collaborative networks.

- *4th Quadrant.* This fourth area of the model concerns the response of the public sector to the needs of a modern and interconnected society. In this context, the necessary infrastructure for secure digital or electronic signatures is implemented, while mechanisms are provided for secure access of citizens and businesses to electronic services such as electronic voting and integrated electronic transactions.

2.3.3 The 5 Levels of e-Government Development

e-Government services are classified into five levels depending on the level of integration of the information systems that support the project of providing digital services. Initially, the division was limited to four levels, but in recent years there has been a need for a fifth, more comprehensive, level. The diagram below (Fig. 2.7) presents the five phases of e-Government, where it is understood that as the services provided are completed, the complexity of the systems that support these services increases along with the need for substantial changes in the structures of public sector organisations.

- *Level 1*—Informational Services (Information): Only information material is provided on how to provide the service. The information concerns the required documents that must be submitted by the interested party, the bodies involved, the order of carrying out the transactions included in the service, etc.
- *Level 2*—Communication Services (Interaction): Provide information material on how to process the service as well as official material (application forms, certificates, etc.) which users can "download" to their computer, print it and use it during their transaction with the carrier on a physical level.
- *Level 3*—Interactive Services (Two Way Interaction): In addition to information, they offer online forms to complete and submit online. Since they include

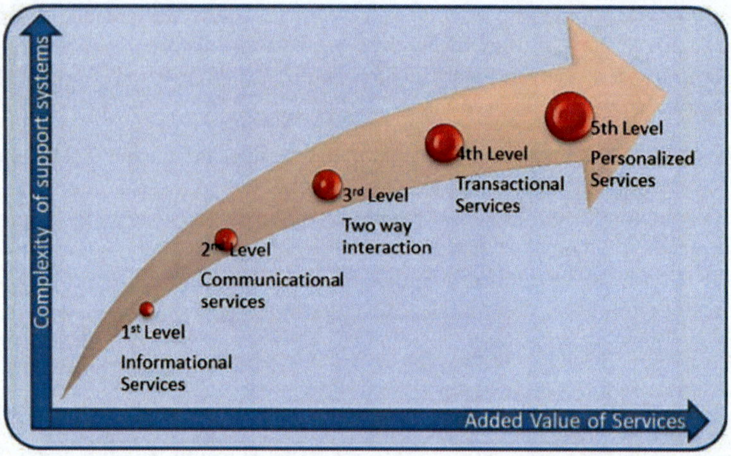

Fig. 2.7 The levels of development of e-Government

online submission of data by the user, they require a mechanism for identification, authentication and protection of the data sent by the user of the service.

- *Level 4*—Transactional Services (Transaction): They support functions where the user completes the transactions included in the service (e.g. payment of VAT). The fact that an online service enables the completion of financial transactions implies the possibility of a complete replacement of the respective contractual service.
- *Level 5*—Personalised Services (Personalisation): They provide pre-filled forms to the user to the extent permitted by the legal framework and inform him about the services that concern him depending on his social and economic profile. The state takes preventive actions in order to promote the quality of service and the degree of user-friendliness, while some services are performed automatically, relieving the citizen or business of the corresponding actions. This level only makes sense for certain services, and expresses the following two dimensions:
  - The proactive service delivery, i.e. the state takes preventive actions to upgrade the service and its friendliness to the citizen. For example, the timely notification of the citizen in case he has to take any action, the pre-completion of data in the user's requests to the State, etc.
  - The automatic provision of services, where the state automatically provides specific services without requiring the citizen (or business) to request them. Also, while until now the direct transaction of the citizen (or business) with the State was examined, now the existence of intermediaries that trade with the State on behalf of the applicant is taken into account (Stage 4B).

2.3.4 The Model of 13 Levels of Digital Service Integration

According to a more specific analysis concerning specific planning for the implementation of e-Government in public sector bodies, 13 distinct stages are mentioned which, from the smallest to the largest, become more and more complex but also more complete in terms of digital services. These levels can also be considered as the necessary steps to complete e-Government (Fig. 2.8).

Level 0—Internal infrastructure of the body

This level is the "starting kick" for the introduction of the body in the process of e-Government. It mainly involves the creation of appropriate ICT infrastructures that will support the overall project. In particular, actions such as:

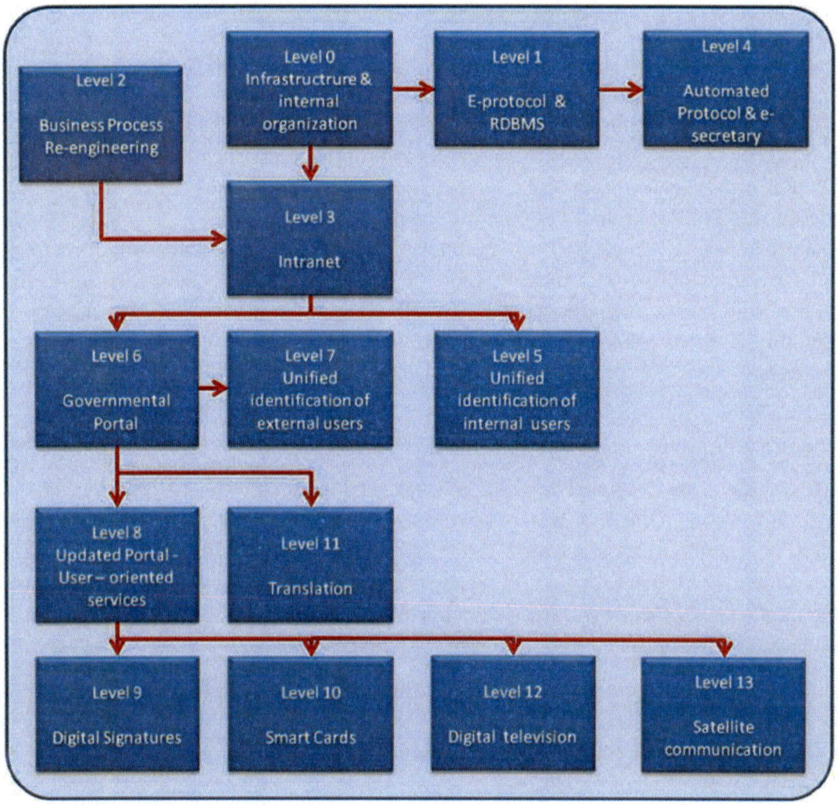

Fig. 2.8 The 13 levels of digital service integration

- Supply of hardware and software such as office automation applications etc.
- Network infrastructure and interconnection of existing and new equipment.

Level 1—Electronic register

It concerns the transition from the manual method of registering documents to their electronic identity. At this level, it is considered necessary to eliminate incoming–outgoing books and maintain electronic databases where a unique register number is manually assigned to each document so that the search for documents becomes easier and any human errors are avoided (duplicates, etc.). The next stage is considered level 4.

Level 2—Business Process Re-engineering

At this level, specific actions are needed that are not related to the supply or implementation of IT infrastructure but to the process of administrative reform. In particular:

- Recording of all the procedures of the institution
- Recording communications and interactions both internally and with other bodies
- Study on the possibility of consolidating interdependent procedures or eliminating unnecessary procedures
- Redesigning the flow of documents
- Specifications, design, and implementation of applications that need modifications
- Preparation of the organisation chart of the institution for the transition to the next stage. It is at this level that any differences from the usual procedures or applications of other bodies are recorded, thus creating the various application levels.

Level 3—Intranet

The transition to this level requires the successful completion of level 0 and the operation of the infrastructure. Employees of the organisation, through an Internet-based environment can share applications or documents of their department or other departments which are located on a server (local server). Each employee maintains and uses a specific code in order to access either the applications or the institution's databases. In particular it may concern:

- Electronic document handling (elimination of hard-copy documents).
- File management and shared document storage.
- Monitoring the availability of employees (calendar).
- Assignment of tasks (to do list).
- Scheduling of internal meetings.
- Execution of documents and other service obligations.
- Library available online.
- Use of employee phone books.
- News, announcements, proclamations.

- Other applications necessary for the completion of daily tasks adapted to the needs of each service (e.g. in town planning, building permit management and control of arbitrary buildings).

Level 4—Automated register

A basic requirement at this level is the automatic assignment of a protocol number by the system to each document, without user mediation. In practice, this application is the gateway to the documents managed by the institution as a whole. Its construction should enable the system to record, monitor, control and inform all internal and external actions performed by those responsible for receiving and serving citizens, thereby improving working conditions and upgrading the services provided. The institutional provision is necessary for the adaptation of the internal procedures of document handling so that there are no implementation problems.

The registration of the documents should be done in a central point and then with the internal assignment in a specific part/employee of the institution and through the internal network the course of the case related to the specific document will be monitored. An important parameter of the application is to ensure communication between the software modules which should be done securely using encryption.

Also important for the next levels is the interoperability of the automated register software used by each body, so that it is possible to inter-exchange documents, as well as the compatibility of register systems between the bodies. Our experience from the Greek reality shows that different procedures are observed between the operators for registration and handling of documents, resulting in the inability of a common automatic protocol software to operate.

Level 5—Unified identification of internal users

Study, design and implementation of access control applications to the capabilities and functions of the institution's systems using a registration code (login/password) for each employee separately.

Level 6—Portal

Creating a "portal" is the first most essential step towards a "one stop shop" type service. Through government portals, citizens can search for information about the public body without the need for registration or authentication. Indicatively it may contain:

- Announcements, news.
- Proclamations, competitions.
- Legislative texts (laws, presidential decrees, circulars, etc.).
- Online libraries with documents that are not provided free of charge.
- Press Releases.
- Instructions to the citizen for processing a transaction.
- Auctions.
- Presentation of a body to its citizens and its supervised bodies.

- Links to other e-Government portals/sites and sources of information of interest, related to the object of the public body itself.
- Information search.
- Staff telephones for the public.
- Citizens help desk.
- Map of the portal.

Level 7—Unified identification of external users

Research, design and implementation of an access control system for any external user wishing to get electronic services. It concerns applications that are already in electronic form and were used with the physical presence of the interested party on the site of the institution.

Level 8—Portal upgrade with applications tailored to each user

Update and enrich the portal with applications that require weak to very strong access control of external users depending on the desired service. The purpose of the portal is to enable different categories of users to access information and services according to the access rights granted to them. The users of the portal can belong to various general categories such as:

- Citizens
- Business representatives
- Employees of public bodies and various organisations
- Administrators.

The portal should provide a central point for data accumulation and interconnection with various applications and systems located in different physical locations and used by different users. Here the issue of secure management and storage of personal data is a first priority. Thus it is necessary for the reliable and efficient operation of the portal to define a security policy according to specific and commonly accepted security models.

The services provided by the portal concern all the above categories of users and include the following [4]:

- Publishing information and linking to existing websites.
- Search and retrieval of information.
- Search for information and submission of data for further processing.
- Sending e-mails to the public body with personal information in the content of the text.
- Receiving e-mails from the employee in charge of the public body with personal information.
- Electronic completion of applications where no signature is required.
- E-learning.
- Participation in government events by personal vote through the "portal".
- Exchanging views with other citizens.
- Complaints submission.

- Electronic completion of tax return and VAT.
- Financial transactions.
- E-commerce.
- Paid online libraries.
- Electronic circulation of medical records.
- Emergency response services.

Additional functional features that will be provided concern:

- Multilingual support. The content of the portal should support languages other than the official state.
- Search engine capable of ranking and indexing structured and unstructured data.
- The search engine is capable regardless of language used.
- Services such as electronic discussion spaces (on-line and off-line) part of which will be addressed to people with disabilities.
- Unified and automated user service through multiple communication channels (web, e-mail). Possibility to expand the communication channels (e.g. in sms and fax).
- Portal content management tools for content maintenance. Content management is about creating, managing and delivering content.
- Ability to create personalised pages (personalization) for each user, that will contain their information and preferences.
- Security services, security policy.
- Open architecture is able to expand according to the increase of users and the traffic of the node.
- Monitoring of online behaviour and traffic of users and statistical tools for analysis of all information circulated through the node.
- Easy node management by easy-to-use central control tool with graded access and security.
- Ability to monitor the progress of processing requests and relevant information of the applicant through e.g. e-mail system.

Level 9—Digital Signatures

In contrast to the traditional way of handing over documents with the applicant's or the draftsman's own signature, electronic documents, although not at risk of robbery on shipments or natural disasters (fire, flood, etc.) present new forms of risk associated with innovative methods of forgery, alteration or even destruction. To avoid the risk of unwanted modifications, special encryption methods and algorithms were used which led to the so-called digital signatures. The digital signature is not, as the term may misleadingly imply, the electronic imprint of the handwritten signature, but it is essentially a "locked" abbreviation of an electronic text. This abbreviation could be documented as a fingerprint of the electronic text. Basis of the process, for creating a digital signature, are cryptographic algorithms, which use different keys to "lock" and "unlock" an email.

The digital signature performs mainly two functions at the same time. One is confirmatory, i.e. the recipient can be sure that the received message belongs, without intermediate modifications, to the sender and the other is confidential, i.e. the recipient trusts that he only will read the electronic text and not any unwanted third parties.

The use and implementation of digital signatures by the general public requires the establishment and operation of respective organisations and procedures that will operate under state control but in an independent environment.

Level 10—Smart cards

It is an advanced and secure form of authentication in the context of personal credit cards and IDs, as it contains its owner's encrypted personal data. The holder of such a card will be able to access the various government portals while being able to trade with the public. These cards have the ability to redefine the data they hold in their memory and can be used in advanced applications. They store unique information about their owner that serves to certify him/her, such as a police ID and information related to advanced applications facilitating the transaction of smart card holders with the Government, automating procedures that would require, done in the conventional way, much more time, bureaucracy and queues at public service offices. Depending on the specifications of each body, a number of applications with smart cards come up, which should also be classified at different levels depending on their functional dimension.

Level 11—Compilation

Translation of the applications that are available to citizens and businesses in different languages, due to the multiculturalism that characterises almost all advanced economic states but also due to the operation of economic associations such as e.g. the European Union market.

Level 12—Digital TV

Access to the "portal" through digital television. The highest stage of e-Government, where every citizen will be able to access these digital services through the receiver of his digital television with procedures similar to ordinary television. The great advantage of such applications lies in the immediacy of television among the average citizen but also in its wide spread across the range of social groups and economic levels. Perhaps it is the answer to the problem of the electronic gap that characterises our time.

Level 13—Satellite Communication

Informing government officials and exchanging views via satellite systems (GPS). This advanced stage is the way high-ranking government officials are informed about what is happening in places of interest, through satellite systems. In addition, it enables the sending of brief information to government portals for faster information of citizens as well as conversation both between executives and with citizens on key government issues.

In conclusion, from the presentation and analysis of the most important models of service development in the context of e-Governance and e-Government, we are led to substantial conclusions about the way in which the modernization of public administration is planned. Specifically, these models include the goals and priorities for e-Government, while the achievement of the goals is based on implementation actions and the use of ICT tools.

Comparing the strategic action plans of the most important initiatives that are in progress, important similarities emerge but also differences in terms of goals and priorities that have been set. The set of strategic plans sets as objectives:

- Modernization of the Public Administration through the reduction of its operating costs, but also the simplification of public procedures.
- Provision of electronic public services through one-stop portals.

A common development model of e-Government and Governance, as shown by the analysis of the above strategic action plans, can be summarised in the following process:

1. Status recording and selection of public services.
2. Creation of one-stop portals, based on the architecture of the respective portals from E-Commerce. The operation of the portals is based on complex information systems, which operate separately from the traditional Public Administration.
3. Installation of mature ICT solutions, which guarantee reliability in transactions. The mature solutions come from ready-made products, which are corresponding e-Commerce applications.
4. Grouping of public services according to human events and business needs (life events, business situations).
5. Offering electronic means for the communication of citizens and businesses with public bodies.
6. Provision of public information and services through portals.
7. Transformation of public transactions into electronic form.
8. Provision of electronic forms for the submission of applications for public services, for those which cannot be transformed into electronic form.

2.4 Towards an Electronic Public Administration: The Roadmap

The environment in which the public administration is called to operate and be transformed towards its electronic appearance, presents significant challenges and opportunities. Perhaps the most important issue for governments is the optimal satisfaction of citizens with the provision of innovative services by increasing the productivity of services combined with the maximum savings. At the same time, the observance of specific specifications and the minimization of the delivery time of the services will contribute to the quality upgrade of the deliverables. To this end, the administration is required to transform from the simple application of laws, circulars and instructions to an administration to achieve goals and measurable results, according to the principles of the New Public Management. It is now necessary to change the way the public administration operates and to redesign its procedures and organisational structures.

The effort for this transformation is facing various difficulties, mainly of an economic nature. This is because, on the one hand, governments in their efforts to attract investment are pressured to reduce the tax burden on companies with a direct consequence on the state tax revenues and on the other hand, they must intervene to mitigate the effects of the globalisation of the economy. Also, due to the participation of states in various international organisations or associations (e.g. European Union), national administrations are forced to adapt their policies and institutions so that interaction between them is easier.

The changes required are undoubtedly linked to the use of ICT and new management tools that have already been used successfully in the private sector. IS creates new data and new opportunities for growth, prosperity and quality of life. Its development is based on the evolution of ICT, which affects the human resources, since they create:

- New working methods in both the public and private sectors.
- New skills.
- The need for continuous learning on the part of employees and adaptation of the education system.

ICT contributes to improving the quality of life of the citizen, by providing improved services in the fields of trade, information—communication, service, health, support of "intelligent transport" and environmental protection. The use of ICT will facilitate:

- The creation of a modern democratic state.
- The required modernization of the Public Administration.
- Improving its efficiency and effectiveness.
- Generally the fulfilment of the demands of society in a framework of transparency and democratic participation.

The further development of ICT, aims to create a networking environment in public administration, to convert existing information into digital form, to create and maintain databases within the framework of security and sound management of personal data, to support the decision-making process with the use of appropriate information systems, in the provision of information and in electronic transactions with citizens and businesses, through the integrated interconnection of information systems and interoperability across the public sector.

The transition to an "e-state" requires on the one hand the transformation of functions in both the public and the private sector and on the other hand innovative actions that take advantage of the possibilities of ICT. In addition to the development of e-Government and the transition to e-Government, the promotion of e-information,

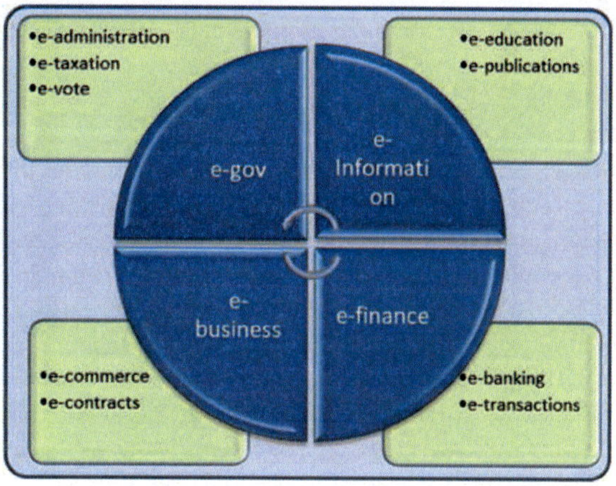

Fig. 2.9 The structure of the e-state

the development of e-economy and e-business, are the key components of the overall argument (Fig. 2.9).

The possibilities of the Internet for the support of distance learning and electronic publications that may include opinions, studies, announcements etc. have already become apparent as well as the possibilities for further developments. For example, in the field of education, we have witnessed the establishment and operation of "open universities" which are based on Internet communication and electronic publishing. In particular, in the field of electronic publications, the current situation can only be described as explosive. The electronic publications of studies, articles, opinions etc. are growing rapidly worldwide and fall into all scientific categories. An important parameter in further development is the protection of intellectual property and ensuring and verifying the validity of sources. The interconnection of educational units of all levels in a national network initially—through an educational portal, and then with other international portals, is an important step in disseminating ideas and knowledge and ultimately establishing e-learning. Certainly the road to e-learning and at the same time interactive education is long and we are still at the beginning.

In the field of business and commerce, the development and spread of ICT in the private sector and the globalised economy have paved the way for innovative actions, such as e-commerce. E-business applications in various sectors have already made their presence felt by changing the classic form of commerce, offering opportunities to increase competition between businesses, reduce monopolies and generally improve market conditions. Electronic communication between suppliers and professionals worldwide, the ability to promote products and orders via the Internet as well as other services, have already given significant benefits to the businesses involved (business-to-business) but also to the citizens (business-to-citizen). Citizens-customers receive new innovative services such as "online" orders, real-time information on the availability of goods, ability to compare prices and quality of products, etc. thus saving money, time and hassle. It should also be noted that the emergence of these new private sector services has prompted citizens to demand from the public administration the delivery of public services to the standards of the private sector.

Related to e-business and e-Government is the perspective of e-economy. As for the economy of a country, it is generally accepted that, at the macroeconomic level, the use of ICT to an even greater extent, will lead to improved productivity and as a result increase the competitiveness of businesses and the economy as a whole. In some sectors, however, increased productivity due to this use may lead to additional jobs while in others to increased unemployment. In this context, the adequacy of ICT-related professional qualifications in the workforce is a major challenge for the overall perception and management of the labour market. On the contrary, the lack of ICT skills is a major risk that will hinder further development. It is noted that this situation is particularly acute in Europe due to declining demographic trends and the declining interest of young Europeans in scientific studies [10].

At the microeconomic level, the e-economy imposes significant changes on market organisational structures. Relevant technological developments directly affect the structure and life cycle of companies. This is mainly due to the fact that with the use of ICT:

- The cost of access to information is reduced due to the elimination of distances from the source of information and consequently the increase of business competitiveness.
- The cost of setting up a small business is often reduced, potentially introducing additional competition.
- Opportunities are offered for new cooperative means of delivery of goods and services, which leads to the improvement of quality and efficiency.
- Innovative products and new services appear.

It is a fact that in the last two decades traditional businesses have adapted to the new market conditions or have been replaced by others, along with the creation of new businesses with innovative products or services. Especially in the last decade this process has accelerated significantly. Products are becoming more and more "extended products", which include a significant component of services. Internet-oriented businesses are beginning to acquire the characteristics of traditional businesses, such as warehouses and chain stores. On the other hand, traditional retailers are beginning to provide some of their services or goods online, adding new distribution circuits and new supply strategies.

The further development of e-commerce, e-procurement and contracts is directly affected by growth in the banking sector. Electronic banking (e-banking) and the possibility of electronic payments are directly linked and affect the situation in commerce. The possibility of transferring the bank itself via the Internet to the computer screen or via wireless networks to the mobile medium (e.g. mobile phone, i-pod, palmtop etc.) of the trader serves in the direction of saving time and money for both parties—banks and customers but also in reducing the hassle of queuing in bank branches and organizations.

Despite the large penetration of ICT in almost all areas of business activity, its impact is heterogeneous. The domains that manage more information that can also be in digital form—e.g. digital goods, information services, financial and economic services etc. contribute to the emergence of new economic models and increasing market competition. Unlike in areas where access to information is not the most important issue—e.g. heavy industry, construction, etc., the impact will tend to be more progressive.

The emergence of new economic models is a feature of the electronic economy. A significant number of these models have failed, as have numerous businesses whose Internet-focused and general-purpose activity is "dotcoms". Others, on the other hand, have proven to be viable, especially in the area of business-to-business (B2B) e-commerce. A more advanced degree of maturity in the entry of businesses into the e-economy would, perhaps, be the answer to the mistakes of the past. Businesses need to use proven solutions and techniques following best practices of sustainable economic models, mainly business-to-customer (B2C) companies, whose potential should be exploited.

The further development of the e-economy requires specific steps—phases which include the following:

- Promoting the cultivation of entrepreneurship.
- Improving skills related to the use and effective utilisation of ICT capabilities in the e-economy.
- Empowering businesses to meet the needs of the global market.
- Improving the operating conditions of the markets and in particular the conditions of the European common market.

In particular, entrepreneurship must be supported by skills related to business issues on the one hand and ICT issues on the other. So it is necessary:

- To organise targeted training programs to improve skills based on developments in the field of ICT
- To implement research programs that combine topics from both the ICT and the socio-economic sciences
- To substantially support small and medium-sized enterprises for the further access and development of e-commerce.

It should be noted that an important parameter for the further development and establishment of e-Government is the provision of information that circulates electronically. Both the public and the private sector manage a plethora of sensitive personal data or financial data of businesses and citizens on a daily basis. Significant steps have already been taken in this direction, but securing information circulating either via the Internet or via wireless networks is an issue that must be constantly reviewed and addressed in accordance with common international rules and standards. In this direction, "smart electronic cards", electronic signatures and other technologies have already appeared, but they do not have the required scope of application, mainly on the part of the citizen.

To make the most of e-Government, barriers to the availability of e-Government services in the Member States must be identified and actions taken to accelerate their development. Related to these barriers are the concepts of e-inclusion and e-accessibility. In particular, e-inclusion [11] is linked to the non-spread of the information society to all, i.e. a society that guarantees to all, at reasonable prices, the same opportunities for access to ICT and the same availability. In particular, e-inclusion aims to implement systems that will allow people with disabilities and the elderly to have easy access to information society services. On the other hand, e-Accessibility refers to initiatives aimed at ensuring that all citizens have access to information society services. The aim is to remove the technical, legal and other obstacles that some people may face when using ICT services.

To address the problems mentioned above, the modern public administration in the context of administrative reform and New Public Management should develop specific features including the following [12].

- The public administration must be result-oriented and not clinging to the simple application of formal rules and procedures. It is necessary to consolidate the role of management as a producer of measurable results and that the current institutional framework should be used as a tool to achieve the desired result for the public and expansively for the benefit of efficiency.
- It is necessary—especially in times when recessions follow one another and economic crises bind governments—that public spending is guided by the production of public products with added value to society. It is therefore necessary to limit time-consuming and anti-financial procedures, unnecessary actions, etc.
- In the ever-changing environment where socio-economic needs change, the public administration must detect the needs of the public and adapt its action accordingly in order to produce new innovative products.
- It is important to adopt specifications for the desired results of public activities both for those that are the final deliverable and for those of the intermediate stages. This will ensure the compatibility of the outputs with the requested public products.
- The transition to management with a staff role is linked to the planning and scheduling of public administration actions, monitoring and evaluation of results as well as feedback to the planning staff, which should consist of both parties—political and permanent staff of the public administration.
- The use of cost-benefit analysis tools will help in the environment of the economic crisis on the one hand to maximise the quantity and quality of the results of public action resulting in maximum social acceptance for public spending and on the other hand to minimise the required resources for public production.
- Improving efficiency through the measurement of the quantitative component of public activity, requires the adoption of a system of performance indicators and its application in all cases of production of public products.
- In the new conditions of the open world economy, the regular practice of the public sector to produce the final deliverables or intermediates with its own forces is no longer sufficient. The organisational structure of public sector actors, in the narrow sense, does not allow maximising efficiency in terms of economy and efficiency. Thus it is necessary, in addition to the structural and operational

reorganisation of public administration, the search for solutions from the free market. The provision of deliverables based on the adopted specifications and within the framework of the existing planning, the monitoring and evaluation of the results and all the executive action may concern the public sector, but in each case it must be decided if outsourcing the production of final or intermediate products and services to entities, in either the wider public or private sector, is preferable in terms of economy.

- In the new complex environment that is formed by the interaction of the public and the private sector but also by the involvement of various public sector actors for the production, the effective and seamless communication between them as well as the coordination of their actions to avoid waste and unnecessary actions becomes necessary. In other words, the synergy and complementarity of the actions of the entities involved is discussed.

2.5 Critical Issues for e-Government

From our experience so far with regard to the actions and policies of national governments towards the completion of e-Government, we have documented and analysed the following critical points on which further effort should be focused.

A. *Access for all*

A prerequisite for the successful course of e-Government is to ensure that everyone has access to e-Government services. Various factors can lead to unequal access to online services and lead to the so-called "digital divide". Education and training are essential in order for the citizens to acquire the necessary knowledge of new technologies, so that they can make full use of the services offered by e-Government. Also important, in our view, is the issue of access to online services, for people with disabilities or the elderly who either cannot or have not been properly trained. In these cases the adaptability of the information systems to the needs of the special groups can make a positive contribution. Finally, providing incentives, financial or other promotional activities, will help the project to have a greater impact on the public.

B. *User trust*

Public services should only be offered electronically under conditions that guarantee users absolutely secure access. In this context, the confidentiality of personal data, the security of digital transactions and communications are paramount issues for which maximum protection must be provided. To achieve this, technologies that improve privacy in e-Government must be promoted. More generally, network and information security, the fight against cybercrime and operational security are prerequisites for the success of e-Government.

The tremendous progress of ICT, the increasing application of electronic transactions, the need for computerization of both businesses and the state and the resulting massive databases, new communication networks such as the Internet and the rapid transmission of information worldwide, provide increased possibilities for collecting, storing, interconnecting and processing data of a private and sensitive nature, putting new risks to the privacy of citizens. The transmission of personal data through open public networks, their storage in publicly accessible computer systems, combined with the increased possibility of interception of this data, complicate efforts to protect and respect the privacy of citizens, one of the fundamental principles of any democratic society.

The threats faced by any e-Government venture are significant. Therefore, the requirements of security mechanisms are correspondingly high, able to guarantee the authenticity of the identity of the traders, the integrity and confidentiality of the content of each transaction and the non repudiation of participation and completion of the transaction. Meeting security requirements is covered by a family of technologies, such as smart cards, Public Key Infrastructure (PKI), electronic forms combined with digital signatures, digital certificates etc.

Very often, in order to transact with the state, it is necessary to possess and present an identity element. It is obvious that in order for the state to be able to provide a series of transactions electronically, a national system for issuing and certifying electronic identities will have to be implemented. Under such a system, the state will issue electronic IDs in the form of smart cards, i.e. cards with a built-in microprocessor. Electronic IDs will contain digital certificates, which will be used to create digital signatures and will certify the identity of their holder, when trading in a public key infrastructure (PKI) environment. Through electronic identities, it will be possible to verify the authenticity of citizens' identities (authentication) and maintain the integrity and confidentiality of electronic transactions, in which citizens participate. Electronic transactions will be made using a variety of access points, such as personal computers, digital televisions, mobile phones, ATMs and more, through private or public data networks, such as the Internet.

C. *Public procurement*

Public procurement is an area in which the use of ICT can offer particular advantages in terms of efficiency, transparency and accountability of the parties involved. Traditional public procurement procedures are proven to be time consuming and complex, require a lot of resources and often conceal procedures from the general public. Specific tactics associated with the use of ICT that can improve the timing and effectiveness of related tenders as well as save resources while increasing accountability and public confidence are:

- The initial public consultation and wide disclosure of the terms of the relevant tenders.
- Fully automated procurement and monitoring procedures.
- The organisation and monitoring of an electronic register of suppliers.
- The organisation and monitoring of a register of goods/works and the corresponding invoices.

- Automated procedures for confirmation of deliverables, in addition to the established committees for monitoring and acceptance of projects.
- Organisation of an open public repository for free access to contracts.

D. *Interoperability*

Interoperability means the way in which systems, information and working methods are interconnected. The interoperability of information systems therefore makes it possible to integrate service delivery into a single-gateway portal—whatever the number of different systems or administrations involved. Interoperability, however, is not just about connecting to computer networks. It also concerns organisational issues, such as the need to ensure cooperation with partner organisations, which have a different way of internal organisation and operation. In addition, the creation of pan-European e-Government services also necessarily requires the conclusion of agreements on common standards and specifications. Most Member States are already addressing this challenge by adopting national "e-Government interoperability frameworks". These actions are complemented, at European level, by the development of a European interoperability framework.

Our view is that while adhering to common systems development standards, as defined in the interoperability framework, can help in the exchange of data and information between the different systems of public bodies, it does not solve the problem of existing systems and especially base systems and in particular database systems developed in earlier times without complying with the new standards imposed by the interoperability framework. Often, the development of new systems is unprofitable in terms of resource economy, so the exchange or retrieval of information requires, as we will develop in more detail in the following chapters, the adoption of innovative tools in the field of Semantic Web.

E. *Human resources and the role of leadership*

One of the most important factors for the success of e-Government is the preparation, information and training of public sector human resources, as change and innovation have proven to be slow or in any case slower than in the private sector. The human resources are the ones that will finally take advantage of the opportunities of technological solutions and will finally enter a new trajectory of work, mentality, perception and detachment from their traditional way of working. The issues that employees are called upon to address are:

- The adoption of the customer-centric concept in order to satisfy the citizens through their every action.
- Their continuous and focused training in new technologies and their ability to interact with other departments and people.
- Increasing the participation in the targeting of the respective entity in order to share the new perceptions for the work and the offer to the society as a whole.

However, in this difficult path, the employees themselves will have to receive the practical support of the state and of the leading figures within the organisations that will motivate them on the path of reorganisation of the Public Sector. The official and political leadership must act in this direction by implementing specific policies such as:

- Focus on how technology can influence and shape Public Sector strategies.
- Using technology to develop innovation and not just to automate processes.
- Adoption of best practices by countries considered to be pioneers in e-Government.
- Improving the flow of funding for the integration of technological solutions.
- Protection of privacy and security.
- Forming communities within the workplace with a common recipient the development of technology and the economy.
- Using technology to create equal opportunities.
- Preparing for Digital Democracy.

F. *Steady flow of funding*

The role of each government in ensuring a steady flow of funding and funds in projects and actions that help the development and adoption of e-Government, is the Achilles heel of projects related to e-Government.

The difficulty of assessing the benefits arising from investments—mainly financial—in projects for the reorganisation of the Public Administration, is often the main reason for not placing stable funding. On the other hand, the poor absorption by the respective government of each country of Community funds regarding projects in the Public Administration, the non-planning and forecasting and therefore placement of funds in the national budget of each country for the development of e-Government, the non-promotion of e-Government as the main weapon for the profitability, productivity and competitiveness of companies and as a consequence reluctance and reduced private initiative for private investment in Public Administration, are the main quantitative reasons for the lack of development of e-Government.

So, whether it is a matter of poor absorption of EU funds by the national government, or of a lack of national resources or of a reduced private initiative, the question is one: As much as there is a willingness to adopt what e-Government proposes on the part of civil servants and citizens, this cannot be continued without the necessary assistance from the State. This assistance cannot be other than investing funds in projects that assist the Public Administration.

G. *Reconstruction of the state*

There has been a lot of discussion about the need for change in the Public Administration itself in order to be a successful path to a fully electronic state. The necessary restructuring of the state, on the one hand, should not be based only on the automation of processes, on the other hand, it should focus on the promotion of innovative services by "building from scratch" the required processes with the ultimate goal of providing optimal public products.

It is necessary to study all the processes of an organisation and decide which ones can be merged, which ones should be abolished, which ones should be automated and which ones should be improved and how. The goal will be the most efficient-effective operation of the organisation in the context of the Redesign of Business Processes. There are two relatively new concepts in the heart of the whole process: The "front-office" and the "back-office" of public bodies. The front-office of public bodies is what the citizen perceives as a "person" or "representative" of the public administration and is related to the final product—service provided to the public. Undoubtedly the quality of the front-office is affected, or even determined, by the mode of operation and the consistency of the back-office. The concept of back-office means all those processes, services and agencies that must interact and combine in order to produce the desired public product and deliver it to the public.

The demand until recent years has been to improve the quality of the services provided, i.e. the front office. This logic, although on the right track, is not complete. Significant care must also be taken to improve the back-office, i.e. the processes and their structures of services and agencies. Related to this issue is the redefinition of the responsibilities of ministries and public bodies. Co-responsibilities of different services are often observed even within the same body, resulting in the most difficult transition to the electronic provision of these services. Finally, the re-examination of the required supporting documents when the citizen arrives at the appropriate branch (front-office) of the public will have significant benefits for the implementation of the much-discussed "one stop shop". Particular attention should be paid to the essential information that the person concerned should submit to initiate the procedure, while the supporting or other relevant documents that each service will seek ex officio in the back office operation should also be specified. In the context of a successful e-Government, the interested party should be able to be informed at every stage of the process in any convenient way (e.g. e-mail, mobile phone message, etc.).

H. *Transnational collaborations*

Although every European country has already developed a national strategy for implementing e-services in public bodies, we are still far from having a pan-European common view. The integration of G2G government systems that allow transactions between public bodies at the transnational level has a long way to go before it becomes a tangible reality. The main factors include:

- The complexity of the systems already developed.
- The time required for public bodies to change and align.
- The economic gap between European countries.
- Differences in administrative culture between European citizens.
- The inadequacy of governments to coordinate their national strategies with a commonly defined implementation.

Coordination at European level is not easy. European and national organisational units must balance their national, regional and local identity and autonomy, and also try to meet the central challenge of a homogeneous knowledge society. European countries must be together in a winning strategy. In addition, their strategies at national, local and regional level must not only be homogenised across borders, but also include European directives.

G2G cross-border e-services are also a big challenge. Such services can help increase national and international cooperation and security. In fact, recent national and international security concerns are becoming increasingly commonplace. The escalation of international terrorism and cybercrime requires international cooperation and real-time information exchange. Due to legislative, cultural, linguistic and other differences, as well as the lack of a strong central direction, the conclusion of cross-border agreements is very rare. Some individual attempts have been made in recent years, but many programs have failed.

I. *Data digitization—Knowledge management*

The provision of e-Government services required from the first steps the digitization of documents, the development of databases and the implementation of specialised ICT solutions. Nowadays, most of the documents related to current cases have been digitised in most of the public bodies. In this context, the amount of data managed by public bodies is constantly increasing and even at an exponential rate. This phenomenon has been exacerbated in recent years by the adoption of Open Government policies, which require the disclosure of public information using ICT. At the same time, the search for information and the management of knowledge generated by traditional methods in such an environment presents dysfunctions and will soon become inefficient, mainly due to the huge volume and variety of data combined with different information systems that do not interoperate in a large part of them.

However, there are areas of activity such as, justice and security in Greece, where digitization has not been completed and e-Government still has several steps to complete. Also, much of the public information is not yet accessible via the Internet. This is either because it is quite difficult and anti-economic as an action, or because it is not a priority for the administrations of the organisations. For this reason, administrations must adopt innovative methods and applications for managing vast databases and intelligently searching for information. The more specific issues related to this issue are developed in the following chapters.

2.6 Back-Office Systems Development—Critical Issues for the Greek Case

A lot of discussion has been made on e-Governance and the need for changes in Public Administration in order to succeed in the transition to e-Government. The objective is to improve the quality and also to expand the services provided to citizens and businesses, following the example of private sector companies. This issue is connected to the, so-called, Front-Office of the Public Administration, that is, the person or representative of the public administration that any citizen deals with. Improving the front-office is a point which governments consider as very important and emphasise its benefits to the everyday lives of citizens by applying e-Government solutions. Thus, the focus now is on the citizen who deals with the administration and on how to serve him/her optimally. Special programs and individual assessments are underway to monitor the progress of the Front-Office project. Moreover, e-Government encourages direct communication between citizens and policymakers enhancing democratic participation. Citizens enjoy the possibility, using innovative tools such as online forums, virtual discussion rooms, and electronic voting, of submitting questions directly to policy makers and expressing their views on the policies carried out for the public.

Considerable attention should also be paid to the other side of the issue of e-Government that has to do with the support to the administration. Effort, therefore, has been made to improve the Back-Office as well, i.e., the procedures and structures of institutions that support the delivery of public goods and affect the quality of the front office. The performance, efficiency and speed of the back-office will affect the overall image of public administration.

Importance similar to the one attributed to the image that citizens will form an e-Government should also be given to the appeal and to employees involved in the administration. If they have been offered modern, efficient and smart applications then maximum acceptance and support of the project will be achieved. Modern tools will enable faster and more effective solutions to everyday issues faced by the employee on the exercise of his duties. Particularly in the environment of the economic crisis that plagued the previous years in almost all the advanced states of Europe, the reduction of costs in the back office administration and faster service delivery is of great importance. The latter is directly related to the so-called paperless office. The transition from hard copy of documents produced to electronic documents and electronic procedures as well as electronic administrative transactions must be at the heart of policy. With such processes we can achieve the maximum degree of transparency and better control of the services involved.

In a political, economic and social environment in which requirements for public administration are constantly increasing, the main goal for a system of servicing citizens and businesses is to provide integrated services and information that meets their objective standards of quality and efficiency. The public needs to know in advance both the services and capabilities of the system provided and the standardisation followed so as to promote a sense of confidence in the abilities of Public administration and to the maintenance of proper procedures.

The system must fully meet the real needs of the public, for providing information and handling of administrative affairs in conjunction with the continuous review of procedures followed to ensure that the trade is more flexible and adaptable to continuously changing requirements. By the correct use of ICT, transactions between the public and the public administration will be easier, faster and cheaper for both parties.

The improvement of the image of the State, the upgrading of relations between the state and social partners, and the transparency of operations, are directly related to the improvement of existing services, providing integrated electronic services to citizens and businesses and the transition to e-Government. The provision of advanced electronic services, direct access to digitised information are the key objectives of e-Government whose main pillars are the electronic connection of public administration (e-administration) and interoperability between different information systems. To this end, the introduction of ICT in public administration is not the only prerequisite. Other prerequisites include the promotion of appropriate education, awareness and retraining of the employees, as necessary reforms to obtain a positive response from official executives and citizens.

In this domain, we mention the following related work. In [12], the essential contributions of e-Government are presented, including e-administration, while in [11] several aspects of developing an integrated Regional E-administration system are presented, by adding "pieces" successively developed under the frame of two PHARE Projects. In [13], the challenges and barriers of implementing e-Government by the investigation on NEIS of Korea are presented. In [14], issues that refer to the Manipulation and Transparency Control of an ICT-constituted e-Administrative Protocol via Digital Watermarking for LDCs are discussed. In [15], the authors contribute to the knowledge of the various issues raised by local e-Administration and propose an analytical framework for the evaluation of potential local online service offerings by the examination of how local public administration has been developed in France. In [16] a framework is proposed which is capable of personalising both the access and the behaviour of an e-office, based on ACE software. In [17], the authors deal with the typology of Citizenship management using ICTs, namely, e-administration, e-Government, e-Governance and "The Learning City". Finally, the authors in [18] present several aspects of building an integrated regional e-administration system. This system is a document-oriented data collecting network that ensures the link between the Timiş County Council and the local administration and it is based on Lotus technology.

2.6.1 e-Administration—The Current Back-Office in Greece

In achieving e-Government, special emphasis has been given so far to the implementation of infrastructure support as well as to software development in the front-office sector. Thus, as in all advanced countries, modern and powerful computers and peripherals have been installed and used throughout the Greek administration. Networks of public administration and broadband networks have been created and government websites and e-Government portals have been developed in order to create the image of a modern and well-organised public administration focused on the user. Services have also been institutionalised which operate to the standards of one-stop-shop, but these are not authorised for the whole of government services [19]. Of course, a lot of actions and steps still need to be taken for the provided services to reach their highest possible level of development.

All previous however, although initially accepted with great expectations for their subsequent evolution, have not had so far the desired results. From the point of view of the public, the new e-services, where implemented, have the greatest acceptance and penetration, while a large part needs further development or even revision. Noteworthy is the fact that people seem to trust more the one-stop-shop type of services, rather than the online services provided until now. It seems that people prefer the personal touch and service from the employee rather than through an online self-service application. We can attribute this either to a lack of experience in the use of these services from the side of the public or to the design of these services, which has the characteristics of personalization. The public seems to expect more, new and innovative online services, more intelligent and interactive applications, according to its own needs and skills as well as faster delivery of public goods.

On the side of the public sector (within the system of e-Government) expected cost savings and faster delivery of deliverables within the parties involved does not seem to have currently been achieved. Direct consequences of this situation are recorded at prevailing view for the overall project to both employees and the public. One dimension of this problem is the need for structural changes in the state and its agencies and the need to develop innovative new applications in the back-office.

Software applications developed are, in many cases, insufficient and depreciated to substantially assist the employee supporting the internal procedures. In particular, the existing systems are characterised by the following:

- Older independent software applications do not support interoperability and standards that are developed on a case by case basis and occasionally overlap. We observe a phenomenon in various institutions: there are similar applications with similar functions that handle the same type of data but do not interact. As a result we do not really achieve the interconnection of all public institutions. The consequences are higher government spending, waste of time when processing data which perhaps another public entity has already processed and incomplete or difficult control of the information whenever needed.
- Main emphasis is being given to the development of databases, which results in the same data being stored in many different computer systems. This fact in addition to the dimension of wasting storage space has also the perspective of potential errors and conflicts at the cross checking of data due to double entries.
- Delay in application of universal electronic signatures. Although established through the adoption of relevant legislation, the validity of the electronic signature has not yet been possible to be used by nearly any of the public employees. This issue is crucial for the transition to another model of back-office function based on simpler processes and the reduction of printable documents resulting in lower costs, shorter processing times and higher transparency.
- Insufficient development of information systems to assist search of administrative information. The search of the necessary information when not covered by specialised depositories of certain organisations or when these are not recently updated is achieved through the common Internet search engines without the possibility of specialised access.
- Procedures oriented to the, so-called, hard copy rather than electronic form on documents. This issue is linked to the existing institutional framework and the need to adopt new forms of administrative acts.

- Insufficient development of applications for recording and browsing processes and workflows. Given that such applications are specialised and non-commercial, their growth has come to a secondary role.
- As a consequence of the previous three issues, searching for the necessary provisions, circulars or guidelines but also the data required for the processing of a case, often requires sending documents between departments involved, resulting in wasting time and increasing spending.

The result of this situation is that employees in the back office of public administration, in many cases have to work just like they used to work several years ago, without being able to yield the expected. This problem is associated with the personalised provision of services that does not only concern the citizen or business, but also the employees of the government themselves, i.e., government-to-government (G2G) services.

2.6.2 The Current Situation for the Personnel of the Public Administration in Greece

A major component of designing a software system is always the user. The interaction with the machine is crucial to its successful use, to the employee's self-fulfilment and also as a means to achieve the targets set within the organisation. Taking into account the preview requirements, the manufacturer of the system should create it, knowing in advance the actual needs and skills of the users. In this sense, the features of the human resources of public administration in Greece are very interesting.

According to available statistics for the entire workforce of the Public Administration [20], its main features are summarised as follows:

- Tendency of the percentage of University graduates to increase. As graduates of technical colleges and universities have become more familiar with the use of computers and new technologies in relation to older employees, applications installed or used may have greater acceptance than in earlier eras. This however has the dimension of stricter evaluation of existing applications from officials resulting in their obsolescence or even their complete withdrawal.
- Tendency of the staff to age. This fact should be taken into account when designing new applications since older employees face particular problems with new applications.
- Low percentage, though increasing, of women's participation in senior management positions (general directors and directors).
- Small proportion of disabled employees. The staff of this category usually has typically ancillary tasks such as serving calls in call centres or clerical staff tasks. However, there are individuals with appropriate qualifications which could by using appropriate software to work in cutting edge sectors.

- Tendency of mass retirement of employees in conjunction with the severe restriction of new hires currently applied in the public sector. Coping with the economic crisis and the ever decreasing salaries, many employees prefer early retirement from the service. The phenomenon of mass retirement from work in the public sector complicates the already difficult situation. This however is not without damage. These employees, having gained experience and knowledge, retire and do not leave behind them this valuable heritage. This makes it an imperative issue for the development of software that will help employees in acquisition and management of knowledge.

The personnel of the Public Administration, despite the efforts made in educational and training programs as well as by applying institutional measures, is still characterised by inherent, traditional dysfunctions such as the following [20]:

- The complexity of settings.
- The uncertainty of settings and the contradictions between settings.
- The fragmentation of production settings.
- The existence of outdated settings which have not been adapted to modern technological data and new socio-economic requirements.
- The absence of codification of existing settings.
- Incomplete "official memory" of organisations.

Some of the consequences of the above dysfunctions are:

- The burden to the government itself to carry out unnecessary paperwork.
- The difficulty in searching for appropriate procedures for handling cases.
- Delays and inefficiency in the delivery of public goods.
- Difficulty in ensuring transparency in the acts of public administration and fighting corruption.
- High cost of transacting with public services.
- The disadvantage of citizens towards public services they trade with, especially those without a high skill, or knowledge on managing their affairs, such as low educated people, immigrants, etc. and small businesses as well.
- The burden on the cost of living and quality of life of the citizens.

2.6.3 Description of the Support System

The current research examines the possibility of developing a single integrated system of support for the back-office of public administration within an organisation's e-administration. The objectives of this system include:

- Provision of the exact information and services needed by the employee depending on his/her position in the organisation and especially on his/her particular skills. Obviously not all employees have the same needs and the same skills. However a common application for all would prevent disputes and conflicts when processing data and contribute to a more cohesive function of the organisation.

- Minimisation of bureaucracy and production of hard documents. This system must introduce innovations in order to be the trigger for the institutionalization of new procedures and ways of working in the public sector.
- Maximisation of the organisation's efficiency and effectiveness to achieve faster and more valid information search in accordance with the objective above.
- Providing tools for acquiring and managing knowledge. Public sector bodies have a huge amount of data and information while employees have transformed this wealth into knowledge. It is therefore necessary to develop specialised technologies to manage this knowledge.
- Raising awareness of staff for effective system support. Employees who use a system are the ones who will note its defects and faults, or suggest improvements. They are also the ones to bring it out as a successful tool or reject it. There are cases where applications are constructed and manufactured according to the original requirements of employees but in practice have been proven inadequate or inconvenient resulting in their removal.
- The interconnection and interaction with the systems of all public sector bodies under the current structure of the pyramid (Fig. 2.10). According to this, there are two dimensions of interaction. One dimension concerns the hierarchical relationship between public bodies and the other, the direct relationship between public bodies with the public (citizens and businesses). In each dimension, back-office systems support interaction between bodies, and provide the requested data. These data are supplied to the intermediate handler who is the ultimate provider of services to the interested parties.

In order to determine the necessary structure of the system to support the operational function and the process of delivery of public services, system components have been classified according to the type of component implementation and the parties involved (Fig. 2.11). These categories are:

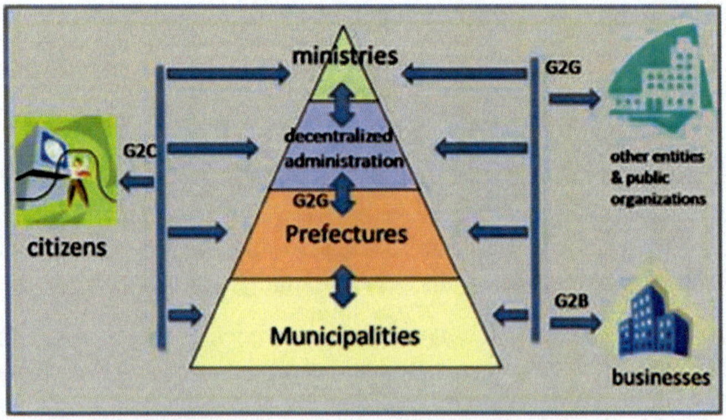

Fig. 2.10 The structure and interactions of the Greek public sector

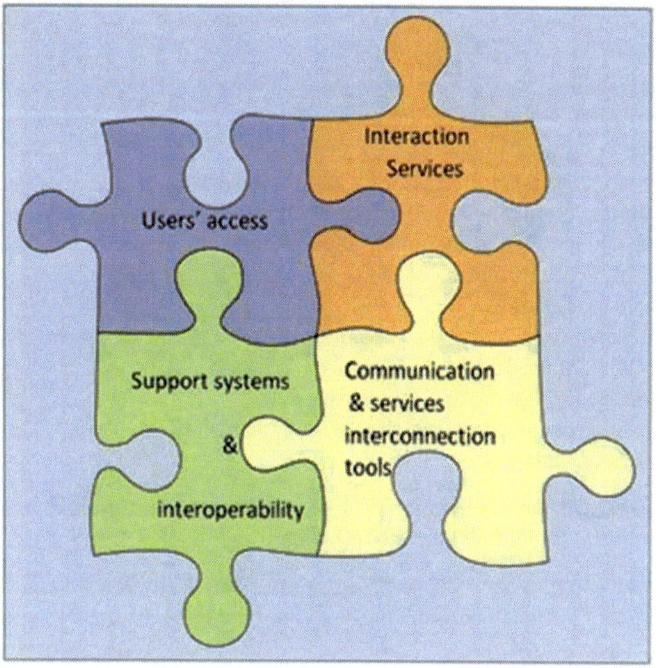

Fig. 2.11 The parts of the back-office system

- Users' access, meaning the user's point of contact with the system. This will be a channel of communication within the internal network of the organisation (intranet) which will also allow connection to other users of the administration outside the institution. This channel may be in the form of an internal portal. The above described system will be personalised according to each employee so that every user can be recognized according to his/her personal and unique features, his/her position, his/her responsibilities in the institution, and his/her record of system use. In this way, any employee will transact with the system and will have the most suitable choices and services. At the same time, the integration of employees into groups, based on specific user prototyping, which will handle specialised services will facilitate the processing of similar cases from employees working in different institutions.
- Interaction Services. This includes all the services offered to users in order to facilitate information. These are the most critical components of the system. The impact that they will have on employees will determine their success and further improvement of services of back-office. The main objective is personalised services and information based on innovative technologies such as the Semantic Web. These services cover all processes in the back-office: those hitherto made in the traditional way and are provided electronically. These services will include depending on the position and duties:

  - Specialised search engine featuring personalised information access. It is an extremely important function for saving time and increasing efficiency and effectiveness.
  - Information on the processing stage and prediction for the final delivery of the service.
  - Recommendations for any further actions needed.
  - Legalisation of documents produced with the use of electronic signatures.
  - Check for legality of incoming and outgoing documents. This control may relate to the confirmation of the issuer and to the validity of data legalisation of each document.
  - Handling of electronically signed documents in subsequent levels of hierarchy until the final production of administrative acts.
  - Automatic processing of documents and legalised acts.
  - Electronic procurement—Resource Management.
  - Personnel Management.

- Support systems and interoperability. By this we mean the systems that support the above services and include the following:
  - Software applications for managing information which has been developed based on open standards in order to support interoperability with relevant applications of other institutions.
  - Dynamic templates based on common standards for all public institutions.
  - A list with the works provided by the institution as well as workflow diagrams.
  - A list with the duties and responsibilities of the employees in order to ease the work distribution and to check the legality of administrative acts.
  - Help desk with personalised access and Semantic Web features.
  - Hardware.
- Communication and services interconnection tools. By this we mean the support of safe handling of information both within the organisation and outwards, in addition to communication networks and applications.

2.6.4 Benefits of the System—Perspectives

As already mentioned, an integrated public service system in e-Governance, is mainly based on effective information systems. These systems on one hand support the internal informational public administration (back-office) and on the other hand are the interface between the Administration and the public. The key benefits of successful development include the following:

- Reduction of operating costs for the state.
- Improvement of productivity and efficiency.
- Maximisation of the transparency of the procedures within the administration.
- Better service for the citizens who are the final recipients of services.

- Support of decision-making.
- Support of designation and realisation of policies and programs.

2.6.5 Critical Issues for the System Development

From another perspective, the issue of improving supporting structures is associated with redefining the responsibilities of ministries and public organisations since several shared services are frequently observed even within the same institution resulting in a more difficult transition to electronic delivery of these services. At the same time the huge amount of laws and circulars runs the risk of overlapping and creating conflicts between them. Part of the reforms for the restructuring of the management concerns designing automation processes and searching for innovative services by "building from scratch" the required procedures. In this case the design should be done in order to save time and money. This could be achieved through the adoption of new standards of organisation of services, such as the so-called paperless office. The adoption of "green" models of organisation and operation of public services that are focused on reducing operating costs such as costs for paper, printing costs and electricity might be an important step during the crisis. For this reason, it is necessary to study all the processes of an organisation and decide which ones can be either combined or eliminated or automated or improved. Also the study on which of these can be converted into fully and exclusively electronic is particularly useful. All these can be achieved within the Business Process Reengineering (BPR) and the, so called, "administrative reform". It should however be taken into account that because of the existing institutional framework, changes and innovations in the public sector are promoted slowly or in any case slower than that in the private sector.

Moreover, one of the major factors for the further development of a new powerful back-office, is the preparation and training of human resources in the public sector. Our workforce is the one who will eventually recover or not the opportunities of technological solutions and decide whether to enter a new phase of work, attitude, perception and unmanning from the traditional way of working.

The questions asked by officials to address are:

- The adoption of a customer-oriented concept whose ultimate purpose will be to satisfy the citizens within every act.
- The continuous and focused training in new technologies and the ability to interact with other departments and people.
- Increased participation in the targeting of the vector to become partakers of the new perceptions of work and contribution to society.

However, in this difficult path employees should receive the active support of the state and the superiors within the bosom of the organisations that will motivate them in the way of reorganising the Public Sector. The service and the political leadership must act in this direction by applying specific policies such as:

- Focusing on how technology can influence and formulate strategies for the Public Sector.
- Using technology for the development of innovation and not simply to automate processes.
- Adopting the best practices from countries considered pioneers in e-Government.
- Improving the flow of financing to complete technological solutions.
- Protecting privacy and security.
- Configuring communities within the workplace, the members of which share common views on technology development and the economy.
- Using technology to create equal opportunities.
- Preparation for Digital Democracy.

Finally, an equally important success factor in developing an integrated back-office system is to secure funding. This is directly linked to the overall strategy of the government to finance policies to improve public administration. Nowadays, finding funds to support e governance is of fundamental importance for the future of the project. The main problems associated with this funding include the following:

- The difficulty of evaluating the benefits that can arise from investment, mainly economic-projects for the reorganization of the Public Administration.
- The inability to absorb EU funds concerning projects in Public Administration.
- The weakness of programming, forecasting and thus placing funds in the national budget of the country for the development of e-Government.
- The economic crisis of the last eight years.

References

1. European Commission. (2016). EU e-Government action plan 2016–2020, acceleration the digital transformation of government. *Communication from the commission to the European parliament, the council, the European economic and social committee and the committee of the regions.* https://eur-lex.europa.eu/legal-content/EN/TXT/PDF/?uri=CELEX:52016DC0179&from=EN
2. European Commission. (2006). i2010 e-Government action plan—Accelerating e-Government deployment in Europe for the benefit of all. *Communication of the commission of 25 April 2006.* https://eur-lex.europa.eu/legal-content/EN/TXT/?uri=LEGISSUM%3Al24226j
3. Osborne, D., & Gaebler, T. (1992). *Reinventing government: How the entrepreneurial spirit is transforming the public sector*. Addison-Wesley Publ. Co.
4. OECD. (2003). *The government imperative.* OECD. https://doi.org/10.1787/19901054
5. Osborn, D., & Gaebler, T. (1992). *Reinventing government.* Penguin Press. ISBN 0-452-26942-3.
6. Burn, J., & Robins, G. (2003). Moving towards e-Government: A case study of organisational change processes. *Logistics Information Management,* 25–35. https://doi.org/10.1108/09576050310453714
7. e-Business forum, Electronic Government and Electronic Services of the State to Businesses (2006). http://www.ebusinessforum.gr/

8. Koh, C., & Prybutok, V. (2003). The three ring model and development of an instrument for measuring dimensions of E-government functions. *Journal of Computer Information Systems, 43*(3).
9. Marche, S., & McNiven, J. D. (2003). e-Government and e-Governance: The future isn't what it used to be. *Canadian Journal of Administrative Sciences*, 74–86.
10. European Commission. (2001). Impact of the electronic economy on European businesses. *Communication of the Commission.* http://europa.eu/legislation_summaries/information_society/n26040_el.htm
11. Robu, N., Stoicu-Tivadar, V., Silea, I., Stoicu-Tivadar, L., Berian, D., Albu, A., & Vlasiu, G. (2005). Managing the development of a regional e-administration network with academic expertise. In *3rd International Conference on Computational Cybernetics* (pp. 319–324). https://doi.org/10.1109/ICCCYB.2005.1511594
12. Wang, H., & Hou, J. (2010). Main contributions of E-governance. In *International Conference on Computer Design and Applications.*
13. Kim, Y. S. (2006). Challenges and barriers in implementing e-Government: Investigation on NEIS of Korea, ICA0T. In *8th International Conference Advanced Communication Technology* (pp. 1635–1640).
14. Alam, S. B., Rafi Sazzad, A. B. M., Nazmus Sakib, Md., Kabir, H. M. D., Shahnaz, C., & Fattah, S. A. (2010). Manipulation and transparency control of ICT constituted e-Administrative protocol via digital watermarking for LDCs. In *2010 IEEE International Conference on Information Theory and Information Security.*
15. Attour-Oueslati, A., Dufresne, D., & Longhi, C. (2007). *The development of the local e-Administration: Empirical evidences from the French case.* Springer.
16. Rykowski, J. (2005). Using software agents to personalise access to e-Offices. In *Challenges of Expanding Internet: E-Commerce, E-Business, and E-Government, IFIP International Federation for Information Processing* (Vol. 189). Springer.https://doi.org/10.1007/0-387-29773-1_7
17. Willoughby, M., Gomez, H. G., & Angeles Fernandez Lozano, M. (2010). Making e-Government attractive. *Service Business, 4,* 49–62. https://doi.org/10.1007/s11628-009-0083-5
18. Ahsan Rajon, S. A., & Ali Zaman, Sk. (2008). Implementation of e-Governance: Only way to build a corruption-free Bangladesh. In *Proceedings of 11th International Conference on Computer and Information Technology.*
19. Official site of the Citizen Service Centre of the Greek Public Administration. www.kep.gov.gr
20. Ministry of Interior, Greece, Operational Program: Administrative Reform 2007–2013 (2007).

Chapter 3
Semantic Web: The Evolution of the Web and the Opportunities for the e-Government

Abstract This chapter presents the wide dissemination and use of Internet technologies, the large number of applications that have been developed at the level of Web 2.0 as well as its evolution to Web 3.0. It also presents the perspectives from the use of the technologies and tools of the Semantic Web. More specific references are made to the historical evolution of Web generations, the basic technologies and tools used in Web 2.0, the development levels and the basic technologies of the Semantic Web, while a comparative presentation of the three web generations is attempted. Particular emphasis is placed on applications that can be developed with Semantic Web technologies and tools, and the advantages and challenges for the Semantic Web are documented. Also, emphasis is placed on the expected opportunities for governments in the more specialised field of e-Government from the development and use of innovative—smart applications that take advantage of the technologies of the semantic web.

Keywords Semantic Web · Web generations

3.1 The Internet as the Foundation for Service Providing

The World Wide Web (WWW), commonly known as the "Web", today comprises the most widespread service of the Internet for the search and the exchange of information. It is used in almost all activity areas of businesses, public organisations, and ordinary citizens. Its rapid evolution and integration to the services of the Internet in conjunction with the range of the tools it provides has led many times to its identification with the Internet itself.

The Internet, which should not be confused with the Web, is the most modern technological foundation through which various independent internets have been connected, regardless of their geographic location or the technology through which they have been implemented. The Internet is defined as "the network of the internets". The basic principle for the function of the Internet is the possibility of communication between any two connected computer systems, which, other than usually being connected to different networks, also feature different technical characteristics. The

S. Theocharis and G. A. Tsihrintzis, *Semantic Knowledge Modelling via Open Linked Ontologies*, Artificial Intelligence-Enhanced Software and Systems Engineering 4,
https://doi.org/10.1007/978-3-031-20585-9_3

communication is based on common standards which follow the client–server model. According to it, the client-computer requests a service and the server-computer provides that service. After the request by the client-computer, the server-computer is activated and processes the request in order of priority, since it usually receives multiple requests each moment from various points and accordingly fulfils it or performs a series of other actions necessary for its fulfilment. The basic advantage of this system regards the client-computer, which doesn't require computing power in order to fulfil its various requests. It only requires a minimum computing power in order to project the results it will receive from the server-computer. By contrast, the server-computer which responds to the requests of the clients requires more significant computing capabilities and must be capable of storing, retrieving, and managing information. In short the benefits to the users of the Internet can be listed as such:

- Effective use of computing power.
- Reduction of the costs of maintenance and upgrades.
- Increased versatility.
- The possibility of creating systems that support various different environments.

All the basic services provided by the Internet, such as the World Wide Web (WWW), Email, File Transfer (FTP), discussion groups (newsgroups, forums) and the remote terminal access service (TELNET), operate based on the client–server model.

From another point of view, the Internet can be considered as more of a concept than a tangible entity. This is because the interconnection of systems is achieved through the use of common protocols, which are implemented at the hardware and software level. The basic idea of the Internet is that the computer must respect the Internet Protocol. This is done with the appropriate software installed on the computer that is required to connect to the Internet. Any type of system that respects the Internet Protocol can be connected with systems that do the same.

With existing technology, each computer is connected "hierarchically" to the Internet, that is, not "directly", but through its own network. This is usually either the Internet Service Provider (ISP) network which it utilises, or the corporate network which in turn connects to the ISP network (Fig. 3.1). The network consists not only of computers but also of connective devices such as hubs, routers, etc. Also, in addition to computers, other so-called "smart" devices such as mobile phones, tablets, televisions, printers, home appliances, etc. are now connected to the Internet.

The fundamental principle of the Internet is equal access and free communication of everyone with everyone. But this was not its original design. The primary idea of interconnecting/linked systems is recorded in the period of 1960–1970 in the form of a small closed military network by the US as a response to the programs of the Soviet Union. Thus ARPA (Advanced Research Project Agency) was created to develop defence technology for the US [1]. In 1966 began the design of ARPANET, the first broadband network. The goal was to create a network that could work reliably, even if some of its nodes were down. In the original form of this network, few computer nodes were interconnected in various parts of the United States, including universities through low-speed telephone infrastructure. In the next two years ARPANET

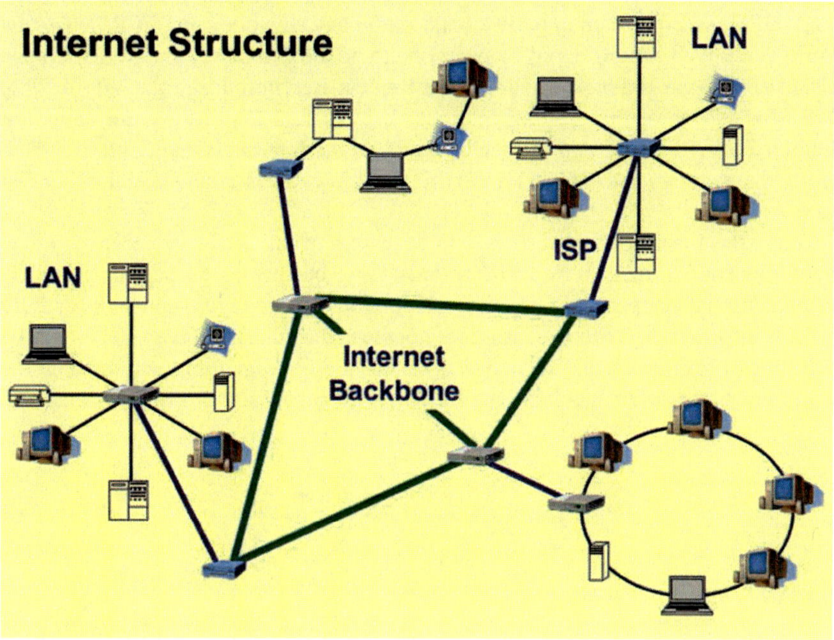

Fig. 3.1 The Internet structure

permitted the connection with nodes outside the US while other networks were created by universities and other organisations. Communication problems due to incompatibility of the functions of the developed networks were not long in coming. The solution was provided by the development and adoption of common communication protocols such as the TCP and IP protocols and their final form TCP/IP. With the maturation of the idea of interconnecting all networks in a single "great" global network, ARPANET is divided into two parts, one of which will remain open using the TCP/IP protocol and the other closed in order to meet military needs.

But an infrastructure is judged by its usability. The first substantial attempt to exploit the potential of the Internet was recorded in 1971 with the development of the basic network service, the creation of the first e-mail. Following that was a "boom" in the development of communication and information exchange standards along with the development of related services. One of the first such services developed during the first years of operation (1984) was the Domain Name System (DNS), a global system for identifying connected nodes based on their email address. Another important step in the development of the Internet was taken by the US National Science Foundation (NSF), which created the first online university backbone, NSFNet, in 1986. This was followed by the integration of other important networks, such as Usenet, Fidonet and Bitnet. The term Internet became widely used during the time

ARPANET was connected to NSFNet and meant any network using TCP/IP. But the World Wide Web is considered by everyone to be the definitive service that helped further spread the Internet. In 1989, Tim Berners-Lee introduced the World Wide Web as a service of the existing Internet [2, 3] (Fig. 3.2). At the same time, the shutdown of ARPANET and the actual launch of the Internet is taking place, which during the next two decades and up to this day has evolved at an exceptionally fast pace [4].

Today the Internet is open, free, and accessible to all while owned by no one. Many educational and governmental organisations, businesses and service providers own and/or manage parts of this infrastructure, but practically no one owns or fully controls the Internet. Of course, this does not mean that all uses of it are allowed. The spreading of fake news, dissemination or distribution of digital material without the permission of the author, publication of sensitive personal data and other criminal acts

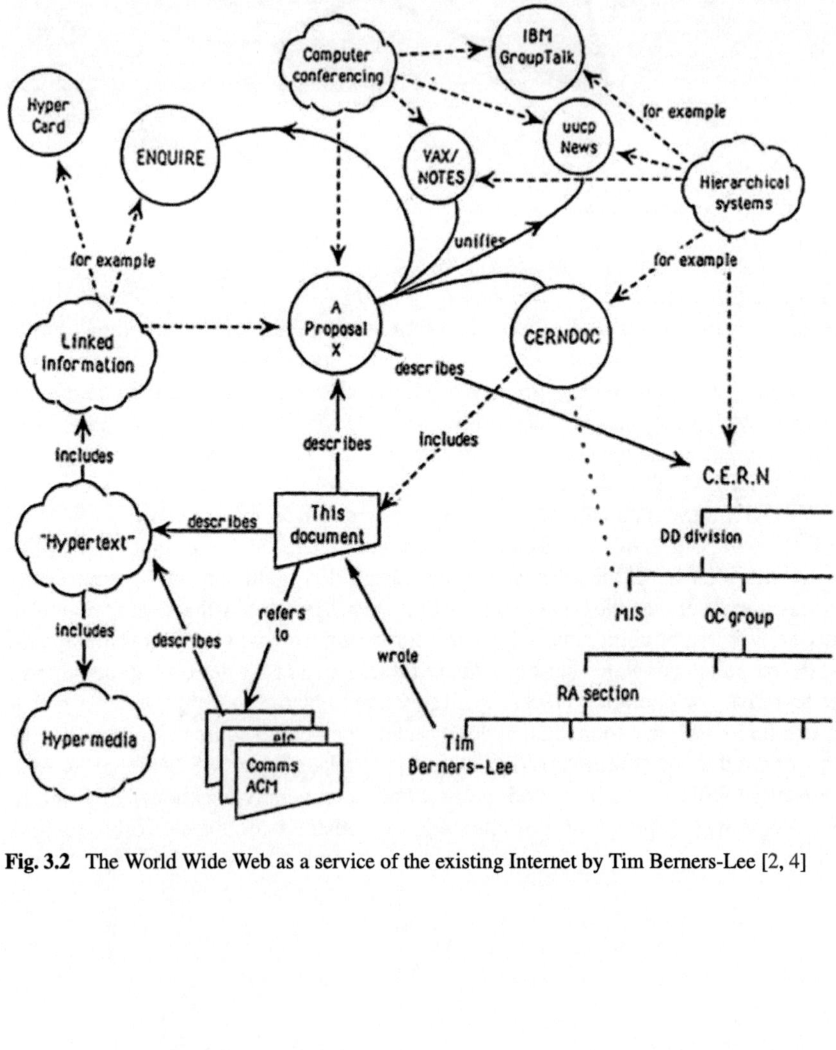

Fig. 3.2 The World Wide Web as a service of the existing Internet by Tim Berners-Lee [2, 4]

are prosecuted on a case-by-case basis by the responsible law enforcement authorities. For the organisation and proper operation of the Internet there are organisations that monitor and define the standards of communication and operation, and assign IP addresses and domain names. Some organisations associated with the development and operation of the Internet are as follows:

- IAB (Internet Architecture Board)
- IANA (Internet Assigned Numbers Authority)
- ICANN (Internet Corporation for Assigned Names and Numbers)
- IESG (Internet Engineering Steering Group)
- IETF (Internet Engineering Task Force)
- IRTF (Internet Research Task Force)
- ISOC (Internet Society)
- NANOG (North American Network Operators' Group)
- NRO (Number Resource Organisation)
- W3C (World Wide Web Consortium).

According to many, the success of the Internet is due to the multiple functions and services that are provided through it but also to its constant upgrades. In adjunction to the ever-growing ICTs, it has become an immense hub of knowledge and information but also a means of communication. Various forms of information can be searched and retrieved for the purpose of processing them for knowledge production, education or even entertainment. The information that is circulated and shared includes, in addition to administrative, financial and scientific elements, traditional art forms (such as film and music) that through digital technology take on digital form (data files). Therefore a concentration and sharing of knowledge concerning a more general digital culture that is directly related to the Internet can be observed. At the same time the possibilities provided by the Internet to achieve communication between users which can be synchronous or asynchronous should be mentioned.

One concern that emerges and troubles the global community regards issues of ownership, administration and management of the Internet but also a more important one related to the control, prefecture and possession of the information and knowledge that circulates through it. Every Internet user has the opportunity to share the information that he produces or owns with other users. There is no direct control over the information "uploaded" to the Internet by a hierarchically superior user or organization. The volume of information on the Internet nowadays is immense and tends to increase. Nevertheless, certain information can be easier or more difficult to access by the user. The Internet has made it possible to gather a large volume of information and has significantly affected the way it is made available, but not so much the way it is produced and managed.

3.2 The Web as an Internet Service—Historical and Technological References

As mentioned previously, the World Wide Web was originally developed as a tool for collaboration and exchange/sharing of information between High Energy Physicists—CERN [4]. From there it spread rapidly to other fields, and evolved into its present form. Its success and acceptance is on the one hand due to the ease of access and sharing of information and on the other hand due to the possibilities it provides for free communication between individuals via the Internet.

Technologically, the Web as an information system manages data and information based on a specific model utilising nodes, anchors and hyperlinks (Fig. 3.3). Website nodes are the carriers of the content while anchors are areas of the content of the nodes where links start or end. The connectors connect/link two nodes. Hyper-text is a text with links, but it can also be considered a way of organising files through which one can extract new information similar to the citation system of scientific articles or the citations of a dictionary. By using electronic files, these links can lead to other files and so on. The process of looking through from one page to the next is called "browsing" and is achieved through specialised software called "browsers". Websites are interconnected/linked through the technique of computer networks, thus comprising the WWW.

The WWW "conceals" details and information regarding the implementation of the websites from the regular user focusing instead on the display of information and links implemented through hyper-text. Behind the apparent simplicity, there is a scientific way of thought to determine the design, the "protocols" and the rules that must be followed for the development and exploitation of the World Wide Web.

During its first steps in order to encourage the adoption of the system, access was allowed to the existing information, without them having to be "compiled" to a special format. This was ensured through the connection to the CERN Computer Center as well as to the already well known Usenet "newsgroups". All this information became readily accessible through simple navigation software that could be run on all operating systems. This initial system, along with an information server and library, implemented the essential services for developers to develop their own software. The system was introduced in 1991 to the community of High Energy Physicists, through the CERN library program, so that all Universities and Research Laboratories could start using it [4]. It soon became fully available through the Internet, especially to the community of individuals working on Hyper-Text Systems. At the beginning of 1993 there were about 50 known information servers. At this stage, there were essentially only two types of navigation software. One was the original advanced version, but also complex and only available on NeXT systems. The other type was line-mode software, which was easy to install and run on any operating system, but was of limited power and user-friendliness. At the end of 1993, there were over 5,000 well-known servers and the WWW accounted for 1% of global Internet traffic (a percentage that was considered very high at the time!). By the end of 1994, there were already 10,000 servers (2,000 of which were commercial) and over 10,000 users. By 1997, the Web already had 650,000 servers and about 1,000 new ones were being introduced every day. In recent years the number of websites served by the web is constantly increasing while the overall picture is shown in the chart below (Fig. 3.4) [5].

Web technology is constantly evolving to meet new needs. The latest data, which have been introduced, regard issues of security, multimedia and the purchase of goods. The key to Web development is the effort to keep it open to all users and avoiding any privatisation of the entire system. This is the main purpose of the International World-Wide Web Consortium (W3C), a body made up of various institutes and companies [6].

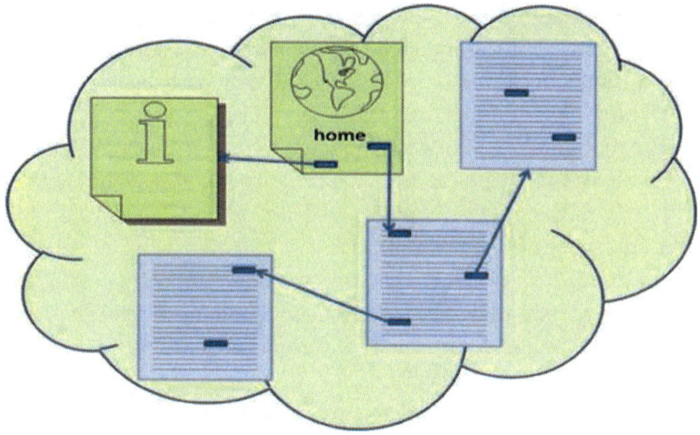

Fig. 3.3 The model with nodes, anchors and Hyperlinks used for WWW

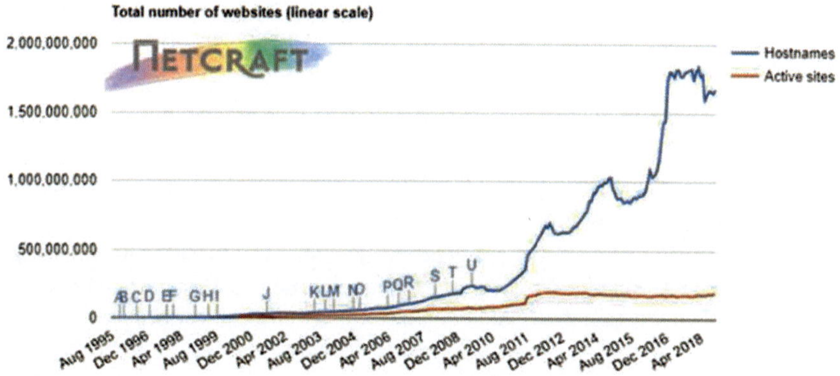

Fig. 3.4 The evolution of the number of websites served by the web over time

3.3 From Web to Web 2.0

Nowadays, the cycle of the first generation of the World Wide Web, the so-called—a posteriori—Web 1.0, has already been completed, while the second generation, Web 2.0, has already been fully developed and there is now talk of its further evolution. In its initial implementation, and with the main purpose being informing the public and easy indexing, the Web was static. Its design was intended to provide a fast and easy way to share the results of experiments conducted by research teams at CERN. For this reason, HTTP (HyperText Transfer Protocol) and HTML (Hyper Text Markup Language) were adopted. Following that, new programming languages such as HTML and JavaScript were developed, as well as tools for creating websites and applications for reading them, the now well-known browsers. The idea quickly spread outside the academic community, and resulted in the exchange and sharing of hundreds of thousands of pieces of information. Every party interested in publicity, such as Universities, corporations, and public organisations, created and published a website, to make their presence known. Creating web pages through HTML, however innovative, was also a tool available to few. The common user would have to retrieve information from ready-made web pages or learn to code in HTML and buy the corresponding equipment, if they wanted to set up their own website. Additionally, the web pages were static, that is, once they were published they were not updated, except perhaps through some announcement section and again through HTML coding. The Internet user was a passive recipient of information who could not interact and influence the content of the websites.

The original "static" state changed around 2004 with the adoption of a new Web Service logic by Tim O'Reilly which was presented as the new generation of Web, Web 2.0—or "interactive Web". The essential difference does not only concern the new technologies that have been gradually integrated but also the evolution of the original Web as a means of communication between users.

During this period the performance of personal computers increases significantly in conjunction to affordable prices for both simple consumers and businesses. Along with the development of programming languages and corresponding software tools, an opportunity was presented for the upgrading of existing websites and also for the development of new innovative services such as e-shops and e-banking.

The term Web 2.0 implies that there has been a technical evolution of the World Wide Web. In reality, however, the changes mainly concern the way websites are constructed and used, but also the emergence of innovative electronic communication methods using new programming languages and technologies. For many, Web 2.0 is not a new version but essentially rather an advertising effort to further promote the original Web [7]. However, a Web 2.0 website displays significant differences and innovations compared to those of Web 1.0. For example, it allows its users to interact and collaborate within a virtual community (a virtual social network) creating content themselves. In contrast, a user of a web page of the original World Wide Web or Web 1.0—as it is subsequently referred to—simply "passively" reads its content without

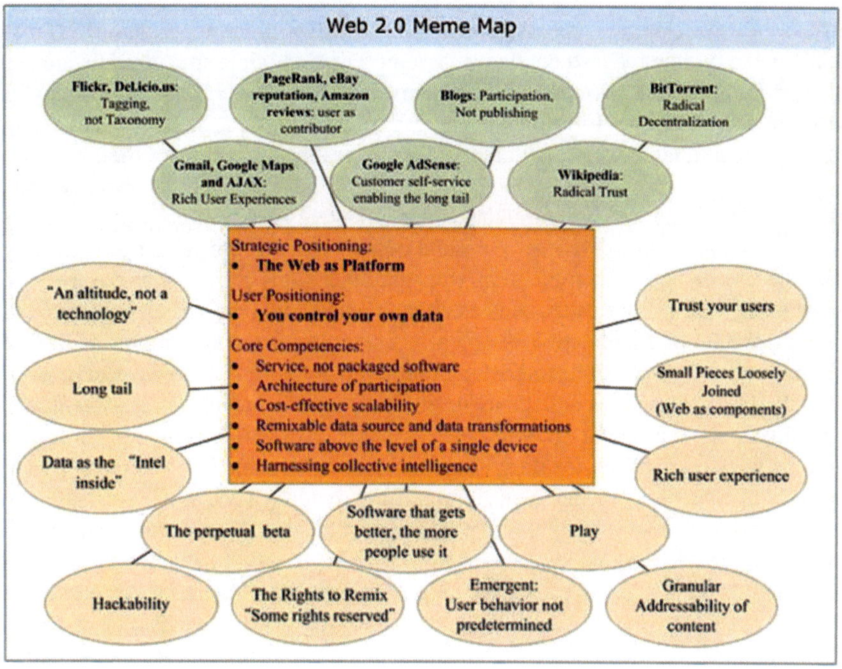

Fig. 3.5 Web 2.0 mnemonic map [7]

being able to edit it. A Web 2.0 site encourages the user to interact, leave comments, sign up, create an account or profile, and upload content themselves (Fig. 3.5).

Of particular scientific interest are two new trends which, in our view, could complement each other. We are referring to the evolution—extension of Web 2.0 to the Semantic Web or otherwise known as Web 3.0 but also to the possibilities that the now advanced Internet of Things can offer. In this work we focus on web 3.0 and the technologies that support it and in particular the opportunities that can emerge for the benefit of Public Administration and e-Government.

3.4 The Road to Web 3.0 and Web X.0

Nowadays, while the predominance of Web 2.0 in the field of information and communications has become indisputable, there is already talk of its evolution and expansion into the next version—generation. The current design of the Internet is anthropocentric, in the sense that most of the content on the Web is designed to be perceived and managed by humans. In the early days of the Web, HTML was used with great success to develop web pages, as the basic means of utilising its potential. With the development of ICT and the emergence of new devices such as tablets, smart phones etc. It was deemed necessary to extend HTML with the adoption of XHTML and HTML5, which focused on the visual representation of the structured document adapted to the device. Nowadays, computers can efficiently analyse web pages in terms of appearance and routines using traditional browsers, but in general, they do not have a reliable way to automatically process the semantics of their content. However, there is a specific trend associated with the attempt to attribute semantics in electronic documents through semantic elements—tags already provided via HTML5 on the existing Web. At the same time, the notation/marking of the pages with the use of tags in combination with the other tools provided at the level of the existing World Wide Web, such as sharing, evaluation, authoring, etc., support the cooperation and interaction between its users. Instead of relying on dedicated editors who are responsible for adding and commenting on content, user knowledge sharing communities collectively contribute to web content. These systems are evolving rapidly but are not as coordinated as they used to be when teams of dedicated creators were responsible for designing and creating content. For this reason, new methods of content organisation and structure are required, such as the attribution of semantics to the content and not to the structure of the documents with the help of a corresponding labelling system.

In particular, tagging means assigning keywords as metadata to resources to describe or evaluate the content of those resources. Keywords are tags. For example, a user might apply the tag "websphere" to a page that provides information about "IBM WebSphere" products. Tagged content enables users to better describe and rank relevant resources. Adding tags to content makes it easier for users to find and emphasise the importance or quality of content, which in turn benefits other users in the knowledge sharing community. The tag provides better search results, as the search criteria are descriptive and users do not have to scan the content to match them to the keywords.

A user usually assigns more than one tag to a single resource. For example, a user might associate a book with the tags: "portal", "Web2.0", and "computer_science". This allows for many ways to find the book, like placing multiple copies of the same book on different shelves. Tagging has become one of the most popular techniques that allows users and entire user communities to sort, organise, and structure content autonomously. By adding tags to the content, users add valuable compilation information and even light semantics to the web content. Tagging allows non-expert users to develop folksonomies that sort the content available on the system.

Also relevant to tags are tag clouds that help users find and retrieve resources by displaying aggregate views of tags that users have made on specific resources. Tag clouds provide different visual representations of all the tags that users have made. For example, depending on the configuration settings, tag clouds can display all available tags or the frequency of display of specific tags. Tag clouds usually emphasise the popularity or importance of specific tags by increasing the size of the tags or changing the colour of the tags.

The notion that these new elements/tags were created to make sense of electronic documents or blog posts does not express exactly what is implied. This is because the "meaning of the concept" and, above all, who the recipient is, are not fully defined. The fact that the "address" field that we find in a shape/model is understood by humans does not mean that it is perceived in the same way by software agents and that the equivalent expression "location" can be used in another shape or model is automatically assigned. This effort of HTML5 purely regards the organisation and structure of the document and not the attributing of the semantics of its content.

If the Web were focused on the meaning of data it would be easier to search for information. If the search engine understands what the user is really looking for, instead of returning the most appropriate keyword matches to the given phrase, the results would be more relevant. For example, in the question "*what is the procedure for filing a complaint in case of official corruption in Greece*" instead of returning websites that refer to services of the European Commission or the General Secretariat for Anti-Corruption or Protection of Whistleblowers the expected response would be the reference to websites related to the submission of specific elements/data or information in a relevant form or the reference of appropriate instructions (Fig. 3.6)

This is just one of the many possible uses for semantic tagging/annotating. Perhaps the concept of semantic tagging will be clearer if we apply it in another context—perhaps in an online library. Let's say that a public organisation like the National Printing Office had a collection of numerous different types of documents such as Laws, Presidential Decrees, Ministerial Decisions, Circulars, other legal texts, etc. The semantic tag/annotation could help both visitors and filing staff to find information faster and the results to be more relevant to the search if there was a proper tagging on the type of document, the signer or the responsible body.

Through this technology the implementation of a new idea is being sought, a Semantic Web with semantic tagging. The basic concept revolves around the ability of machines to understand humans automatically. In this context, the goal of the Semantic Web is precisely to provide the means for the description of the meaning of the data to be understood by the machine. However, the Semantic Web is not really about the Internet and HTML, but rather a new process of managing data and specifically describing the meaning of data in a formal and clear way. These tools as well as their use in the context of e-Government will be elaborated on in the following chapters.

The next generation of Web, also referred to as Web 3.0 or Semantic Web, is expected to be a smart Web that will integrate advanced Artificial Intelligence technologies into existing services that will no longer focus on documents but on information. The goal is for Web 3.0, as an extension of the second generation of the Web, to do all the work for the benefit of the user. For example, after having already collected data about the user's habits, it will be able to project what is best for him before he even asks for it. It will implement in an automated way the personalised provision of digital services. The Semantic Web focuses on the meaning of the content. As a result of the developments in Artificial Intelligence, the search will now be done with

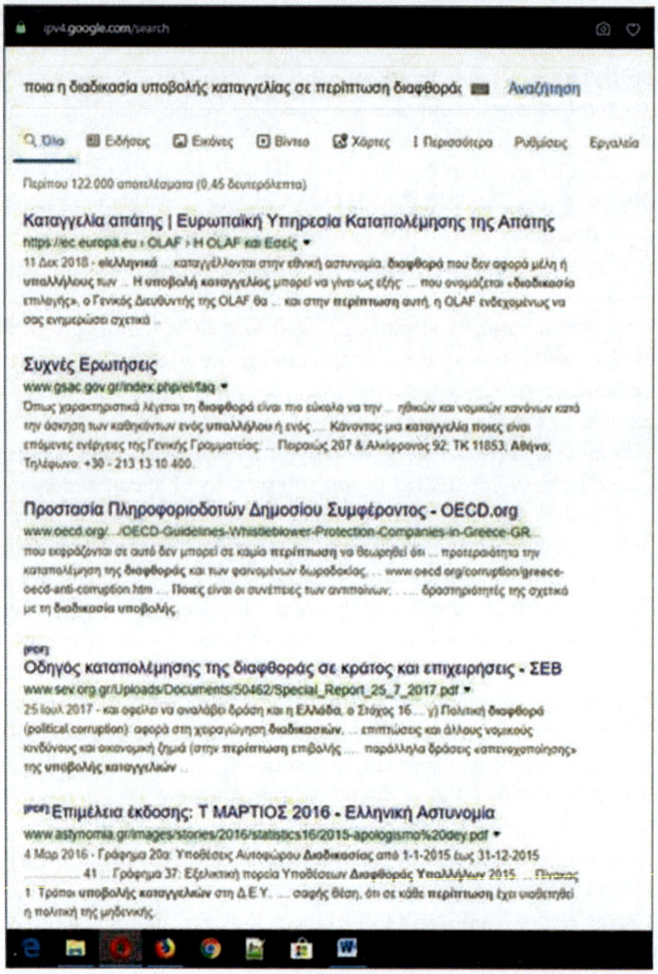

Fig. 3.6 Results from submitting a specific query to a popular search engine

regular expressions and sentences instead of keywords. Web 3.0 attempts to associate meanings and concepts with corresponding entities/documents, thus turning them into information recognizable by both humans and machines.

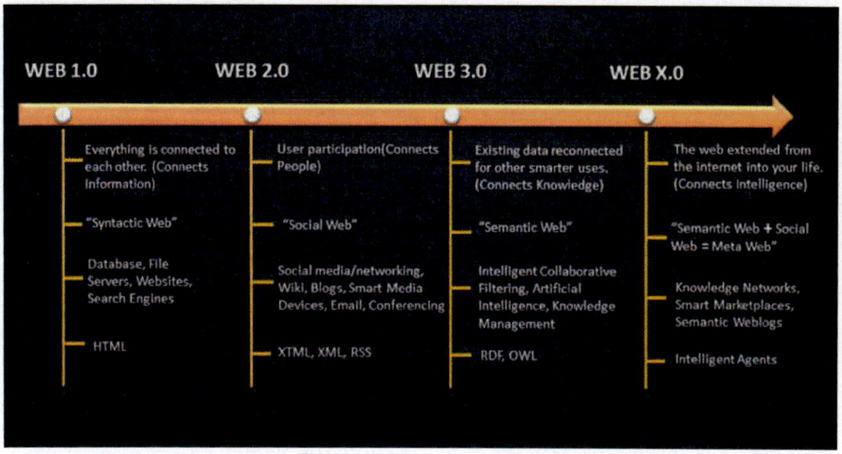

Fig. 3.7 From Web 1.0 towards Web X.0 [8]

The next step is expected to be the so-called Web X.0 (eXtended Web) as a final synthesis of existing Internet technologies. It includes and integrates the current Web, the Semantic Web, and advanced means of transmitting information such as 3D and virtual reality. The Web is now expanding beyond the Internet and into everyday life. Schematically, if the original World Wide Web (Web 1.0) connects data and information, Web 2.0 connects people to each other, Web 3.0 connects types of knowledge, and Web X.0 connects types of intelligence (Figs. 3.7 and 3.8).

3.5 Web 2.0 Versus Web 3.0: A Comparative Analysis

3.5.1 Web 2.0 as a Collaborative Interactive Platform

The term Web 2.0, as mentioned above, is used to describe the commonly accepted "second generation" of World Wide Web Services and applications that has been around since 1989. The introduction of the term is mainly related to the current view of the Internet as a service in contrast to the initially prevailing view of the Internet as a software application. Web 2.0 technologies expand Web Services with an emphasis on user interaction by supporting collaboration and content sharing. These are technologies that focus on the user, who can now intervene and create web content through interactive applications. Social media (e.g. YouTube, flickr etc.), social networking platforms (Facebook, Twitter, LinkedIn etc.) wikis, and blogs are considered among the most characteristic and popular applications of Web 2.0. Through specialised tools, such as search, tagging, citation links or authoring, users

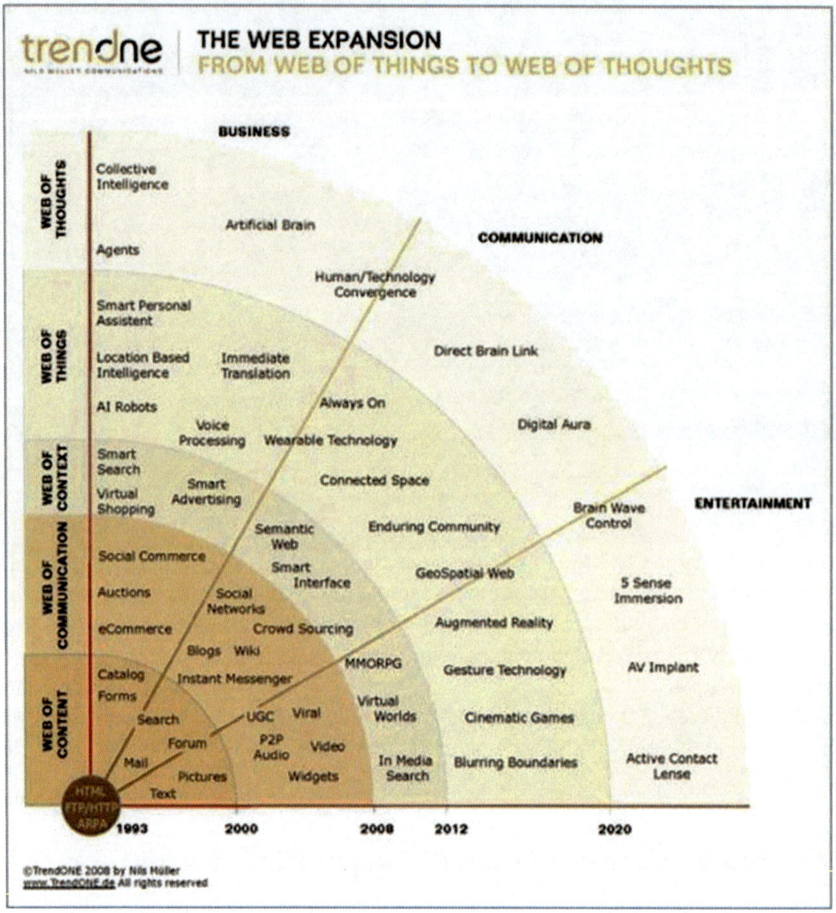

Fig. 3.8 The Web expansion [9]

can create digital content (podcasting), update it whenever needed, share or delete the content they have created. These applications strengthen the role of the user in the creation and circulation of information.

Web 2.0 applications serve basic user needs including:

- Interactivity between user and application.
- Dynamic web content.
- Collaboration between the users.
- Contribution to the members of the communities.
- Participation through community development and social networking.

Web 2.0 has created the conditions for new service delivery practices in almost all areas of interest to businesses, government agencies and the average user:

- In the social sector, it provided the technologies to develop and establish in a minimum amount of time social networking through social platforms such as Facebook, twitter, etc. along with the publication of cultural content from social media such as YouTube.
- We consider the effect of Web 2.0 on education to be catalytic. Tools such as wikis and distance learning platforms, virtual rooms, applications for collaborative text writing, etc. are considered innovative and have been adopted and used for educational purposes in almost all scientific areas and levels as far as on professional levels.
- Informing the public has changed significantly since the time of Web 1.0. Nowadays, the information is achieved in real time mainly through Really Simple Syndication (RSS) channels. These channels provide an economical solution for news producers since the printing of news is no longer required while at the same time eliminating the time of broadcasting. The audience which the news reaches has expanded both spatially and in terms of other qualities such as social, living, and education level.

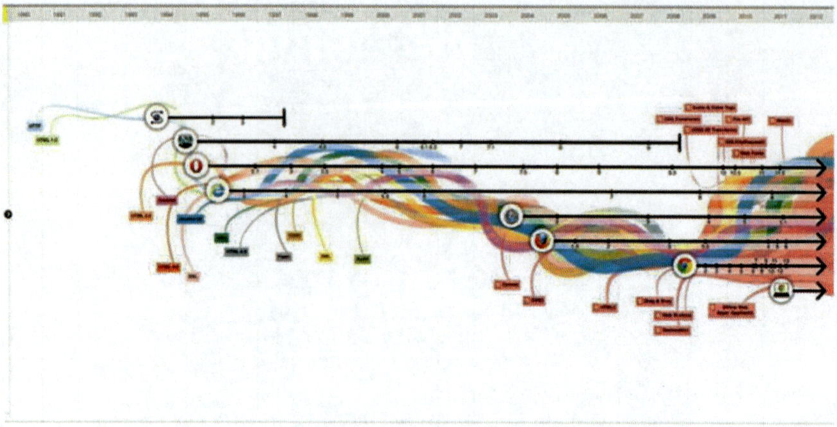

Fig. 3.9 The evolution and the interaction between web technologies and browsers over time [10]

- In the field of governance, as explained in the previous chapter, Web 2.0 was one of the pillars for the completion of e-Government and the provision of innovative public services to the public.
- Applications supporting e-commerce, e-procurement for the public sector and e-banking are directly linked to the introduction of Web 2.0.
- In the field of managing information that accumulates through search engines and the utilisation of collective intelligence they are considered to be the essential advantage of this new Web.
- The users become creators and shape the content: text, image, audio, video, they categorise it, evaluate it, rank it, for example news they consider important. With the new network services, users collaborate and exchange data online.

Over time web technologies have evolved to give web developers the ability to create new generations of useful and immersive web experiences. Today's Web is a result of the ongoing efforts of an open Web community that helps define these web technologies, like HTML5, CSS3 and WebGL and ensure that they're supported in all web browsers [10]. The colour bands in the visualisation (Fig. 3.9) represent the interaction between web technologies and browsers, which brings to life the many powerful web apps that we use daily.

3.5.2 *Web 2.0 Technologies*

3.5.2.1 Web Services

The establishment and spread of Web 2.0 was primarily linked to the upgrading of the functions of the Internet beyond mere machine interconnection and exchange of information. This upgrade included the development and use of innovative web-based applications known to users as "Web Services". Web Services have become a new way of creating applications, providing access to remote functions that are distributed over the Internet. They are now a widespread model of software application development that utilises the capabilities of the World Wide Web.

Web Services cover the need of applications hosted on remote systems to interact with each other over a network. This model allows the publication of business functions on the Internet and universal/open access to these functions. The Web Services model simplifies application development and interoperability while allowing users to create their own Web Services or discover other services through browsers.

Highly defined Web Services are any part of software or application component that is available over the Internet and that can perform tasks and interact with the user. In particular, Web Services are a technology that allows applications to communicate freely between each other regardless of platform and programming language [11]. A Web Service is a software interface that describes a collection of functions that can be accessed over the network through standard XML messages. An interactive Web Service group defines a Web Service application. Also, according to the W3C [12], a Web Service is a software system designed to support interoperable machine-to-machine interaction over a network. It has an interface described in machinable format (specifically WSDL). Other systems interact with the Internet service in the manner specified by its description using SOAP messages, which are typically transmitted using HTTP with XML serial analysis in conjunction with other Web-related standards. In other words, a Web Service can be thought of as software that knows how to interact over the Internet with other software that may have been developed with different technologies.

Among the functions the Web Services perform are the following:

- They report and describe their presence, i.e. they describe themselves using common protocols in order to enable other services to understand the functions they perform
- They are perceived by the other services through their reference in appropriate directories
- Combine information and functionality from multiple Web Services to run a business process.

The basic functions can be organised into three categories depending on who performs them. At the same time, three distinct roles can be considered, meaning three "types" of entities that undertake the execution of a specific type of action. These roles are: the service provider, the client-requester and the service broker/register (Fig. 3.10).

A Service Provider is the entity that provides specific software applications in the form of services. Service providers have the power to publish, revoke, and renew their Internet services. They are responsible for the selection of applications that will be provided to the users of this service, the level of security and accessibility as well as pricing. The responsibility of the Provider that the specific service is visible to all Internet users is considered important. For this reason, its cooperation with another

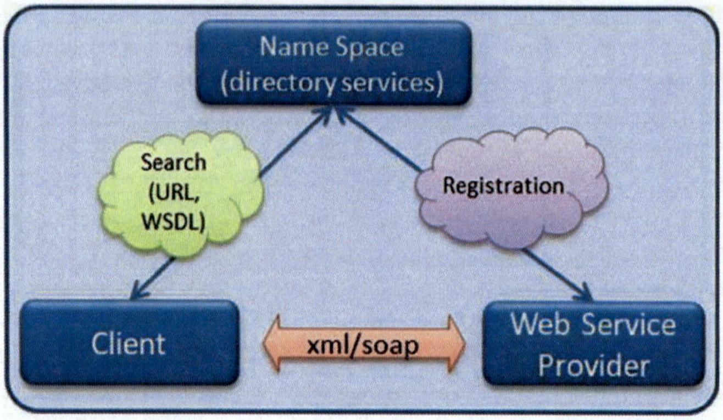

Fig. 3.10 Web Services components

building unit, the Directory Service, is required. The Service Provider describes its service and provides some access information to the service directory. At the same time it categorises the service in the directory and defines what kind of agreements are required to be able to access the specific function. From a business point of view, the provider of a service can be considered as the owner of the services while from a technical point of view it is the platform used to implement the applications in the form of services.

The Client—requester of a service is the entity that has a functional need and which it tries to satisfy by searching for the appropriate service on the Internet. From a business point of view it can be a company that wants to outsource a need to another company while from an architectural point of view it could be an application that is simply looking for another application to use. The requester can be anyone, from a user accessing a service through his computer, to an application or another service. The requester searches for the desired services through the service intermediary and then commits them to the provider.

The Service Broker (Name Space or Directory Service) acts as a registry of services. It is an entity that provides a directory—a repository of service descriptions and can accept and answer questions about the existence of services that meet specific specifications. The providers describe their services in this directory with the help of metadata while the applicants look for the services they want there and retrieve both their description and the necessary information in order to be able to secure it from its provider. In essence, it is the connecting link between the two previous entities. Typical examples of such directories are the Universal Description Discovery Integration (UDDI) [13].

A. *Stack of technologies-standards of Web Services*

Internet service providers, applicants, and directory services interact based on the use of common standards and technologies for service description, communication, and data engagement. The use of templates allows developers to create services independently of the platform and implementation language. The main standards related to services are: Simple Object Access Protocol for communication between services, XML language for uniform definition, representation and data exchange, Web Services Definition Language for service description and Universal Discovery Description and Integration for registration of service descriptions in indexes so that users can easily discover them (Figs. 3.11 and 3.12).

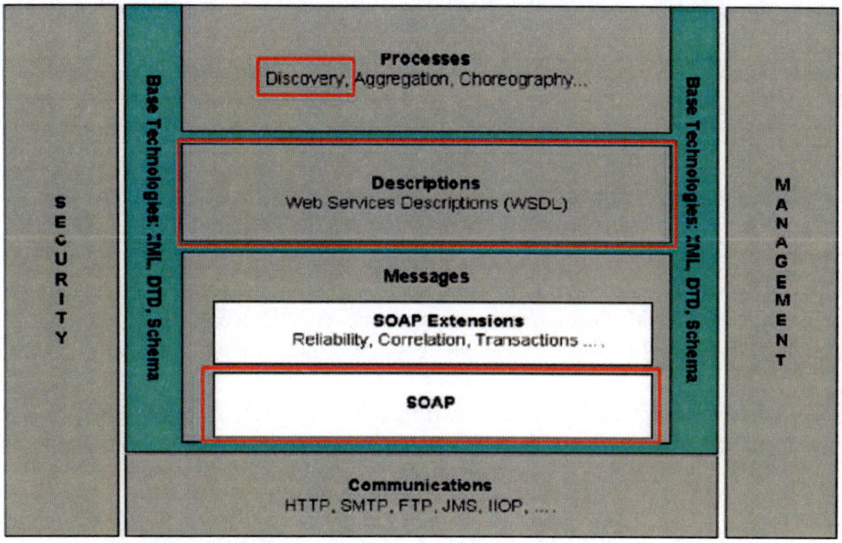

Fig. 3.11 Stack of technologies-standards of Web Services [14]

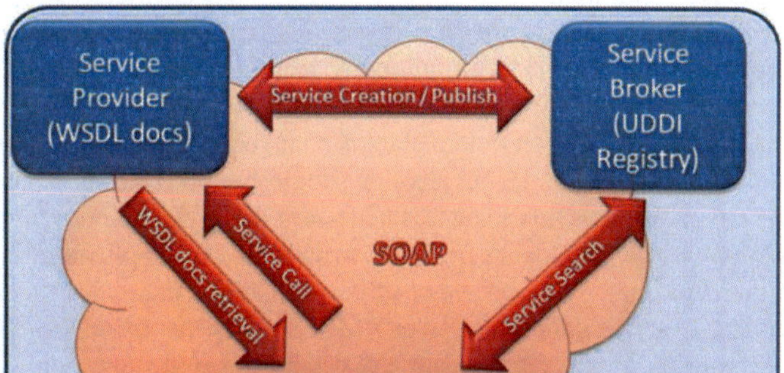

Fig. 3.12 The interactions of the basic technologies/components of Web Services

- Simple Object Access Protocol (SOAP) is the method by which one can send messages to different software modules. It's similar to the way communication with a search engine works. It is the channel used for communication between an application-Web Services provider and an application-client. Although called a protocol, the simplicity of SOAP is that it does not specify any new transfer protocol. Instead, it reuses, among other things, Hyper Text Transfer Protocol (HTTP) or Simple Mail Transfer Protocol (SMTP) for transferring data as messages. This use of HTTP or SMTP as transport protocols ensures that server applications can communicate with client applications using the Internet as a base. It is the use of SOAP that enhances the capabilities of services.
- Extended Markup Language (XML) is a great technology for transferring structured data over the Web. Data is for XML what text is for HTML. When Web Services are responsible for handling data in a reliable and automated way, HTML documents are not appropriate. By providing data in XML, Web Services can process this data in numerous useful ways. The separation of content and presentation of XML information is ideal.
- Web Services Definition Language (WSDL) is how different services are described in UDDI. This corresponds to the actual search engine. The following figure shows the hierarchy of technologies used to create, publish and find Web Services.
- Universal Description, Discovery and Integration (UDDI) is the universal search base for locating services. It is analogous to the indexing service of a search engine, in which all web pages are listed and related to keywords.

B. *Advantages of Web Services*

The advantages of developing and using Web Services include the following.

- *Interoperability.* The main problem that is usually encountered in distributed technologies that are developed in different logic, concerns the interaction of applications. An application calling another remote one requires strict connection between them by passing the appropriate parameters to the processes. In the event that the definition of an object class is changed or a procedure is modified, the application that acts as an accelerator should be modified to be compatible with the new situations. Obviously it is necessary to provide relevant instructions and directions to the developers from the providers of the application in order to achieve the necessary modifications to the applications that call it. Conversely, Web Services describe data typically in XML in a way that is independent of the implementation platform. This ensures that the interaction between a provider and a client-requester is completely independent of platform and language. This interaction requires a document in WSDL to define the interface and describe the service in conjunction with a network protocol, usually HTTP. Based on this interaction, there is talk about loosely coupled applications.
- *Encapsulation.* Requesters and service providers are essentially dealing with the interfaces that exist for their interaction. Thus, the requester does not need to know how the provider implements its service and vice versa. This element of

encapsulation, that is, the perception of services as if they were "black boxes" with a specific behaviour, greatly facilitates the work of developers who no longer deal with the details of the applications they request. Thus the complexity of service development is significantly reduced.

- *Simple communication protocol.* The communication protocol used by Web Services is SOAP which is simpler than the older technology protocols used by distributed environments.
- *Easy communication.* With previous technologies, collaboration between different companies was more demanding as distributed technologies used non-standard ports to communicate with each other. This required adjustments by opening holes in the firewalls, which was often not accepted by the security rules set by a company, as there was a risk to the security of its systems. This did not allow for dynamic collaboration due to the fact that it required a manual process for a company to communicate with its partners. Since Internet services mainly use HTTP as a transport protocol and most firewalls allow access through port 80 as a standard gateway for HTTP, this leads to easier and more dynamic collaboration between systems.
- *Easier and faster application creation.* Businesses and organisations that use Web Services can create new services without the investment and delays required by traditional models. This is mainly due to the ability to create new services by reusing and combining existing applications.
- *Application independence.* Previous collaborative applications were characterised by a lack of flexibility as a result of both the tight connection between the applications and the static commitment between them during their design. In this way any change in one of the "sub-applications" could cause the overall application to collapse and need to be reprogrammed to become compatible with the new data. The Internet services model supports the disconnection between services, and the "on-site" development of applications and services by dynamically discovering and composing services available on the Internet. Loose connection and dynamic engagement help a service that uses another service to meet a need to replace it at any time with another that best suits it.

C. *Web Services applications*

Web Services are being developed to perform a variety of functions from simple requests to complex business processes. The first Web Services were intended to perform a basic IT service which one can very easily integrate in their applications, such as current stock prices, weather forecasting, public transport timetables, etc. Since most of this information is already available on the World Wide Web, Web Services are expected to make access to it easier and more reliable and, above all, more automated.

Publishing existing applications as Web Services will allow users to build new, more powerful applications that will use the original ones as building blocks. Many examples can be given of applications already provided as Web Services in all areas that use the Internet as a means to achieve their business goals. For example, public organisations in Greece can already access the central procurement application and be informed about the current prices offered for public bodies, supplier profiles, etc. The central database is updated by the relevant central procurement service while through the Web Services the users of this service can be informed in real time. Also, a service is already provided to various public organisations, for example the Citizens' Registry that operates under the supervision of the Ministry of Interior and provides specialised access to bodies such as the Ministry of Finance, the Ministry of Education, access through Web Services to central databases where each body has classified access according to its field of activity and their respective responsibilities.

3.5.2.2 Service-Oriented Architecture

Every service, in order to function efficiently, to be maintained, and to evolve, must be based on a specific structure. It is also advisable to follow a specific, commonly accepted model on which all the functions and protocols of the respective service will be developed. The model that has prevailed and is used to date for the development of Web Services is SOA-Service Oriented Architecture. In this model there are separate building blocks, which can be distributed over a network, combined and reused to create business applications. The building blocks include the provider, the directory service and the requester.

SOA is based on three key technologies: WSDL, SOAP (a protocol that allows messaging in XML) and UDDI (a protocol that allows services to be published and retrieved). What makes this model stand out is the ability to discover services that match the requirements of an application, the ability to negotiate their usage contracts and their accessibility from anywhere and at any time.

An SOA service has the following features:

- Manages business processes such as calculating a bid, delivering email, managing a technical issue such as accessing a database.
- It may have access to another service. With the right runtime technology, it can access a traditional program and respond to different types of applicants, as with Web Services.
- It is relatively independent of other software or programming languages. Changes to the requester require little or no changes to the service. Changes in the internal logic of a service require little or no changes to the applicant. The relative independence of the service and other software is called loose pairing.
- It can handle interactions within the organisation that provides it but also between the organisation and any suppliers, partners and customers.
- Standard service contract: Defined through one or more service description documents.
- Subtraction: A service is fully defined by service contracts and description documents. They conceal their logic, which is embedded in their application.
- Reusability: Designed as components, services can be reused more efficiently, thus reducing development time and associated costs.

- Autonomy: The services have the control of the logic they incorporate and from the point of view of the service consumer there is no reason to know their application.

The advantages of adopting SOA as well as the case-by-case support infrastructure for the development of respective services are shown in Table 3.1.

Table 3.1 Table: SOA features, benefits, and infrastructure [15]

Feature	Benefits	Supporting infrastructure
Service	Improved information flow Ability to expose internal functionality Organisational flexibility	
Service re-use	Lower software development and management costs	Service repository
Messaging	Configuration flexibility	Messaging program
Message monitoring	Business intelligence Performance measurement Security attack detection	Activity monitor
Message control	Application of management policy Application of security policy	PDPs and PEPs
Message transformation	Data translation	Data translator
Message security	Data confidentiality and integrity	Encryption engine
Complex event processing	Simplification of software structure Ability to adapt quickly to different external environments Improved manageability and security	Event processor
Service composition	Ability to develop new function combinations rapidly	Composition engine
Service discovery	Ability to optimise performance, functionality, and cost Easier introduction of system upgrades	Service registry
Asset wrapping	Ability to integrate existing assets	
Virtualization	Improved reliability Ability to scale operations to meet different demand levels	
Model-driven implementation	Ability to develop new functions rapidly	Model-implementation environment

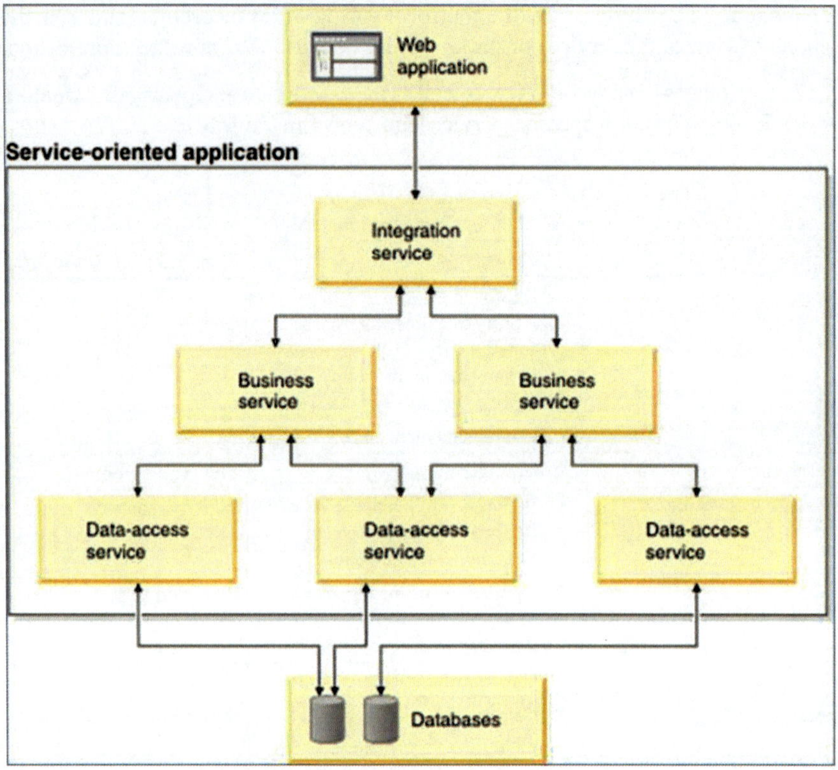

Fig. 3.13 SOA basic architecture [15]

Basic architecture

A service-oriented application is an application that consists primarily of services that are often structured hierarchically (Fig. 3.13).

The top level includes one or more integration services, each of which controls a flow of activities, such as the processing of a user request. Each integration service invokes one or more business services.

The second level consists of services that perform a relatively low level business project. For example, an integration service may invoke a number of business services to verify the details provided by an agent of an organisation. Answering these services may trigger the call of other services or the answer to the user.

The third level consists of data access services, each of which handles the technical task of reading and recording to data storage sites, such as databases and message queues. A data access service is commonly used by the business level, but easy access to the services allows for different uses. For example, a requester for such a web application may have access to a data access service to assign initial values to a form.

The focal point is flexibility. Some integration services provide different features to different applicants, and some cite other integration services. In addition, a requester may have access to different types of services from a service-oriented application. The requester may have access to a consolidation service at one point and a business service at another.

3.5.2.3 HTML (Hypertext Markup Language)

HTML is not a programming language but rather a markup language on which the World Wide Web was originally based. IBM initially used it to achieve the desired display of text on its various computer systems. It was first reported in 1991 by the W3C through a file entitled “HTML tags” [16]. This document described the 20 elements that made up the original and relatively simple design of HTML. Aside from the hyperlink tag, the rest were heavily influenced by SGMLguid, an SGML-based documentation creation format created at CERN. Berners-Lee saw HTML as an implementation of SGML. Thirteen of those original elements still exist today in HTML 4 [17]. This was formally defined by the Internet Engineering Task Force (IETF) with the publication of the first proposal for an HTML specification in mid-1993 [18], which included a Document Type Definition (DTD) of SGML, which defined the relevant grammar. This draft expired after six months, but it contains something remarkable: the recognition of the NCSA Mosaic tag for embedding images within text, which reflects the IETF’s philosophy of successfully incorporating prototypes into standards [19]. Something similar was contained in Dave Raggett’s competing draft, “HTML+ (Hypertext Markup Format)” from his end, which suggested the standardisation of some already implemented features, such as tables and forms [20]. Following the completion of the draft HTML and HTML+, in early 1994, the IETF set up the HTML Working Group, which in 1995 completed HTML 2.0, with the intention of being the first specification on which future implementations would be based on. HTML 2.0 was published as RFC 1866 [21], and contained ideas from the draft HTML and HTML+. The number 2.0 was simply intended to distinguish the new version from the previous drafts. Further development under the supervision of the IETF was delayed due to conflicting views. HTML standards have been complied with since 1996, with the consent of software developers, by the World Wide Web Consortium (W3C) [22]. However, in 2000 HTML also became a global standard (ISO/IEC 15445: 2000). The latest HTML specification, HTML 4.01, was published by the W3C in 1999, and in 2001 its errors and omissions were also published.

HTML is used to describe both static and dynamic web pages. Its key element is the links it can create, i.e. specific sections of the page content which when selected by the user lead to another part of the document or to another document that may be on another computer of the Internet. It also uses a number of tags to format text. HTML tags work in pairs (for example <h1> and </h1>), with the first being called the start tag and the second being the closing tag. Between tags, web designers can place text, images, tables, etc. HTML provides means for creating structured documents (that is, documents consisting of the content they convey and the content formatting code) by defining structurally important elements for the text, such as headers, paragraphs, lists, links, quotes, and more. Command scripts can also be embedded in languages such as JavaScript, which influence the behaviour of HTML web pages and turn them interactive from static. Another key feature of the language is that an HTML file can be created either with the use of specialised tools or with the use of basic copywriters provided by all operating systems.

Websites are "read" with the help of special applications called browsers. The purpose of a web browser is to read HTML documents and compose them into pages that can be read or listened to. The browser does not display HTML tags, but uses them to interpret the content of the page. When a browser opens an HTML file the various elements are translated into appropriate features resulting in the desired appearance and functionality of the specific page. Web browsers can also refer to CSS formatting styles to define the appearance and layout of text and other material. W3C, which creates and maintains HTML and CSS templates, encourages the use of CSS instead of multiple HTML elements for content presentation purposes.

3.5.2.4 HTML 5 (Hypertext Markup Language 5)

HTML 5 is the 5th version of HTML. It is the new standard template for HTML, XHTML and HTML DOM. HTML5 was developed in collaboration with the World Wide Web Consortium (W3C) and the Web Hypertext Application Technology Working Group (WHATWG). WHATWG worked on web forms and web applications, while W3C, which created and manages the HTML and XHTML templates, was involved in developing the new XHTML 2.0 template. In 2006 they decided to work together to create the new template, HTML5. The main goal of the 5th edition is to support the latest types of multimedia while maintaining the logic of the original version regarding human and computer comprehensibility. HTML5 is essentially an attempt to combine HTML and XHTML to define a common, unique markup language that can be written in both HTML and XHTML syntax. In previous versions of HTML, there were no universally accepted names for headings, footers, etc. Developers used their own created names for configuration elements. This made it very difficult for search engines to index the contents of the web pages. Thus, HTML-5 introduced semantic elements. This consistency helps search engines and developers. There is a large list of HTML-5 semantic tags that can be used. In this list we can typically mention the elements `<header>` for defining a header, `<section>` for defining a document section, `<article>` for defining an article, etc. However, as will be described below, this is not a Semantic Web but an attempt to attribute semantics to the description of HTML documents.

The semantic version of HTML refers to the use of HTML tagging to enhance the semantics or meaning of information in web pages and web applications instead of defining its presentation or appearance. Semantic HTML edits traditional web browsers as well as many other user agents. It includes detailed editing models for using more interoperable applications. It extends, enhances and explains available tagging for web pages. HTML5 offers more web design capabilities such as more options for building web pages, email fields, contact fields, etc. It also supports direct embedding of video or audio on web pages, without the need for additional. It should be noted, however, that not all browsers support the new HTML 5 features yet, so when new tags are used, they are treated as plain text by certain older versions of popular browsers.

3.5.2.5 PHP

PHP is a scripting language that was originally designed to generate dynamic web pages that interact with databases. PHP is now a widely used, general-purpose language that is suitable for web development and can be embedded in HTML code. It runs on web servers and with the appropriate settings it can produce the desired web pages on the user side. It can be deployed on web servers and on almost any operating system and platform. PHP offers scripts for object-oriented programming as well as a wide variety of functions for objects that make them much more flexible and easy to use. PHP uses a combination of interpretation and compilation providing developers with the best possible combination of performance and flexibility.

The advantages of PHP include the following.

- Enhanced performance
- Connection with many different database systems
- Built-in libraries for many common Web processes
- Low costs
- Ease of learning and use
- Object oriented support
- Portability
- Availability of source code.

3.5.2.6 MySQL

MySQL is an open source relational database management system characterised by the ability to efficiently and time-effectively manage large volumes of data. Uses Structured Query Language—SQL, the best known language for managing data in database systems. The advantages of MySQL over competing trading systems include:

- Enhanced performance
- Low costs
- Ease of configuration and learning
- Portability
- Availability of the source code.

3.5.2.7 XML

XML is a markup language for documents that contain structured information. It was developed by an XML Working Group under the auspices of the World Wide Web Consortium (W3C) in 1996. It was established by John Bosak of Sun Microsystems with the active participation of an XML Special Interest Group organised by the W3C [23]. A markup language is a mechanism that defines structures in a document. The structured information includes content and some clarifications about the role that content plays. Almost all documents have the same structure.

It was designed with an emphasis on simplicity, and usefulness on the Internet. It is a text data format, with strong unicode support for all languages of the world. Although XML design focuses on text, it is widely used to represent any arbitrary data structures that come up in Web Services. It can also be used by any individual or group of individuals or companies who want to share information consistently. XML complements and does not replace HTML. While HTML is used in the formulation and display of data, XML represents the relevant meaning of the data. In HTML the tags are predefined while XML allows users to define the tags and the structured relationships between them. XML was designed to meet numerous different needs by giving documents a greater level of flexibility in style and structure than was previously the case with HTML. It offers HTML designers the ability to add more elements to the language. By 2009, hundreds of XML-based languages had been developed, including RSS, SOAP, and XHTML. This language solves many of the problems faced by web designers, and will be used for many years to come because it offers effective and dynamic solutions. Default XML-based encodings are available for most office suites, including Microsoft Office, OpenOffice.org, and Apple iWork.

3.5.2.8 AJAX Technology

AJAX is featured as a key component of WEB2.0 applications [24]. It's not a new programming language—it is not really a language, but rather a set of Internet technologies. Although the technologies used have existed for a long time, during the post-2004 period they were linked as a network of technologies that enabled the development of innovative e-services. As the name implies—which derives from the initials of the words Asynchronous Javascript XML, this technology is related to existing technologies and templates such as Javascript (for page layout and interactivity while combining all other technologies), and XML (for the format of moving files). It should be noted that XML encoding has in practice been made obsolete by JSON which has more advantages as it is a lighter template and is supported by JavaScript and all modern programming languages. It also integrates technologies such as XHTML and CSS (for dynamic display and page stylization), Document Object Model (DOM) for dynamic display and interaction, and the XMLHttpRequest object for asynchronous data retrieval.

It can be found in typical WEB 2.0 applications such as Flickr, YouTube, Gmail, Google, Facebook etc. Its use marks the beginning of an unprecedented period of user interface innovation, with the creation of web applications that on one hand provide content to the user and on the other hand are developed as local applications and in the context of services provided instead of classic software applications. Such services, for example, were used in Gmail, which combined web strengths such as ubiquitous accessibility, database usage, and search services.

With its use, faster interaction with the websites was made possible, since the exchange of data through a server and the updating of the page by the user is now done without the time-consuming process of reloading it or submitting a request by the user as older, classic websites do. The basic idea behind the application is to load the website independently of the browser at speeds that are not particularly affected by the low speeds that may be supported by the network infrastructure. This is mainly achieved by using the well-known XMLHttpRequest object through which the user can send and receive information in various formats such as XML, JSON, HTML and text files. In practice, the update is done asynchronously with the exchange of data of small size between client and server in the background, without the user taking note of it. So for example on different pages one can interact by "clicking" or just moving our mouse over a point or filling in a field of a form often with an indication based on search history and the update can be done without exiting the page. Its basic function is shown in the following figure (Fig. 3.14).

In classic websites, all the html instructions that the server sends to the visitor's (client's) browser when asked to see, the content of the page is sent only once and after the end of the sending the server-browser communication is shut down. In case any changes occur on the page and in order for the visitor to be made aware of them, they must refresh the website by reloading the browser. In contrast, with the AJAX technique, it is possible for a web page to be updated asynchronously by exchanging

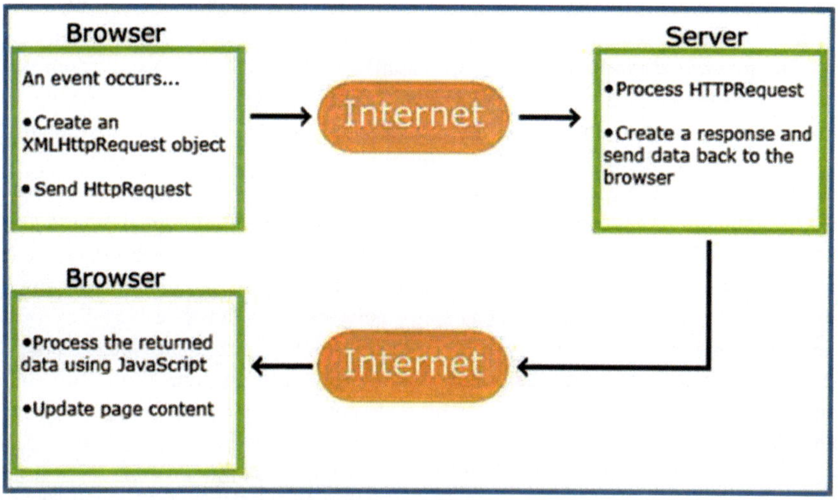

Fig. 3.14 The basic function of AJAX technology [24]

small volumes of data with the server in the background, allowing parts of the web page to be updated (a div for example), without updating the entire page! The AJAX technique is very often used on pages that contain registration forms on a website or forum.

3.5.3 The Semantic Web: Making Data Meaningful

Under the guidance of Web1.0 developer Tim Berners-Lee and the World Wide Web Consortium (W3C) [6], the Semantic Web expands the capabilities of the World Wide Web through the use of templates, markup languages, and related editing tools. Specifically in 2001 it was said by Tim Berners-Lee that "*The Semantic Web is an extension of the current Web, which gives a clearly defined meaning to information, enabling computers and humans to work better together*" [25].

This endeavour seeks to create a universal medium for the exchange of information based on the meaning (or semantics) of the content of Web documents in a way that can be understood by both humans and computers. As a result, it is expected that automated production and connection of knowledge from existing information to the Internet will be made possible.

The Semantic Web as an extension of the World Wide Web, according to the vision of its creators and developers, is expected to make the web more useful by using new technologies that will make this vision a reality: a common language for representation of data that can be understood by all types of software agents, ontologies—as sets of statements—that translate information from different databases into common terms and, rules that allow software agents to justify the information described by these terms. Data formatting, and ontology and logic management software are expected to function as a major application on the World Wide Web, analysing all raw data stored in digital databases as well as all text, image, video data and communications on the web. As more groups develop and publish their own taxonomies, Semantic tools allow them to link their designs and translate their terms, gradually increasing the number of people and communities whose web software can make them understand each other automatically.

The Semantic Web is expected to contribute to the more intelligent access and management of information circulating on the Internet through new technologies and the development of corresponding new applications. The overall vision is to move from the existing Internet of static pages to a network of dynamic service providers (Web Services) that automatically discover the information sought after, trade goods that the user intends to buy or gather information from different sources and combine them into homogeneous forms with the ultimate goal of sharing and interacting with other "common language" systems.

The semantic integration of the web is expected to give another qualitative impetus to the already existing infrastructure in various fields such as e-Government, e-commerce, distance learning, etc. with the support of the communication between different systems (interoperability) but also the saving of time and energy on finding, classifying and processing the information.

However, the Smart Web scenario has received a host of critical comments insofar as its effectiveness and performance go, and many remain sceptical. This is because nowadays the web tends not only to intervene, and it does so obviously, at all levels of user interaction with systems, but it also collects data frequently without the user being able to object. There is already a great deal of concern about the protection of users' data—personal or not—by webmasters and popular social networks worldwide. Legislative initiatives such as the General Data Protection Regulation of the European Parliament and of the Council of the European Union formulate a single legal framework at a European level with the aim of updating and modernising existing data protection regulations. Another example that highlights these concerns is the fact that in October 2018 members of the European Parliament's Committee on Civil Liberties, Justice and Home Affairs said that EU agencies should be allowed to conduct a full audit on a social networking website in order to assess the level of protection and security of the personal data of its users. MEPs have taken note of the improvements on privacy made by the site following the "Cambridge Analytica" scandal, but clarify that the company has not yet conducted a full internal audit. They suggest that the company make "substantial changes to its platform" in order to fully comply with EU privacy rules [26].

3.5.4 The Levels of the Semantic Web

According to the creator of the Semantic Web, Tim Berners-Lee, the Semantic Web includes a series of levels, that is technological levels of functionality [27]. The stratification of the required technologies for the support of the Semantic Web is analysed in 7 levels as shown in the following figure (Fig. 3.15).

1st level

The HTTP protocol used in the existing network is the cornerstone of data transfer and the Semantic Web, while Universal Resource Indicators (URIs) support ontologies, and Unicode encoding supports universal access.

2nd level

In the next fundamental level the Extensible Markup Language (XML) also plays an important role as it allows the writing of structured documents with user-defined vocabulary using tags in them. Although XML allows for a rudimentary and arbitrary structure of data, by itself does not fully offer the possibility of rendering and disseminating semantics of data through the Semantic Web. XML does not impose any semantic restrictions, however it does set rules and guidelines for designing text formats that make it easy to structure data in such a way that it is easy for people and programs to read and edit. As a result, XML Schema is a language that limits the structure of XML documents. Specifically, it defines the "grammar" for creating

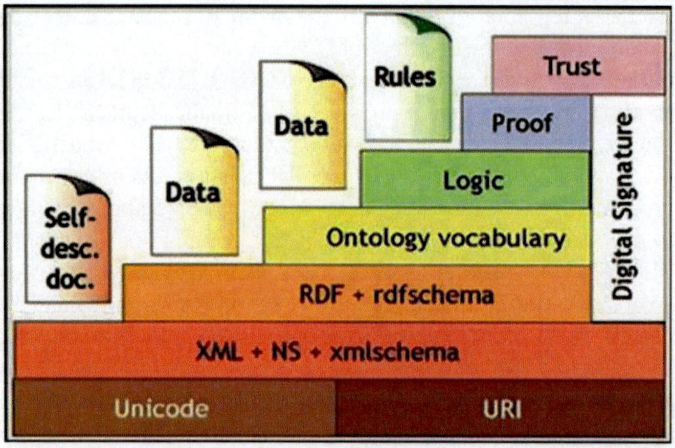

Fig. 3.15 The stratification of Semantic Web technologies [27]

valid XML files. Files can refer to different namespaces to distinguish between two tags that represent different things but have the same name.

3rd level

The third level is based mainly on the RDF model while it is the level at which the performance of semantics in the data is based. The RDF model is based on XML and concerns the description and processing of metadata while providing the possibility of interoperability between applications that exchange information on the Internet. Uses URIs to identify online resources as well as a tagged pseudo-graph model to describe the relationships between these resources.

Through the RDF Schema—the semantic extension of RDF, design principles are offered for the organisation of web objects in hierarchies while it is considered a key tool for the development of ontologies of the next level. Key concepts used here are classes, properties, relationships of subclasses and sub-properties, and constraints of domains and rows while providing a simple reasoning framework for inferring resource types.

4th level

At this level a common representation of ontologies is established, so that the terms used in the data level are defined and correlated with each other. There are different languages for the representation of ontologies such as RDFS, DAML + OIL, OWL which have similarities and differences with the ability to describe and define classes and properties. In all cases a set of concepts is included, a hierarchy, and correlations between the concepts.

5th level

At this level, through logic, automated reasoning and inference based on information structured on an ontology are supported. The use of standard rules also enables and strengthens the pseudo intelligent decision-making process by computer machines.

6th level

At the level of proof, the results deduced from data in the Semantic Web can lead back to the assumptions that caused them. For example, if someone sends to page A the proof that they can use it, then that page should be able to check and verify the existence of this proof. In general, statement makers should be able to provide evidence of the validity of their statements which will be verifiable at the computer level. At this level it is not necessary for the reader (computer machine) of the statement to look for the necessary evidence for the validity or not of the statement, but simply to be able to check the evidence provided by the creator of the statement.

7th level

At the level of trust in combination with digital signature technology (which is already implemented) the reliability of the information that is circulated, processed and inferred on the Semantic Web is determined, in an automated manner in combination with the use of digital signatures and certificates. Digital certificates provide evidence that a particular person has written (or agrees to) a document or proposal. In this way, it can be known who has created (or approved), for example, some RDF proposals of interest, and one can set up a sentence editor, regarding which "signatures" can be trusted and which cannot.

3.5.5 The Basic Technologies Used in the Semantic Web

A prerequisite for the operation of the Semantic Web, is the access of computer systems to structured collections of information and sets of logical rules that they can utilise, to perform the automated drawing of conclusions. That is why we are talking about the "representation of knowledge", a topic that has piqued the interest in the field of Artificial Intelligence in recent years and that is fully developed in the next chapter.

Among the technologies used by the Semantic Web are the following:

- Universal Resource Identifier (URIs): strings that uniquely identify an entity (a Website, a property, a person, a thing, etc.)
- Extended Markup Language (XML): allows users to add arbitrary structure to their documents without specifying the semantics of this structure
- RDF (Resource Description Framework) technology: used for data representation and knowledge sharing on the Internet
- OWL technology: used to create and distribute ontologies, supporting advanced web search, software agents and knowledge management.

Each of the above technologies is based on those mentioned before it. Thus e.g. RDF is based on XML and uses URIs. These technologies are combined to provide descriptions that complement or replace the content of documents on the Web, as will be shown below. These machine-recognizable descriptions allow meaning to be added to the content, making it easier for computers to automatically search for information. The meaning of the content is expressed through the RDF, which encodes sets of triples that represent approximately the subject, the verb and the object of a sentence. These triples can be written in XML. An RDF document states that some entities (websites, people or things) have certain properties with specific values (e.g. element X is required for document Y). This way of representing can express most of the data that computers understand. The subject, the verb and the object are identified by a URI, just as with web pages. So anyone can add a new entity or a new property.

A key component of the Semantic Web are ontologies. Ontologies typically define the concepts and correlations of concepts for a field. They contain definitions of object classes and relationships between classes, as well as rules for drawing logical conclusions. Through ontologies it is possible to combine data from different sources, which share the same ontology. Also, terminology problems are solved as the meaning of the terms that appear on a page can be defined with pointers to the ontology. Ontologies can improve Web performance by increasing the accuracy of information searches, as they search for information that refers to a specific concept and not to a keyword. They can also be used to relate a site's information to corresponding knowledge structures and logical rules. Common ontologies help exchange data and meanings between different web-based services.

There are many web-based automated services that do not use semantics, but other programs (such as software agents) cannot detect such a service on their own for a specific function. This can only happen when there is a common language for describing services on a network so that the various agents can advertise their services on a yellow page service.

Once the desired agent has been identified, the two communicating software agents can understand each other by exchanging ontologies. The Semantic Web provides this flexibility. By discovering new ontologies, software agents gain new possibilities for drawing logical conclusions.

The unifying logical language of the Semantic Web enables the integration of concepts that can be defined by anyone (via a URI) into a Web. This way software agents will be able to semantically analyze people's knowledge by providing a new form of useful tools.

3.5.5.1 Metadata

Metadata, as data for data are named (or according to many, information for information) is a fundamental concept in the creation of the Semantic Web. Metadata is the information that describes a set of data and is the essential solution to the ever-increasing volume of information circulating on the Internet.

Metadata serves the user multiple ways. They are the means by which a specific data set is identified, while by recording the content, quality and characteristics of a data set, it is possible to check the suitability of the data set in relation to the user, without the user needing to access the data themselves.

However, the increase of moving information as well as their diversity, undoubtedly leads to the increase of metadata for their description. This fact dictates the need for the evolution and more specific formatting of metadata. So the W3C has proposed dividing metadata into three zones:

- The first zone contains the relevant unstructured data, which are practically automatically extracted from the information sources. These data are of low clarity semantics, do not support field research, and do not allow the user to have an objective assessment before retrieving the information.
- The second zone includes data that contains a level of description such that the user is able to evaluate the usefulness of the source of information without the need to connect to the source.
- The third zone includes descriptive formats of a higher level, which can be used to locate and find, but also have a role in describing objects.

Three key elements are essential for the effective use of metadata:

- Semantics: meaning a set of commonly understood terms to describe the content of information sources
- Syntax: meaning a commonly recognized syntax template for linking terms to meaningful metadata sentences
- Interoperable elements: meaning a framework that allows the exchange and recombination of metadata sentences between different applications and objects.

3.5.5.2 Agents

By agents, we mean programs that perform some functions autonomously, usually without the direct supervision of the user, and produce results after the end of these functions. These programs usually browse the Internet, process the information they find and are already used for search, sorting and data selection functions.

Their main characteristics include the following:

- Autonomy. Their behaviour is based on goals, without necessarily being based on the existence of external events to the system that cause it.
- Collaborative behaviour. One agent can work with another to achieve a common goal.
- Reactivity. An agent can perceive events in his environment and react accordingly.
- Knowledge based communication. An agent can communicate with others with the help of a communication language and not with the use of common protocols or standards.
- Ability to draw conclusions. The possibility of the agent acting with the abstract description of the data of a problem is mentioned.

- Mobility. An agent can act independently of the platform.
- Personality. An agent has human character traits.
- Adaptability. Increasing the agent's experience leads to learning and improving his abilities.

Regarding the possibilities of the agents' programs, the "intelligent agents" are also mentioned. They are programs that display some form of artificial intelligence and, having the ability to learn, based on their experience and the data they have, can make decisions. Semantic Web technologies will increase both the number and capabilities of agents while smart agents will change navigation and information collection with the help of ontologies.

3.5.5.3 Web Services

According to the W3C Web Services research team [6], Web Services are defined as a resource—not necessarily within the World Wide Web, but also at the level of an organisation's local internal network. In particular, it states: "*A Web Service is a software application, defined by a URI, whose interfaces and links can be defined, described and located based on XML, and support direct interaction with other software applications, using XML-based messaging through web-based protocols.*"

A basic principle of Web Services is their implementation with emphasis on interfaces, the point of contact between the service itself and the software that calls for it. The pioneering element of these services is the existence of many independent entities that communicate with others on the basis of open standards and without being strictly interconnected.

The functions that can be achieved by the online services are the following:

- Automatic service discovery. It means the automatic identification of the offered services that cover the restrictions set by the user.
- Automatic network service call. To date, most services require human intervention during their execution, during which the user makes various calls to achieve the desired result. In the case of automatic execution of an online service, the user will formulate the query—request and the smart agent will activate all the necessary procedures automatically. For example, it will send a tax registration number and a bank account number in order to complete the tax clearance process with information that will be on the user's smart card.
- Automatic composition of services. From the introduction of a set of services it is possible to create a new service to achieve a goal that was not achieved by the original services. For this to happen, the requirements and results of each service must be described with a specific methodology. For example, in the case of automatic debit of a bank account, the sub-services, "check account balance", "debit the amount", "update" appear.
- Automatic monitoring of service performance. In case the services "run" for a long period of time, it is useful to provide information about the stage of the service at all times.

Web Services are used in addition to scientific applications in the fields of e-Government and e-commerce. Their architecture is based on:

- The exchange of messages using the SOAP protocol, which leads to the construction of structured packages with data exchanged through applications.
- The description of Web Services with WSDL (XML extension), which provides the description of the protocol, services and structure of the messages exchanged. It also contains all the necessary information for the use of the service.
- The publication of descriptions for their discovery and use. Storing information in UDDI registers leads to the creation of online directories of information, enabling the user to discover a service and communicate with it via a SOAP message.

The next step in these services are the "Semantic Internet Services" which will contain additional semantic information in order to better describe them and integrate them harmoniously into the Semantic Web. In this way the search for such services will become more efficient and more automated.

3.5.5.4 Ontologies

The word "ontology" comes from philosophy, where it is a branch that deals with the nature and organisation of reality. In computer science, ontologies aim to capture knowledge of a field of interest in a basic way and to provide a commonly accepted perception of the field that can be reused and shared between applications and groups. Ontologies provide a common vocabulary for a field and define, with different levels of formality, the meanings of terms and the relationships between them.

Ontologies can also be defined as descriptions of perceptions and correlations that may exist for an agent or community of agents. Ontologies can also be considered as an essential technology to support intelligent search, given that on one hand they allow computers to perceive them as they involve formal semantics and on the other hand they are simultaneously perceived by humans.

According to the W3C, ontologies describe the terms used to describe and represent a field of knowledge by defining:

- Classes, which are general concepts of the field of interest
- Relationships that may exist between the classes
- Properties or characteristics they may possess.

Ontologies are a dynamic field, first explored by scientists of artificial intelligence since the early 1990s, of the Engineering of Knowledge, of the processing of natural language, and of the representation of knowledge. More recently, the concept of ontology is prevalent in areas such as: intelligent information integration, collaborative information systems, information retrieval, e-commerce, and knowledge management. The reason they have become and are becoming so popular is because they promise "a shared and common understanding of a field that can communicate between people and applications". Because they aim for consensual field knowledge their development is a collaborative process of different people of different backgrounds and locations.

As already mentioned, ontologies have been associated with the development of the Semantic Web and have proven to be particularly useful in its further development. And this is without the existence of the necessary infrastructure, at a methodological level, from the field of knowledge representation. These structured representations or patterns of known and accepted events are built today to make a multitude of applications more capable of handling complex and spread-out information. They become more effective when the semantic differences that people take for granted are critical to the purpose of the application. This can mean managing common sense lurking in natural language passages, or experiences stored in specific languages and work repositories.

Examples of the usage of ontology include areas such as semantic network research, creating medical guidelines for patient health, mapping animal and plant names, searching for specific sources of public information, collaborative organisational planning, in-depth security analysis, and the automated exchange of information between trading partners. Ontologies in the WWW are used to classify websites, as well as to categorise various products for sale, according to their features, in various online stores etc.

The development of ontologies for the concepts of government agencies and in the context of the further development of e-Government, offers significant advantages. On the one hand, internally, it provides a formalism in the representation of information as well as knowledge of its structure and distribution, and on the other hand, it enables external applications to reuse the stored data with maximum interoperability, thus offering a high level of services to the citizen and in other organisations.

A. *Types of ontologies—Categorization*

From time to time different classification systems have been developed for ontologies. A simple system of classification—categorization of ontologies is based on the level of generality according to the schema described in the figure below (Fig. 3.16).

- High level ontologies are used to describe general concepts such as space, time, or a fact, which are independent of a particular problem or area.

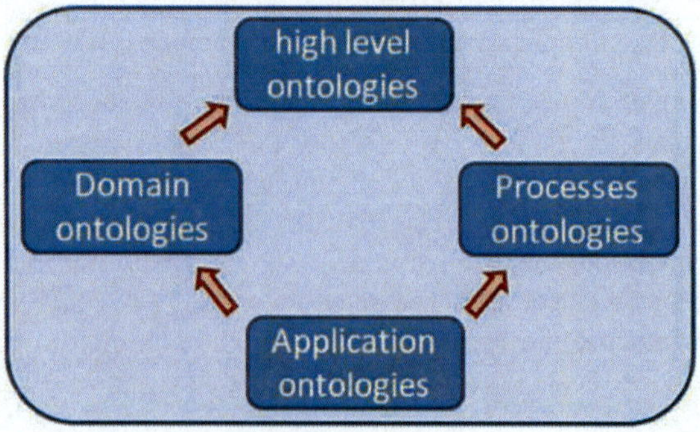

Fig. 3.16 Ontology categorization [28]

- Domain ontologies describe the vocabulary of a specific domain each time, specifying the meanings of the ontologies of the previous class.
- Process ontologies are used to describe vocabulary related to a specific elementary task or activity while also specifying the concepts of high level ontologies.
- Application ontologies are the most specific ontologies. The concepts described correspond to the roles played by domain entities performing a certain activity.

Another categorization of ontologies emerges depending on the purpose for which they are developed [29]. According to this, ontologies are categorised as follows:

- Knowledge representation ontologies: they provide representation entities without specifying what these represent.
- General or common ontologies: they aim to capture general knowledge about the world, providing basic concepts such as time, space, events, etc.
- Top-level ontologies: they provide general concepts under which all terms are related to pre-existing ontologies.
- Metadata ontologies: they provide a vocabulary to describe the content of information that is available online.
- Domain ontologies: they represent knowledge about a specific field, e.g. medicine etc.
- Methodology or task ontologies: they provide terms that refer to specific tasks, e.g. diagnosis etc.

An ontology, according to the degree of formality of its representation, can be:

- Highly informal, i.e. expressed in a natural language.
- Semi-informal, for example formulated in a limited and structured subset of a natural language.
- Semi-formal, expressed in an artificial and strictly defined language.
- Rigorously formal: definitions of terms with strict semantics, theorems and proofs of properties.

B. *Ontology representation languages*

Ontologies are usually expressed in a logic-based language so that they can be detailed, accurate, consistent, correct, and expressive in distinguishing between classes, attributes, and relationships. Some ontology tools can automatically perform reasoning using ontologies, and thus provide advanced services in intelligent applications such as: conceptual/semantic search and retrieval engines, software agents, decision support machines, natural language speaking and comprehension machines, knowledge management machines, intelligent databases, and e-commerce machines. The use of ontologies is a way for the emerging Semantic Web to represent the semantics of documents so that they can be used by web applications and intelligent agents.

The architecture of the Semantic Web consists of the following three layers [30]:

1. The metadata layer: The data model at this layer contains mainly the concepts of resources and properties. The RDF (Resource Description Framework) language is the predominant data model for this level.
2. The schema layer: At this level, web ontology languages that define hierarchical descriptions of concepts and properties are introduced. RDFS (Resource Description Framework Schema) is the predominant candidate format for this layer.
3. The logical layer: Consisting of more powerful ontology languages. These languages offer a larger set of formulated principles that can be mapped to known expressive description logics. Here OIL (Ontology Inference Layer, 2000) and DAML-OIL (Darpa Agent Markup Language-Ontology Inference Layer, 2001) were the two predominant languages. W3C now suggests OWL Web Ontology Language as the official ontology language.

In general, the languages of Ontology representation could be divided into three categories [31]:

1. Traditional languages. Predicate logic first order (cf. Prolog), Logical frameworks (Frame-based logic), Descriptive logic (Description logic)
2. Web-based languages. Their syntax is based on XML
3. Languages were developed to represent specific ontologies and use in specific applications. Examples: CycL, GRAIL, NKRL.

The most common ontology representation languages we come across are Web-based languages, and to mention a number of them: Simple HTML ontology extensions (SHOE), Ontology exchange language (XOL), Ontology markup language (OML and KML), Resource description Framework schema language (RDFS), DARPA agent markup language (DAML), Ontology interchange language (OIL), Ontology Web Language (OWL).

The differentiation and separation of ontology languages is mainly based on (a) syntax, (b) terminology (e.g. Class or concept, Instance or object, Slot or property), (c) expressiveness, i.e. something that can be expressed in one language cannot be in another and (d) in semantics as the same statement can mean different things in different languages.

In addition, an ontology language is defined by some design objectives that describe its general motivations and have emerged from the study of various real cases of their use. The following are eight design objectives for an ontology language suitable for the Semantic Web.

1. *The ability to reuse established ontologies*: Ontologies must be publicly available and different data sources must be able to refer to the same ontology.
2. *The ability to modify established ontologies*: An ontology can change during its lifespan. A data source must specify the version of the ontology to which it refers.
3. *The ability to correlate established ontologies*: Different ontologies can model the same concepts in different ways. The language must provide ways of correlating

the different representations, thus allowing the data to be converted into different ontologies, creating an "Ontology Web".
4. Different ontologies or data sources may be contradictory to each other.
5. *Balancing expressiveness and scalability when creating ontologies*: Language must be able to express a wide range of knowledge, but it must also provide efficient means for its logical processing.
6. *Avoiding unnecessary complexity* that may discourage widespread use and adoption of the language: The language should be easy to learn and contain clear concepts and meanings.
7. *Maintaining compatibility with other standards*: The language must be compatible with other standards commonly used on the Web and also with industry standards.
8. *Support for internationalisation*: Language should support the development of multilingual ontologies, and possibly provide different views of ontologies that are appropriate for different cultures.

3.5.6 Web 1.0–2.0–3.0: A Comparative Presentation

From the analysis of the characteristics of the current version of the web and the Semantic Web mentioned above, specific and important differences are recorded in various areas such as content and its management, software, and technologies and other parameters from which we analyse below the most basic versions as well as summarising the key features of each version.

Web 1.0

Web 1.0 refers to the first stage of the evolution of the World Wide Web and is a content delivery network (CDN) that allows the presentation of information on web pages. At this early stage there were only a few content creators while the vast majority of users were content consumers. Personal web pages were static hosted on web servers running ISPs or free web hosting services.

Cumulatively, the basic technological components of a Web 1.0 site include:

- Static pages.
- Content is viewed from the server file system.
- Pages created using the server page or the Common Gateway Interface (CGI).
- Frames and tables used to place and align items on a page.

Web 2.0

Web 2.0, as the current and most widespread version of the World Wide Web, also known as the Collaborative Social Web, refers to websites worldwide that highlight user-generated content, usability, and interoperability. It did not change the level of technical specifications, however it did change the way the websites were designed and used. Interaction and collaboration between users and creators is achieved in web 2.0 through a system of social media where creator and user communicate in a virtual social network. The web now contains a number of online tools and platforms where people share their views, thoughts and experiences. Web 2.0 applications tend to interact much more with the end user. Therefore, the end user is not only a user of the application but also a participant with tools such as Podcasting, Blogging, RSS, Social Networking, Social Media etc.

Key technology features of Web 2.0 include:

- Free information classification which allows users to retrieve and collect information collectively.
- Support for dynamic content that meets the needs and demands of each user.
- The web browser technologies used in the development of Web 2.0 include AJAX and JavaScript frameworks which provide advanced dynamic features of websites.
- Direct interaction and information flow between the site owner and users with the support of evaluation and commentary.
- Access to Web 2.0 raises concerns and reservations about privacy and data security issues in relation to the original version of the web, highlighting technology measures such as security software to enhance the security of data and of information systems.

Web 3.0

The Semantic Web promises to introduce the "web of data" in a more logical way than the current version of the web by search engines developed in conjunction with indexing mechanisms. With new tools based on logical rules it is possible to automatically arrive at conclusions and new information that is not based on keywords but on concepts. The main aim is to interconnect the information so that it can be automatically retrieved by machines without human intervention. To achieve this it is necessary to document the information through the development of ontologies using appropriate declarative languages such as OWL.

The study and development of the Semantic Web has been linked to the evolution of the use and interaction in the Web environment that involves changing the Web into a database. It mainly concerns the upgrade of the back-end of the web, after a long period of time focusing on the front-end with Web 2.0. With the new view of the web, the goal is opening and interconnecting the data so that they can be freely distributed to users and machines based on specific protocols, while respective new services undertake their personalised display. The main features of the new web 3.0 are the following.

- Semantic Web. The constant evolution of the Web includes the development of web technologies to support the creation, linking and sharing of content through searching and analytics based on the concept of word meaning rather than keyword matching.
- Artificial Intelligence. Combining this capability with natural language editing in Web 3.0, computers can discern information like humans in order to deliver faster and more relevant results. They become smarter to meet the demands of the users.

- 3D graphics. 3D design is widely used in websites and services on Web 3.0. Museum guides, computer games, e-commerce, geospatial environments, etc. are all examples that use 3D graphics.
- Connectivity. With Web 3.0, information is linked and can be automatically retrieved by machines thanks to semantic metadata. As a result, the user experience evolves to a new level of connection that utilises all available information.
- Ubiquitous presence. Content is accessible from multiple applications through any device connected to the web and can perform all of the services provided (Fig. 3.17).

As mentioned above, web 2.0 and web 3.0 show differences in key factors, which we analyse below and are summarised on the table.

- *Content*

The Semantic Web contains real content available to everyone, along with its official semantics. Formal semantics is mechanically comprehensible content, produced in logical languages such as OWL. No current semantics of existing content is attributed to the current web. However the content is readable by the machine, but not understood.

In the original version of the web the content was published by the creator who owned it as opposed to the current version where the content is published by a user and can be edited and shared with others.

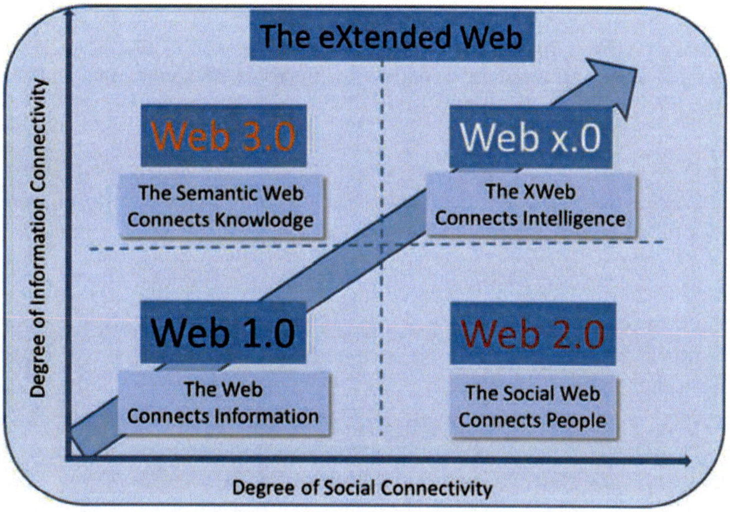

Fig. 3.17 Towards the eXtended Web

- *Conceptual perception*

The current web site manages an astronomical number of web pages based on hyperlinks and tags that their creators have already assigned to specific keywords. In any case, the semantics, i.e. the real meaning of the hyperlink or the tag is not obvious, so the user must visit the new website or analyse the meaning themself in order to understand their relationship. In contrast, in the Semantic Web this limitation is eliminated through ontologies where data is defined with clearly defined meanings and properties, which are understood by machines.

- *Scope*

As mentioned above, the number of websites on the current web currently exceeds 1.7 billion, but only a very small fraction of them are returned in response to a search through existing search engines. This is through the methodology of identifying the keywords within the content of the saved web pages. On the contrary, with the Semantic Web all the web pages are expected to be directly accessible through the semantic search engines which are based on the automatic processing of the concepts through the ontologies that have been developed and officially correspond to the meaning that is of interest each time.

- *Environment*

The Semantic Web is a web of ontologies with data labelled with formal meaning comprehensible to machines. The Semantic Web also has data as well as documents that machines can process, transform, assemble, and even infer and produce new information. On the contrary, the current web is the web of information, providing access to virtually unlimited information in the form of documents, but without automatically understanding their meaning.

- *Utilisation of resources*

There are many web resources that can be very useful in numerous user activities. Resources mean any recognizable "thing", digital, physical or abstract, such as a digital document, a photograph or video as well as a service or person. They are recognized based on Uniform Resource Identifiers (URI). In the current web they are difficult to locate because the metadata that is already in use and assigned to them regarding their meaning or properties is not understood by the machines. The Semantic Network develops a network of related resources which is supported by the Resource Description Framework (RDF) and OWL. With these technologies, Internet resources are automatically understood and accessible by machines.

In summary, the differences between traditional HTML-based web and Semantic Web are shown on Table 3.2.

Table 3.2 Differences between traditional HTML-based web and Semantic Web

Factor	Web 1.0	Web 2.0	Web 3.0
Oriented to	• Trade • Business	• Regular users • Communities	• Knowledge • Machines
Content	• Content ownership-publishing	• Cooperation • Exchange of information	• Real content with semantics • Open to everyone
Condition	Reading	• Reading • Interaction • Execution	• Personalization • Smart customization
Components	• One way services • HTML sites • Portals	• Services • Software packages • Participation architecture • Expandability • Databases • Collective intelligence	• 7-level service framework • URI • XML • RDF • Ontologies • Logic • Proof • Trust
Content organization	Classified in directories	Use of tags	Each thing corresponds to a unique URI
Via	Static websites	Dynamic websites	Linked data
Content management	Content management systems	• RSS feed • Social Media • Blogs • Wikis	Smart apps with an emphasis on personalised customization
Structure	Simple stable structure	Brainstorming	Framework of technologies and standards
Communication between users	One-way communication via email	Two way communication	Two way communication
Software	Software as a product	Software as a service	Software as a service
Technologies-languages	HTML	XHTML	• RDF • OWL • SPARQL
Means	PC	Ubiquitous connectivity	Ubiquitous connectivity
Advantages		• Ease of use • Wide acceptance • Low page creation costs • Use by unskilled personnel • Freedom of speech • Education	• Organisation of world information • Reasoning • Improved search • Personification

(continued)

Table 3.2 (continued)

Factor	Web 1.0	Web 2.0	Web 3.0
Disadvantages		• No protocol updates • Lack of new ideas • Ethics • Privacy	• Not widespread • Few applications support it

3.6 Semantic Web Applications

Semantic Web technologies are already popular and applied in areas such as research and life sciences. Especially:

- In health sciences, these technologies are expected to assist researchers in gathering data on various drugs and diseases that have multiple names in different parts of the world. In this area, the creation of unified medical terminology languages and new innovative services are expected to assist medical staff and guide consumers to reliable health information about their condition.
- In education, the Semantic Web will contribute significantly to learning, mainly in the way of searching for information, in the organisation of results and in the creation of a learning program specific to each and everyone.
- In the business sector, there will be better organisation of companies, better experiences for users in online shopping and better coordination between different companies.
- Also, there already is an effort by companies, researchers and non-profit organisations to produce ontology standards, mainly for the fields above, so that there are common languages and that more data that can be combined for better results.
- Finally, social media and other daily Internet activities of Internet users are expected to be affected, as the methodologies of searching for information using specialised software and respective browsers such as BrowseRDF, Facet, Marbles, SIOC Browser etc. change significantly [32].

Perhaps the most obvious examples, although limited in scope and communication to the general public, are the marking systems developed on the Web, such as:

A. *dbpedia.org.*
   DBpedia [33] is a project designed to extract structured content from information generated by Wikipedia into the World Wide Web. DBpedia allows users to semantically explore relationships and properties of Wikipedia resources, including links to other relevant datasets [34]. Tim Berners-Lee described DBpedia as one of the most famous parts of what is called the linked data project [35].
   This project was started by the Free University of Berlin and the University of Leipzig in collaboration with OpenLink software and the first available data set was published in 2007 [36]. The articles posted in this work are mostly free form

text, but also include structured information embedded in the articles, such as "infobox" tables ("pull-out" tables that appear in the upper right-hand corner of the default view of many Wikipedia articles or mobile versions), categorization information, images, geo-coordinates and links to external web pages. This structured information is extracted and placed in a uniform data set which can be further investigated.

The 2016-04 version of the DBpedia dataset describes 6.0 million entities, of which 5.2 million are classified into a consistent ontology, including 1.5M individuals, 810k positions, 135k music albums, 106k movies, 20k video games, 275k organisations, 301k species and 5k diseases [37]. DBpedia uses the Resource Description Framework (RDF) to represent exported information in triples and consists of 9.5 billion RDF triples, of which 1.3 billion were exported from the English version of Wikipedia and 5.0 billion from other language versions.

From this data set information can be extracted into multiple pages, for example the bibliography of an author can be compiled from pages about the work or about the author. This is possible based on the defined properties that connect objects that are defined based on a unique URI.

One of the challenges in extracting information from Wikipedia is that the same concepts can be expressed using different parameters in infobox and other templates, such as the "birthplace" and "placeofbirth" attributes, which mean the same thing. Because of this, questions about where people were born would have to look for answers in both of these qualities in order to get more complete results. However, the DBpedia Mapping Language has been developed to map these properties into an ontology while reducing the number of synonyms. Due to the great variety of infoboxes and properties used on Wikipedia, the process of developing and improving these matches has opened up to public contributions [38]. In terms of its functionality, DBpedia retrieves information from Wikipedia pages, allowing users to find answers to questions where information can be found in many Wikipedia articles. Data is accessed using SPARQL, an SQL query language for RDF.

DBpedia as it has been formulated provides a wide range of entities from various fields of human knowledge. This has turned it into a hub for linking datasets, where external datasets could be linked to its concepts. The DBpedia dataset is linked at the RDF level to various other open Data datasets on the Web. This allows applications to enrich the data with what is automatically retrieved through the linked data. As of September 2013, there are more than 45 million interfaces between DBpedia and external datasets, such as: Freebase, OpenCyc, UMBEL, GeoNames, MusicBrainz, CIA World Factbook, DBLP, Project Gutenberg, DBtune Jamendo, Eurostat, UniProt, Bio2RDF Data inventory [39].

B. *schema.org.*

Schema.org [40] is a collaborative effort with the mission of creating, maintaining and promoting structured data formats on the Internet, web pages, email and beyond. In addition to the people from the founding companies

(Google, Microsoft, Yahoo and Yandex), there is substantial involvement of the World Wide Web community through public mailing lists such as public-vocabs@w3.org and through GitHub. Since April 2015, the W3C Community Group Schema.org has been the main forum for schema collaboration and provides the mailing list of public-schemaorg@w3.org for discussions. Schema.org related topics are monitored on GitHub [41] where all the necessary components of this activity are hosted. Its key component is vocabulary which can be used with many different encodings, including RDFa, Microdata and JSON-LD. These vocabularies cover entities, relationships between entities and actions and can be easily extended through a well-documented extension model. More than 10 million websites use Schema.org to index their web pages and emails. Many applications including Google, Microsoft, Pinterest, Yandex and others already use these vocabularies to power rich, scalable experiences.
Based on Google, Microsoft, Yahoo and Yandex services, Schema.org vocabularies are developed through an open community process, using the mailing list public-schemaorg@w3.org and through GitHub. A common vocabulary makes it easier for webmasters and developers to decide on a scheme and make the most out of their effort. In this spirit, the founders together with the wider community came together—to provide a common collection of designs. The day-to-day operations of Schema.org, including scheme decisions, are handled by a guidance team, which includes representatives of the founding companies, a W3C representative and a small number of people who have actually contributed to Schema.org. Guidance group discussions are public.

C. *wikidata.org.*
Wikidata [42] is a free, collaborative, multilingual, sub-database that automatically collects structured data to support Wikipedia, the Wikimedia Commons, the other Wikimedia movement wikis, and everyone in the world. Wikidata is a central repository that can be accessed by others, such as wikis maintained by the Wikimedia Foundation.
Content that loads dynamically from Wikidata does not need to be retained in every single wiki project. For example, statistics, dates, locations, and other common data can be found in Wikidata.
The Wikitata repository is based on Semantic Web logic and the RDF format. It consists mainly of objects, each of which has a label, description and any number of aliases. The data are identified in a unique way—ID. Statements describe detailed characteristics of an item and consist of an attribute and a value. Attributes are also determined by a unique feature. For an individual, it is possible to attribute an attribute to a trained instance, and to assign a specific value to the corresponding educational institution. For buildings, it is possible to specify attributes of geographic coordinates with latitude and longitude values. Attributes can also be linked to external databases. An attribute that associates an item with an external database, such as an authority check database used by libraries and files, is called an identifier. Website Special Links link an item to the corresponding content on the client wikis, such as Wikipedia, Wikibooks or Wikiquote. All this information can be displayed in any language, even if the

data comes from a different language. Upon access to these values, the client wikis will display the most up-to-date data.
There are several ways to access Wikidata using built-in tools, external tools, or programming interfaces. For example, Wikidata Query and Resonator are some of the most popular tools for searching and examining Wikidata objects. The tools page has an extensive list of projects to explore. Customers of the application can access data for their own pages using a specific interface while it is possible to retrieve all the information through a special API.

D. *Europeana:*
Europeana [43] collections provide access to over 50 million digitised items—books, music, artwork and more—with sophisticated search and filtering tools to help retrieve exactly what the user wants. Dedicated thematic collections on art, fashion, music, photography and World War I contain galleries, blogs and exhibitions that inform and inspire the user.
The semantic and multilingual enrichment of metadata in Europeana is a key concern, as it improves access to the material, defines the relationships between objects and allows interlingual retrieval of documents. The quality of these enrichments is vital to ensure that highly curated content by providers is properly presented in a variety of languages. In order to ensure that these enrichments develop their full potential and facilitate access, semantic enrichment of metadata is required [44].

E. *geonames:*
This database contains over 25,000,000 geographical names corresponding to more than 11,800,000 unique features [45]. All functions are categorised into one of nine attribute categories and further classified into one of 645 attribute codes. In addition to place names in various languages, stored data includes latitude, longitude, elevation, population, administrative subdivision, and zip code. All coordinates use the World Geodetic System 1984 (WGS84). This data is accessible free of charge through a range of Web Services and a daily database export [46].
The core of the GeoNames database is provided by official public sources, the quality of which may vary. Through a wiki interface, users are asked to manually edit and improve the database by adding or correcting names, moving existing features, adding new features, and more [47]. Each GeoNames attribute is represented as a web source identified by a fixed URI. This URI provides access, through content trading, to either the HTML wiki page or the RDF description of the feature, using elements of the GeoNames ontology [48]. This ontology describes the properties of GeoNames functions using the OWL Language, attribute classes, and codes described in the SKOS language. Via Wikipedia article links that are linked to RDF descriptions, GeoNames data is linked to DBpedia data and other linked RDF data.

3.7 Advantages and Challenges for the Semantic Web

From the development and application of Semantic Web technologies to date, the main advantage over the existing web version is the fact that the information is understood by the computers along with the ability to produce new information through automatic inference methods. This will facilitate the process and response times to information search requests as the new search engines will run through all the web pages with new "smart" procedures.

The advantages from the further development of the current web and the adoption of the technologies of the Semantic Web include the following:

- More effective handling of huge data volumes in traditional databases that are also referred to by traditional search engines. Excess data received after a search will be filtered based on defined semantics, contrasted and compared to better meet the needs of users.
- Automation of the production of internal representations of the concepts and complex information that are managed and stored in the billions of web pages and databases of the current web.
- Transition from traditional databases to knowledge databases in order to effectively support more complex queries for retrieving logic-based information.
- Automatic reasoning processes with the development of corresponding methodologies and tools that support the automatic production of new information as they implement the 5th and 6th level of the Semantic Web, meaning the levels of logic and proof.
- Enrichment of the content of the current web through the development of more accurate data and metadata models.
- Increase of service users as with Semantic Web technologies services can also be discovered automatically by computers.
- Minimization of costs and time for the maintenance and development of electronic services.
- Assistance to the ordinary user with the transition from the model of adaptation of services based on the settings desired by the user to the model of providing personalised services based on their specific needs, which are automatically discovered by the new systems and without the user's effort.
- Facilitation of the exchange of knowledge and the educational process in various fields by supporting the low learning curve.
- Effective support of interdisciplinary activities in various fields by matching online resources and individuals worldwide.

Despite the positive aspects and advantages that the application of the Semantic Web is expected to bring, however there are challenges that need to be addressed in order to make the original vision of web3.0 a reality. These items include the following.

- The vastness of the web is unquestionable. Nowadays we are talking about an immense electronic universe which consists of billions of pages which is constantly expanding. Advancing to the new web, which will understand the precise meaning of the concepts and will automatically produce new information, requires enormous effort and resource consumption in order to access, analyse and semantically mark the entirety of this universe. At the heart of our scientific interest is the automatic semantic labelling of data managed by relational databases, a topic that will be elaborated on in the following sections of this Book.

- The Semantic Web and its innovative solutions do not yet have the recognition and impact that would be expected on the general public of the Internet. This is in part due to the lack of particular interest on the part of businesses and e-commerce in adopting and utilising the solutions offered, due to the need to retrofit existing and costly information systems. From another point of view, many consider the Semantic Web to be a theoretical model, which will not be widely accepted as it is difficult for the general public to understand. This is also related to the view of the "black box", meaning the view that the regular user is not interested in the processes behind his screen but in the results he sees on it. After all, especially in the field of information retrieval, the regular user is already familiar with the meaning and use of keyword phrases and is not interested in switching to another search model. Therefore, the issue of adapting query processing procedures and returning the best results based on semantics is up to the developers and ISPs and not the consumer.
- Semantic Web technologies offer the ability to develop a huge knowledge sharing network through linked data. Often, however, the sharing of knowledge is not desired by companies mainly for competitive reasons and therefore this project mainly concerns the scientific community and the voluntary action of researchers in this field. From the point of view of commercial exploitation, and since the technologies of the Semantic Web are not a profitable tool for businesses, in fact one of the key factors for further development of the Semantic Web is the public sector that owns and circulates huge amounts of data its primary goal is not generating profit.
- At the highest levels of the Semantic Web are proof and trust. A key parameter of the success of the project is the degree of satisfaction of the basic principle that "anyone can say anything to anyone, in the context that the original source makes the specific statement (proof) and that this source is reliable (trust)". Regarding the proof, digital signatures will play an important role in conjunction with encryption and control of access to information. These tools are already running on the current web and it is important that they adapt to the web 3.0 environment. With regard to trust, the parameters that will determine the level of trust of resources must be defined separately. This means that any user of the Semantic Web can determine their trust in any existing source by introducing the concept of "web of trust". So when a user trusts the sources from user A they would automatically trust the sources that user A in turn would trust. This would create a huge hierarchical network that would make it easier for agents to retrieve information based on trust knowledge. At present, the concepts of both proof and trust have not yet been pointed out and they have not yet been integrated into the Semantic Web.

However, these technologies are very important and are the foundation for the proper operation of the Semantic Web.

- Since the development of the Semantic Web largely depends on the ontologies, this is an area which requires significant attention. As we will develop below in a separate section, new systems for the development and management of ontologies are required, as well as the establishment of commonly accepted models and languages of representation.
- At the moment, the content of the web as well as the data of the databases that are compatible with the Semantic Web are very limited in terms of availability. Existing web content will have to be upgraded to Semantic Web content, including static HTML pages, XML content, dynamic content, multimedia, and Web Services. Similarly, web-based relational databases will have to be transformed into knowledge databases to support new smart search engines. To this end, the commentary of the entities and their properties used as web content plays an important role. The ontologies are used to comment on the content of the Semantic Web, and they determine the meaning of the words/concepts and the relationship that exists between them. This commentary relies mainly on RDF especially at the metadata level as well as on RDFS for the representation of more complex schemes in order to establish the formal semantics of the concepts. However, these models are not compatible with the annotation currently used on the web with the corresponding HTML annotation tools/tags. These tools have been proven to be ineffective in the semantic annotation required in web 3.0. Therefore, the creation of Semantic Web content is a serious challenge, as Semantic Web architecture is still "under construction and under constant review" and the available Semantic Web content is little. However, in addition to architectural development, researchers are also focusing on building tools to support the annotation of web content and database data. However, these tools have inherent limitations since on one hand they should be used to comment on existing static pages and on the other hand they should support the creation of new content. This creates the need to create a complete set of commenting services (as middleware) that will govern both static and dynamic web documents that may contain multimedia and other Web Services. The desired effect of content commenting services is to allow different agents to understand the different languages used in the Semantic Web. However, there are specific views that suggest commentary in a single language with the multilevel approach being taken into account in language development [49].
- In the first steps for the transition to the Semantic Web, considerable effort was made to organise the content, store it and provide the basic functions for its search. However, further development of the systems that will support the aforementioned is required. Storing and organising the pages of the Semantic Web is the first problem that needs to be addressed. The "basic" Semantic Web consists of pages that are annotated based on ontologies, whose link structure reflects the structure of the WWW where the pages link through hyperlinks. This hyperlink configuration does not make full use of the relevant semantics of the Semantic Web pages.

Creating appropriate semantic markers can accomplish the dynamic linking of web pages using ontological information and semantically annotated texts.

- The next issue is related to the discovery of information in the Semantic Web, also known as coordination between semantic markers. An architecture-like technology used in the current web should be provided to easily locate content on the Semantic Web taking into account the semantics of Internet resources.
- Semantic Web technologies are expected to lead to the automatic production of information and knowledge through automated processes of understanding the meanings of words and concepts. This project has received criticism regarding the protection of users' privacy and personal data. Therefore, special attention should be paid to securing personal data from their automated processing and user profile creation.
- As in the current web, the issue of multilingualism also exists in the Semantic Web. Semantic Web technologies should facilitate access to information in multiple languages, allowing the creation and access of semantic content regardless of the content provider's native language and user. Multilingualism plays an important role both in the level of ontology annotations and in the user interface. At the ontology level, ontology developers usually develop ontologies using their native language on which annotations will be based. As not all users use the same language this issue is of lower priority. This is because existing multilingual and language resources can be used, such as WordNet [50], EuroWordNet [51], etc.
  At the ontological commentary level, content commentary may be done in a variety of languages. As more users, especially content providers, are expected to be more interested in commenting on content than in developing a new ontology, appropriate support is needed, allowing providers to comment on content in their native language.
  At the user interface level, millions of people would like to have access to relevant content that has annotations in their native language, without even considering the source language. Although at present most of the information on the Internet is in English, any Semantic Web approach should include facilities for accessing information in multiple languages. Internationalisation and tracking techniques need to be adapted and implemented to take advantage of users' personalised access to their native language.
- The issue of multilingualism is also linked to the issue of ambiguity due mainly to human language which is full of particularities that differ from language to language. For example, the word "base" in the Greek language can be used in many different ways depending on the field of interest. For example in chemistry this word identifies the kind of chemical compound, in the field of geometry it defines the kind of side of a planar shape while in common usage it can mean the beginning of a state. It is difficult to teach machines to understand vague user queries without a proper framework.
- The uncertainty which characterises precise concepts with uncertain values. However, different types and methods of reasoning may be utilised to eliminate uncertainty.

- The inconsistency encountered by logical contradictions. The answers to logical contradiction questions may vary and there is often no right or wrong answer [52].

3.8 Benefits from the Application of the Semantic Web

From the original version of the Web, which was a set of websites for presenting content to users and then, in the form of the Social Web through which users could create content and communicate with each other during production. The search and retrieval of information in most of the categories of users emerged as a basic function-service. The same need is observed nowadays where social services are growing significantly, as users participate in social networks, chat rooms, buy products and services online and watch and share multimedia content. From a technological point of view, in the current version of the web, the development of static pages is no longer enough, and the content must now be dynamically updated in order to publish as much updated content as possible, personalised to the needs of the users. Thus, more and more professional websites and pages of large corporations and public sector organisations provide structured data even in the form of open data drawn from relational databases through various templates, so that it is possible to create applications and generate knowledge by using them.

As the size of data and information continues to grow and the way users interact through the web is mainly focused on content management, existing web technologies already have specific shortcomings in this area. However, the Semantic Web, as an extension of the existing web, is expected to offer innovative solutions and benefits related to the basic functions related to content management including search, information integration and data mining. The main goal is to provide a new experience in the field of browsing web pages and through them to information and knowledge. User-computer interaction will be based on the "intelligence" that the computer will acquire as its role is expected to be upgraded in the field of guidance for retrieving information that will be tailored to the user's needs and decision making.

Search is the most basic use of the Internet to find information or Internet resources. In its current form searching is not efficient enough, as it relies on finding pages that contain the search term usually based on a keyword. If the word is contained in a document it appears in the results and in the user, who is responsible for reading and interpreting it, without necessarily being sure that he has found what he is looking for. From the most modern point of view of the web, the interest does not rest only in information but more so in Internet resources. This means, in addition to multimedia electronic content (video, audio, photography, interactive maps, etc.), natural persons or other entities which at the level of the Internet correspond to a unique URI and which are distinguished for their particular properties or characteristics. With the help of web page indexing and the use of ontologies for the semantic definition of entities it will be possible to visit and examine all web pages while the search will focus on pages that are semantically completely related to the concept or entity of interest.

Information Consolidation can simplify and reduce the time required to organise an activity. For example, if one wants to issue a certificate or a professional licence, they must first look for the respective public body, then find the required documents, find the expected time of issue and delivery and finally locate the nearest branch that may accommodate him and possibly the route to follow to get there. In this version of the web these steps are distinct and are usually provided by different websites which the users have to search for themselves and to which they have to submit their keywords each time. The above steps can be simplified and consolidated using Semantic Web technologies and delivered to the user even with additional information that they did not initially request but is relevant to them, such as contact details or alternatives or the next steps that they may be required to follow. An application that automatically finds exactly all the services that interest the user and dynamically utilises whatever is needed to achieve the goals would be very convenient and would save time and potential capital.

Data Mining from a database or a set of data is another activity that interests the largest percentage of web users, especially considering that the web itself is an immense distributed data storage. As an example we can mention a case where someone is interested in renting an apartment in an area that they are not familiar with. Its decision takes into account various parameters, including some that arise from elements that are relatively unchanged such as the existence of supportive structures and green areas or others that are changeable such as crime, weather and the transport network. All the previously mentioned data are stored in respective databases of which some are static and others are updated on a regular basis. The purpose of this example is to develop a web crawler application, which will visit the respective pages, collect only the data of interest and store it in some type of data corresponding to the query. The application, however, may need reprogramming in case some information changes on the pages that have the needed data, so it would be convenient if the application could understand the importance of the content so that it does not need frequent changes to the code.

Fields that are expected to benefit from the evolution of the current web using the technologies of the Semantic Web include Knowledge Management as well as all the fields to which it relates such as business administration and public administration through the evolution of e-Government, e-trade through the development of new agents for the extraction of useful information, e-learning as a development of the now widespread digital—distance learning as well as the field of education in general in various fields, etc.

A. By "knowledge management" we mean all business strategies and practices in order to create, develop, collect and disseminate knowledge as well as the transformation of individuals into collective knowledge [53]. Given the enormous size, diversity and complexity of information nowadays, there are specific limitations to knowledge management with existing technologies to which the Semantic Web can provide innovative solutions. The use of Semantic Web technologies enables the development of knowledge management systems with emphasis on the importance of concepts, the automatic maintenance and upgrading of databases with various tools.

Limitations/areas that new technologies are expected to contribute to include:

- Search for keyword information
- Automatic extraction of information,
- Maintaining and upgrading information due to inconsistency or ambiguity in terminology and retention of outdated information
- Automatic discovery of information by data mining
- Information display tailored to the needs of each user.

B. Search engines are found in all repositories as a provided tool but also individually as independent online services. Traditional search engines, which appear as the most frequently chosen means of searching for information on the Internet, operate on the basis of a "keyword" or a corresponding phrase, returning links to any entry that satisfies the condition set by the user, but without semantic control of the subject. Most of the time links are returned but most of them are not exactly what the user is looking for. This process nowadays has been proven to be inefficient, as the information that the user has to process manually is large in size but also in variety. The time and the required energy on the part of the user, make the production of knowledge through the Internet a difficult task. Even in the case of limitation of the field of interest in specific areas such as e.g. administrative data or searching in specialised repositories such as open data repositories, retrieving information using traditional tools is inefficient and time consuming.

The efforts observed in recent years in the direction of search engine optimization and automated knowledge production have resulted in the so-called "metadata". This data is in fact "information for information" and aims to support an intelligent and automated way of searching for information on the Internet. However, the development and management of metadata is not done automatically but is left to the initiative and responsibility of the data holder or administrator.

Regarding the search for information through the Internet using traditional search tools, we record the current situation as follows:

- Too many results. In most cases, the return of hypothetical "relevant" links involves several thousand or even millions of reports, making it virtually impossible to effectively control and ultimately retrieve the correct information.
- Insufficient results. Many times the keyword results are not enough to satisfy the user so the user has to do a new search with different synonymous terms and then combine the relevant results with the ones they may have collected.
- There are cases with little or no results. This occurs when a user searches for specific information, based on a specific keyword or phrase in a strictly defined field, or searches for media files or a specific piece of text within text. In trying to deal with this situation, many machines work in addition to giving additional

results that may have matched the needs of the user. These cases, however, make the whole process of searching difficult, providing mostly useless information.

- Conceptual search approach. To date, search engines perform simple lexical analyses based on search strings and produce listings of results that require human users to draw conclusions as to which of the data is appropriate for the subject area being researched. In other words, logical analysis on the part of the user is needed to find the relevant conclusions. Often, webmasters, using the logic of lexical analysis, enter common words into the "metadata" section so that their web pages rank high in the list returned by search engines. This, however, often leads to conceptual confusion, as many words have ambiguous meanings and are often used incorrectly, resulting in misplaced web page retrieval by search engines. The most typical example is the word "sex" (when searching in English usually) which in addition to defining "sexual act" is also used to define "gender" as in "male" or "female". Search engines using the "sex" key return a few hundred million pages of content in the first version and very few in the second. This is because websites do not contain enough information about the semantics of their content so the user software can not draw conclusions from the content of the page.
- Use of synonymous terms. So far the use of synonyms or different words as metadata for the same entities has not been effectively addressed by traditional search tools. The data holder and administrator has the responsibility to anticipate the existence of synonymous terms and to include them in the relevant statements. As a result, when searching, the user must anticipate the use of synonyms and do the corresponding searches. For example, we mention the repository of administrative acts which was developed in the framework of the "Diavgeia[1]" program [54]. This repository is characterised, among other things, by the existence of many categories of documents and a large amount of data. In this system, each posted operation also has specific metadata, for which the relevant body is considered responsible. A phenomenon that has been observed is documents belonging to the same category have been marked using different terms such as "order" versus "decision". In this case, the user must know in advance the use of these "synonymous" terms in order to properly adapt the query filters to the provided search engine.
- The development of special repositories does not always have the most efficient and effective support for the search tools provided. Usually no new tools are developed and instead traditional search engines are used with the problems mentioned above.

[1] The "Diavgeia" program concerns the publication in a public repository of the acts of administrative content of ministries and public sector organisations in Greece, as open data under specific licences (CC etc.).

From the above it is understood that although the search for information is supported by the existing technological solutions of IT and optimization efforts are being made nevertheless, the whole process is characterised as inefficient when the search for information is done in order to produce knowledge and gain comparative advantage.

In order to address this situation, we believe that the technologies and tools developed in the so-called Semantic Web can contribute effectively, which is expected to be the smart version and an extension of the existing web. The use of its technologies is linked to the emergence of linked open data and the Internet of Things. In the following paragraphs we analyse the technologies of the Semantic Web and the opportunities it can offer in the management of information and the production of knowledge in the field of e-Government.

C. The field of e-commerce, which is a subset of the broader strategy of "e-business" companies, includes a large set of online processes including the development, promotion and advertising, sale, delivery, service and payment for a huge number of products and services in many fields. At the same time, in the context of a fully competitive environment, methodologies are being developed that concern both the promotion and delivery of products and services as well as the organisational part with the development of in-house planning systems, data storage and management, information exchange and interbank support and cooperation. The tools and methodologies applied today are limited by the capabilities of existing technologies, mainly in the field of automatic information retrieval but also in the personalised provision of information and services to the prospective client during a period when the range of transactions conducted electronically has increased unusually fast with the widespread use of the Internet. The commerce that is carried out in this way, motivating and absorbing innovations in digital money transfer, supply chain management, Internet marketing, web processing, electronic data exchange, system management, and the automation of data collection systems, will benefit greatly from the application of the technologies of the Semantic Web. With the development of "smart" software agents, able to interpret the product information and terms of use of a service, we would have the correct export of this information, pricing negotiations and direct comparison and response to customer needs. It would be possible to have additional information about the reputation of the stores, which will be derived from other sources, such as independent websites with consumer or expert ratings. Even in business-to-business (B2B) transactions, which are perhaps the most important part of e-commerce, significant progress could be made using these new technologies.
D. Information and Communication Technologies have evolved into significant resources for teaching and learning at all levels and types of education in the education community's effort to adopt the basic principles for undergraduate education formulated in 1987 by the American Association for Higher Education and Accreditation, AAHE [55]. The principles, which can be applied proportionally at all levels and types of training, include: encouraging communication

between learner and trainer, cooperation and reciprocity, the adoption of active learning techniques, etc. In this context, the Internet already performs basic functions since, among others, it is: (a) a source of information, evidence and educational software for learners and teachers, (b) a means of communication and exchange of messages between them and (c) a means of expression. The main goal through the Internet and educational software applications is the transition to a new model of education at the centre of which is the learner. With this model the learner is encouraged to play a more active and dynamic role in the learning process, through the sharing of information and knowledge, the development of collaboration with emphasis on the autonomous experiential exploratory process. The most important of the ICT tools that have emerged as the most useful in the educational process include:

- Individual teaching tools that include traditional self-study and internship programs. They are usually designed for use by one person and are good tools for acquiring basic knowledge.
- Informational tools provide students with the necessary material and resources to build knowledge themselves. Examples of such technologies are encyclopaedias and various Internet resources. These tools support the generation of ideas and can provide the user with information based on different perspectives. In addition, these tools also serve as rich external resources, which learners can compare to check the validity of their knowledge.
- Creative tools that are widely used such as word processors, multimedia creation software, spreadsheets, simulations, etc.
- Communication and social networking tools, which proved to be particularly useful and efficient in education with the COVID19 crisis, mainly as a means of distance education. These include video conferencing software, mail and communication management systems such as Zoom, Skype and Google Hangouts, virtual classroom—lab management systems, etc., through which it is possible in addition to teleconferencing and sending messages, educational material as well as the sharing of the instructor's screen in order to exchange ideas and information and finally the production of knowledge.

To date, the use of ICT in education has highlighted specific issues, both positive and negative. The positives include the following:

- It facilitates the development of critical and synthetic thinking, the ability to combine information and ideas and creating new knowledge.
- Learning opportunities and incentives are provided through a variety of strategies and engaging processes related to information management in particular search, processing, evaluation, synthesis and finally the presentation of information and ideas retrieved.
- It provides an environment for exploring and facilitating experimentation and promotes the development of modelling and complex problem-solving skills.
- The participatory, collaborative and collective character of learning is strengthened.

- Network technologies are used as a tool for communication, thinking and upgrading the social learning environment.
- Tools, processes and learning opportunities are also provided to students with different social, ethnic or racial backgrounds and with different learning abilities contributing to the alleviation of social or other inequalities.
- It approaches learning globally, interdisciplinarily, and mitigates the artificial division of knowledge from traditional curricula.

However, specific shortcomings are recorded in an effort to integrate and utilise new technologies in the educational process. From our experience in the field of education we record the main reasons why information and communication technologies have not been utilised to a large extent, especially in secondary education.

- The traditional teaching models in combination with the physical space and the infrastructure of the schools are not able to integrate the essential possibilities of the Internet and the software applications, since they were designed before them.
- The lack of educational software with dynamic characteristics that automatically responds to the individual needs and learning profile of the trainees and students as well as the learning goals set by the respective curriculum are inhibiting factors in the effort to integrate them into the educational process.
- The lack of appropriate skills on the part of teachers in combination with the reduced possibilities of adapting the applications to the skills of the trainers and the subject they are called to teach.
- The efforts to introduce these technologies in the educational process are of a centralised nature, they are usually designed centrally without enabling teachers to participate in the processes of organising, implementing and evaluating the innovation.
- The applications mentioned above are provided by independent sources and not by an integrated environment, resulting in non-interoperability at both conceptual and operational level and difficulties in searching and passing information from one tool to another.

In the context of dealing with the above, there are certain requirements which concern, in addition to the self-evident review of the technological infrastructure of ICT, the corresponding investment in human resources. Regarding the e-learning infrastructure, we believe that the Semantic Web is expected to give a new impetus to the use of ICT in the educational process as it is connected not only with people, but also with knowledge management. Improvements are expected in the operation of two main axes related to e-learning [56]:

- Web-based education: by creating appropriate ontologies that will be accessible, exportable and scalable by all participating institutions (universities, students, etc.). The representation of knowledge with the above philosophy can codify educational resources and form the basis of a content repository where educational material will be presented to the learner through an appropriate e-learning system.

- Educational data management: management information systems in education record data concerning learners and teachers, based on the representation of knowledge each in its own philosophy. Semantic Web tools, such as XML as an intermediate language, can support the exchange and processing of information from all existing information systems with an emphasis on exporting homogeneous information with an emphasis on personalised online services.

At the same time, the development and use of artificial intelligence and personalised service delivery will transform the way humans interact with computers. The main goal is to help users find relevant information more effectively, automatically and independently of their skills through innovative search methods. This information is collected from various sources, such as websites, e-books, blogs, listings and multimedia on various platforms using computers, mobile phones, tablets, etc.

In addition, access to virtual reality games and 3D lab simulations provides new perspectives on teaching, encourages collaboration, role play and teamwork, and makes learning more interesting and different from traditional education. In this area the Semantic Web is expected to provide a better control of educational resources by the user and more integrated knowledge management. Semantic Web-specific tools and applications can provide innovative solutions that can be applied to both educational software and information retrieval systems. By implementing these solutions it is possible to create upgraded, dynamic learning environments, which will contribute to the individualised and more efficient search and retrieval of information and ultimately to the production and management of new knowledge. At the same time, less effort and savings will be required for the necessary further development of relevant skills of the users, resulting in savings.

In summary, specific developments are expected per average ICT (Table 3.3).

3.9 Opportunities for e-Government from the Implementation of Semantic Web Solutions

In its initial stage, e-Government was considered by many to be synonymous with the introduction of information and communication technologies in Public Administration and at a later stage with the development of Public Administration information websites and the digitization and electronic processing of data and their maintenance in local databases. The current design of government websites and software applications used today is mainly anthropocentric, in the sense that most of their content is designed to be accessible and understandable by humans and not by automated computers. The web pages are mainly based on the HTML environment, which serves to describe a structured text with an emphasis on visual representation while not providing special features for providing intelligent services. So far, citizens can use their electronic devices to visit websites and either obtain the information they request or interact with public organisations by submitting relevant electronic applications and waiting electronically for their deliverables. On the other hand, within

Table 3.3 ICT impact in the domain of education [57]

Means	Activity	Examples	Needs met	Benefits of the Semantic Web	Principles of the WEEE being implemented
E-mail	– Dissemination of information – Provision of knowledge	– Announcements – Discussions – Work groups	– The possibility of rapid information dissemination – Contact of trainees and instructors outside the classroom	– Dissemination of personalised information based on the specific needs of the user – Personalised approach based on proposals based on profile analysis	– To encourage contact between students and teachers – Development of reciprocity and cooperation between students – Provide immediate feedback
Non-interactive websites	– Dissemination of information – Provision of knowledge	– Administrative services [e.g. comments, scheduling, announcements]	– Paper saving, administrative tasks – pdating	– Provision of personalised services – Proposals for related services	– Increase of contact

(continued)

Table 3.3 (continued)

Means	Activity	Examples	Needs met	Benefits of the Semantic Web	Principles of the WEEE being implemented
	– Provision of resources	– Indicators on network resources – Notes [text, shapes, images] – Published texts [e.g. student manuals] – Case studies or slide show exercises with superscript/lectures – Additional material/review of exercises – Readings – Homework/solutions to exercises – Standard solved exercises – Basic teaching material – Documentation	– Paper saving, administrative tasks – Facilitation [prof./student] – Renewability – Ease of access – Ease of providing additional learning resources – Manage different learning styles/levels – Distribution of colored material – Increase of student success	– Delivery of personalised content to the special needs of the learner – Effective information search – Access to appropriate websites and information sources – Increase in trust in information sources	– Setting high expectations – High quality material delimits the standards – Providing a rich framework and resources for the work – Promotion different types of learning – Work of various types
	– Posting	– Exhibition of students' work – Development of an evaluation portfolio	– Students are provided with a means of publication	– Increase of the accessibility of publications	– Active learning: creativity, planning – Feedback (in published works) – High expectations (publishing increases quality)

(continued)

Table 3.3 (continued)

Means	Activity	Examples	Needs met	Benefits of the Semantic Web	Principles of the WEEE being implemented
Interactive websites and multimedia	– Presentation	– Enriched presentations [animation, audio, video, virtual reality] – Manuals/online program guides	– Presentation of complex information through multisensory pathways in order to increase understanding and retention	– Providing the most appropriate content based on the specific needs of the learner which are detected during the user's interaction with the system	– Promotion of active learning: apprenticeship model, simulations, design experience. Authentic localised learning situations. Learning based on dealing with problematic situations or case studies – Activation of various forms of learning. Use of various evaluation methods
	– Hypermodal databases/case studies	– Collection of multimedia information/Case studies – Providing questions/answers	– Students can seek answers to common questions as necessary	– More efficient information retrieval process using new search engines	
	– Simulations	– Simulation of biological, physical, or mechanical systems/environments that respond to user input	– Enabling everyone to access complex, dangerous or inaccessible systems		
	– Interactive exams and teaching	– Exercises and practical tests which also provide advice, monitoring the user's progress/Adjustable difficulty	– Help with basic concepts, such as identification, although it can sometimes be tedious for the teacher	– Personalised access	
	– Intelligent teaching	– Activities focused on the task or process, such as the implementation of a skill or proper use of equipment	– Students are able to accomplish their goal with one-on-one teaching for optimal learning through experience	– Performing activities based on the real needs and skills of the users	

(continued)

Table 3.3 (continued)

Means	Activity	Examples	Needs met	Benefits of the Semantic Web	Principles of the WEEE being implemented
	– Scenario-based targeting/guided simulation	– The user enters a simulation environment motivated by a target, such as e.g. to operate the kitchen of a restaurant successfully, keeping satisfied the guests, the staff, and the owners, while dealing with various problems that arise	– Software that directly maps the desired learning outcomes has a high rate of applicability to real world conditions – Ability to assess the way in which students perform in complex situations		
	– Sharing documents	– Students are able to produce collaborative projects, exchange and follow the various stages of writing and editing	– Possibility of collaborative work – Preparing students for collaborative work environments	– More efficient operation of the groups based on the real common characteristics of their members	

(continued)

Table 3.3 (continued)

Means	Activity	Examples	Needs met	Benefits of the Semantic Web	Principles of the WEEE being implemented
	– Online sending and receiving	– Students have the opportunity to submit their assignments to the instructor safely. The teacher is able to comment and return the assignments	– The teacher d ur hotel offers safe ways of receipt/delivery of works – Easy access for students	– Provision of personalised services – Proposals for related services	– Active learning – Provide immediate feedback
	– Score	– Students have the opportunity to view their grades online safely; – They have access to comparative/cumulative reports. Teachers have the ability to enter and manage student grades	– Ease of access		

(continued)

Table 3.3 (continued)

Means	Activity	Examples	Needs met	Benefits of the Semantic Web	Principles of the WEEE being implemented
Computer	– Ease of communication [emphasis on direct communication is crucial for other goals as well]	– Discussion: questions or topics around the content – Expression of opinion, criticism – Learning based on problem solving or case studies scientific research: the presentation of research – Notes – Immediate help/support team – Study groups – Seminars – Small group project design, document exchange, collaborative writing – Specialised part-time teachers – Mentors – Teaching/assessment by classmates and feedback – Discussion—views Report – Support for collaborating students	– Possibility of extending the discussion, outside the contact hours – The reduction/elimination of contact time and travel – Increase reflective learning activities – Opportunity for shy students to participate – Avoiding the problems of meeting small groups – Meet experts outside the classroom – Monitoring the progress of the teams' work – increase contact with students – Gradual support of student develop-ment/communication, support		– Development of reciprocity and cooperation: study groups, problem solving – Promoting active learning: discussion, analysis, problem solving – Emphasis on proper time management: learning becomes as accessible and flexible as possible – Utilisation of various forms of learning

(continued)

Table 3.3 (continued)

Means	Activity	Examples	Needs met	Benefits of the Semantic Web	Principles of the WEEE being implemented
	– Structured learning [focus on the object of study and the dialogue that are stored and can be retested]	– Recording answers to frequently asked questions that arise through the use – Development history of the project – Competitions and other repetitive procedures [e.g. design] – Development of the discussion, reaching agreements on various issues	– Creating a student community – Process evaluation and outcome – Creating a database for future classes	– More efficient operation of communities based on the real common characteristics of their members	

the public administration, in the so-called back-office, a key question arises as to whether the digitization of documents and various data and their storage in traditional relational databases is enough to improve delivery times of services and their more efficient operation than the traditional model of governance. Still, structural reforms and the constant mobility of staff in the public sector have brought major changes and dysfunctions in the back-Office of the administration. Reducing the operating costs of the public sector is vital and linked to the efficient operation of public bodies. This is particularly important given the European economic crisis of recent years. What is required is a corresponding evolution in the overall view of e-Government.

The opportunities for the most effective functioning of e-Government are linked to the fifth level of development that we have mentioned above. With the achievement of this level, the further development of e-Government is sought with the aim of integrating smart e-services for the general public as well as to the public administration bodies themselves. These services are characterised by advanced features such as recommendations to traders, pre-filled applications and smart information retrieval that takes into account the importance of concepts. Their implementation is linked to the adoption of technologies and methodologies of the Semantic Web for the benefit of saving resources and infrastructure. The fifth level of development is particularly important if we take into account the amount of information that is circulated both electronically and physically in conjunction with the number and complexity of transactions with public administration. In the implementation of this level we consider that it is possible to effectively use the technologies of the Semantic Web that we will develop in the following chapters. Our focus is on the possibilities that Semantic Web technologies can offer both in the field of modelling and knowledge management of the public sector and in supporting open government through the development of linked open government data.

We believe that the field of e-Government can be one of the main consumers of innovative services based on the Semantic Web. This is because the modern trend of open government is supported by the publication of "open data" and their further connection to Semantic Web technologies. With the use of Semantic Web technologies, improvements are expected that concern both parts of the overall system, i.e. the so-called "front-office" and the "back-office":

- In terms of serving the citizens, the search for the provided services will be significantly facilitated both electronically through the government portals (development of semantic portals) and with the physical presence at the service point (within the operation of the one-stop shops). Searching for the right service from the relevant body can become more efficient and effective. At the same time, the provision of personalised services and pre-completed forms and applications will make online services more attractive to the general public.
- As for the back-office, the search for information from the involved bodies and the sharing of data based on common terminology and semantics will improve both the internal delivery times of the data and the overall performance of the

institutions, since when all public systems "speak" in the same language any sharing problems are more easily solved.

- The Semantic Web is expected to contribute to the more intelligent access and management of information circulating on the Internet through new technologies and the development of corresponding new applications. The overall vision is for the transition from the existing Internet of static government pages to a network of dynamic public service providers (Web-services) that automatically discover the information or services they are looking for, trade in goods or services that the user intends to request, or gather information from different sources and combine them into homogeneous forms with the ultimate goal of sharing and interacting with other "common language" systems.
- In the field of knowledge modelling in the public sector ontology management tools have to offer particular benefits for creating and managing the knowledge base around public data. We consider this to be a crucial point, as the field of public information is quite complex and the information comes from different bodies that often use different datasets while there is no commonly accepted dictionary of concepts and interrelationships. This issue also concerns the field of simplification of procedures since with the technologies of the Semantic Web and especially the ontologies, their modelling becomes easier and more efficient.
- Particular benefits may arise in the field of management of the various registers maintained by public sector services in the areas of trade, codification of legislation, procedures, etc. using ontologies and knowledge base management tools.
- The tools and technologies of the Semantic Web can assist in dealing with problems stemming from the opening of public data through the "Diavgeia" program in Greece, such as:
  - The existence of a large number of different systems which do not communicate with each other as they do not support the existing interoperability standards.
  - Many operators store the same data in different places, ignoring the existence of duplicates resulting in wasted resources and conflicts when using them.
  - In terms of the structure of organisations, similar—but not the exact same terminology is used to name organic units that have the same object.
  - There is confusion as to the correct completion of the subject areas and the type of decision.
  - The search for a published decision in the existing search engine is based on a keyword or expression either by entity or in all entities but the results are not related to each other.

The changes expected with the use of Semantic Web technologies in the field of e-Government in relation to the current situation through web 2 are recorded in Table 3.4.

Table 3.4 Impact of the web technologies in the domain of e-Government

	Before web	Web 1.0	Web 2.0	Web 3.0
Main goal	Providing public services	Better services	– Openness – Collaboration	– Optimization of resources – Personalised & Smart services
Transaction method	– Physical presence	– Physical presence – Linked governance	– Physical presence – Electronic identification – Open and collaborative governance	– Physical presence – Electronic identification – Smart governance
Place	Counter	Counter	– Counter – Sites – Portals	– Counter – Sites – Portals
Object	Document	Document	– Document – Online application	– Document – Online application – Smart application – Pre-filled forms
Transaction	– Information – Application – Adoption of individual administrative acts	– Information – Application – Adoption of individual administrative acts	– Online Information – Online Application – Adoption of individual administrative acts	– Smart information retrieval – Online application – Online adoption of individual administrative acts – Personalised services
Function	Process oriented	Process oriented	Client oriented	Knowledge management driven
Mean	Hard copy	– Hard copy – PCs	– Hard copy – PCs – Tablets – Smart phones	– Hard Copy – PCs – Tablets – Smart phones
Key ICT area	–	– Infrastructures – Organisation	– People – Data	– Machine intelligence – Internet of Things

(continued)

Table 3.4 (continued)

	Before web	Web 1.0	Web 2.0	Web 3.0
Key obstacle/Risk	Public sector mentality	– Public sector mentality – Education	– Public sector mentality – Infrastructure support – Resources – Interoperability	– Public sector mentality – Education – Data protection – Users maturity

References

1. Connolly, D. (2000). *A little history of the World Wide Web.* https://www.w3.org/History.html
2. Berners-Lee, T. (1989). *History of the Web.* https://webfoundation.org/about/vision/history-of-the-web/
3. W3C. *The history of the Web.* https://www.w3.org/wiki/The_history_of_the_Web
4. CERN. *The birth of the Web.* https://home.cern/about
5. Netcraft. *Web server survey.* Retrieved 3, 2021, from https://news.netcraft.com/
6. W3C official site: www.w3c.org
7. O'Reilly, T. (2007). What is Web 2.0: Design patterns and business models for the next generation of software. *Communications & Strategies, 1,* 17. First Quarter. Available at SSRN: https://ssrn.com/abstract=1008839
8. Extended Web. https://sites.google.com/site/Internetofthings4867/o-pankosmios-istos/ektetamenos-istos-web-x-0
9. Dineva, S., & Nedeva, V. (2018). *Expanding Web and innovation skills for 21st century.* https://www.researchgate.net/publication/327558119_Expanding_Web_and_Innovation_Skills_for_21st_Century
10. The evolution and the interaction between web technologies and browsers over time (2021). http://evolutionofweb.appspot.com/?hl=en
11. Web Services Architecture Working Group, Web Services Glossary. https://www.w3.org/TR/ws-gloss/#defs
12. Cerami, E. (2002). *Web services essentials: Distributed applications with XMLRPC, SOAP, UDDI & WSDL.* O'Reilly.
13. The SOA Source Book. https://www.opengroup.org/soa/source-book/soa/p4.htm#tbl1
14. Bratt, S. *Developing the foundational standards for Web Services, W3C.* https://www.w3.org/2005/Talks/0420-sb-gartnerWS/slide10-0.html
15. IBM Documentation, Service-oriented architecture (SOA). https://www.ibm.com/support/knowledgecenter/en/SSMQ79_9.5.1/com.ibm.egl.pg.doc/topics/pegl_serv_overview.html
16. W3C, Index of elements in HTML 4, World Wide Web Consortium. December 24, 1999.
17. Berners-Lee, T. *Hypertext Markup Language (HTML) "Internet-Draft".* https://www.w3.org/MarkUp/draft-ietf-iiir-html-01.txt
18. IETF and the RFC Standards Process. http://www.faqs.org/docs/artu/ietf_process.html
19. A Review of the HTML+ Document Format. https://www.w3.org/MarkUp/htmlplus_paper/htmlplus.html
20. Berners-Lee, T. (1995). *MIT/W3C, D. Connolly, Hypertext Markup Language—2.0, Network Working Group.* https://www.ietf.org/rfc/rfc1866.txt
21. Lam, J., Alexander, I., Kmiec, M., Ragget, D. (1998). *Raggett on HTML 4* (2nd ed.). Addison-Wesley.
22. Mark Pilgrim, Dive into HTML5. https://diveintohtml5.info
23. O'Reilly, T. *What is Web 2.0, design patterns and business models for the next generation of software.* https://www.oreilly.com/pub/a/web2/archive/what-is-web-20.html
24. W3schools, AJAX introduction. https://www.w3schools.com/xml/ajax_intro.asp

25. Berners-Lee, T., Hendler, J., & Lassila, O. (2001). The Semantic Web a new form of Web content that is meaningful to computers will unleash a revolution of new possibilities. *Scientific American: Feature Article: The Semantic Web.*
26. Lawspot.gr, Cambridge Analytica: Facebook under the microscope of the European Parliament for the data breach case, article published in www.lawspot.gr. Retrieved November 9, 2018, from https://www.lawspot.gr/nomika-nea/cambridge-analytica-sto-mikroskopio-toy-eyrokoinovoylioy-facebook-gia-tin-ypothesi
27. Bernels-Lee, T. *Semantic web architecture.* https://www.w3.org/2000/Talks/1206-xml2k-tbl/slide10-0.html
28. Guarino, N. (1998). Formal ontologies and information systems (pp. 3–15). In *Proceedings of FOIS '98.* IOS Press.
29. Gomez-Perez, & Richard Benjamins. (1999). Applications of ontologies and problem-solving methods. *AI Magazine, 20*(1).
30. Haav, H.-M. (2011). A practical methodology for development of a network of e_Government domain ontologies. In *IFIP International Federation for Information Processing.*
31. Paslaru, E., Simperl, B., & Tempich, C. (2006). *Ontology engineering: A reality check.*
32. W3C Semantic Web, Semantic Web Tools, Category:Semantic Web browser. https://www.w3.org/2001/sw/wiki/Category:Semantic_Web_Browser
33. Dbpedia official site: https://wiki.dbpedia.org/
34. Dbpedia. DBpedia information extraction framework. github.com/dbpedia/extraction-framework/wiki
35. Paul Miller of Talis, Sir Tim Berners-Lee Talks with Talis about the Semantic Web, Talis (2008). http://talis-podcasts.s3.amazonaws.com/twt20080207_TimBL.html
36. https://web.archive.org/web/20140921021528/http://wiki.dbpedia.org/Team
37. The official dbpedia blog: https://blog.dbpedia.org/
38. DBpedia Mappings Wiki: mappings.dbpedia.org
39. Wikipedia official site: https://en.wikipedia.org/wiki/DBpedia
40. Schema official site: www.schema.org
41. Schema repository on github: https://github.com/schemaorg/schemaorg/
42. Wikidata official site: www.wikidata.org
43. Europeana official site: www.europeana.eu
44. Manguinhas, H. *European semantic enrichment framework—Documentation.* https://docs.google.com/document/d/1JvjrWMTpMIH7WnuieNqcT0zpJAXUPo6x4uMBj1pEx0Y/edit
45. Geonames official site: www.geonames.org
46. GeoNames API: https://www.programmableweb.com/api/geonames
47. Geonames forum page: forum.geonames.org
48. Geonames ontology: www.geonames.org/ontology
49. Pandey, G. (2012). The Semantic Web: An introduction and issues. *International Journal of Engineering Research and Applications, 2*(1), 780–786.
50. https://wordnet.princeton.edu
51. https://archive.illc.uva.nl//EuroWordNet/
52. Lukasiewicza, T., & Stracciab, U. (2008). Managing uncertainty and Vagueness in description logics for the Semantic Web. *Journal of Web Semantics, 6*(4), 291–308.
53. Nonaka, I., & Takeouchi, H. (1995). *The knowledge-creating company.* Oxford University Press, Inc.
54. The diavgeia project official site: www.diavgeia.gov.gr
55. Chickering, A., & Ehrmann S. C. (1996). Implementing the seven principles: Technology as lever. *AAHE Bulletin,* 3–6.
56. Kerkiri, S., Manitsaris, A., & Mavridis, I. *Towards an e-Learning system in the Semantic Web.* Department of Applied Informatics, University of Macedonia.
57. Paroutsas, D. *Teaching and learning using the Internet.* https://paroutsas.jmc.gr/pc-learn.htm

Chapter 4
Representation and Knowledge Management for the Benefit of e-Government—Opportunities Through the Tools of the Semantic Web

Abstract The main subject of this chapter is the representation and management of knowledge as well as the opportunities from the development and use of knowledge retrieval applications that exploit the tools and technologies of the Semantic Web in the field of e-Government. In particular, the concepts of knowledge and knowledge management are examined in terms of the Semantic Web with an emphasis on the critical points and guidelines for their application in the public sector and in the field of e-Government. Tools and applications of Semantic Web technologies for modelling and knowledge management are presented with examples of their application in the more specialised field of e-Government. Through relevant tools and applications, e-Government specific ontology is developed, proposed and evaluated for use in terms of open data. Also, technologies and methods and algorithms for knowledge retrieval through querying ontologies using Semantic Web tools are proposed, while examples of modelling and knowledge retrieval through ontologies from the specific fields of official statistics and public sector internal audit are given.

Keywords Knowledge management · Semantic tools · e-Government · Semantic modelling · Ontologies

4.1 Knowledge and e-Government—Competitive Advantage for the Public Sector

The knowledge that can be generated from the processing of information has emerged today as a comparative advantage for a business, a research community or a public organisation. At the same time, it is a factor for the development of the economy in the context of the so-called knowledge economy. A key factor for knowledge production is the effective search and retrieval of information mainly from the Internet in an automated way. The World Wide Web in the context of the free flow of information that it supports, is nowadays a key tool through which the search, retrieval and dissemination of information is achieved in all fields: science, business, government, etc. Unlike the processing of data for the production of information, the production of knowledge has so far not been automated to be fully supported by machines

S. Theocharis and G. A. Tsihrintzis, *Semantic Knowledge Modelling via Open Linked Ontologies*, Artificial Intelligence-Enhanced Software and Systems Engineering 4,
https://doi.org/10.1007/978-3-031-20585-9_4

(computer systems), and is a matter mainly of human intellect and effort. Existing technologies of web2.0 support the user in searching and retrieving information via the Internet with specific tools e.g. search engines, but then this is the determining factor in its further processing.

The improvement of services provided by the public sector is directly related to the reinforcement of support systems of public administration, the so-called "back-office". The situation currently faced by the employees, is characterised by the huge amount of data and information generated and handled, by the inadequacy of existing systems to manage them successfully enough to transform the information into valuable management knowledge. The whole situation is getting worse when employees who leave the service due to retirement or relocation to other departments, do not leave behind them the valuable knowledge they gained during the years of their service. Even in the case that organisational knowledge exists "somewhere" in the institution, its recovery is an extremely difficult and time consuming task. The result is obvious: waste of time, poor quality service, reduced profitability, poor impression to the citizens. Thus, the issue of managing and sharing the knowledge generated within the public services is considered important, while the knowledge itself practically remains undeveloped. Moreover, a key aspect for the successful operation of back-office systems to e-Governance, are properly trained and informed staff, who can use the tools offered by ICT.

The challenge in the near future for the successful operation of the public sector is the creation of mechanisms to discover, draw, share, and exploit the knowledge of each employee to increase productivity and efficiency of services. This part of our book is an attempt to identify issues related to the concept of knowledge and its codification as initial steps in the process of creating a mechanism to manage the public sector. Of particular interest is the possibility of developing a system perspective based on knowledge as essential tools in support of the back-office administration and the investigation for the creation of relevant controls for the further development of e-Government.

4.1.1 Knowledge as a Concept

The concept of knowledge in various fields of human activity has ancient roots revealed in the works of Greek philosophers. Knowledge emanates from the people through their experiences and ideas. Plato, for example, argued that the correct belief can be translated into knowledge if established through logic and reason. Aristotle believed that knowledge of an object requires understanding its causes. Also, the theories of the Pythagoreans about wisdom included the field of knowledge, which, as mentioned by historians of philosophy, is not scientific unless it is mathematical. For the sophists, knowledge depends on two things: possession of capabilities that support the perception of reality and existence of an objective reality which is the object of knowledge. Yet, bodies of knowledge are the human senses which now become particularly interesting.

Knowledge, as a subject of study in Western philosophy, was perceived as something ideal, universal, impartial, and reasonable. Thus, it was considered as a reality with a natural substance that could be reproduced and, therefore, properly used through technology. In recent decades, efforts to manage knowledge have resulted in the development of such tools with the help of ICT. Most of the early models of knowledge management (KM) tools have faced knowledge as an object and not as an element deeply rooted in human cognition and behaviour. The nature of knowledge is more complex than the data and information because it involves the social component. Thus, it is now generally accepted that the active involvement of people is largely required in the supply, information and management of knowledge in information systems [1].

Therefore, for the proper understanding and application of KM, it is necessary to specify in advance the main differences between data, information and knowledge. Concepts such as data, information, knowledge and wisdom are concepts that are closely linked and interdependent [2]. For this reason, often there is confusion in daily practice as to their distinction and use.

From time to time, the scientific community has tried to define what data is, what information is and what knowledge is. These concepts change according to the area of interest, adjacent properties and characteristics. The relationship between data, information, knowledge and wisdom are shown in pyramidal shape. At the base of the pyramid there is the data and hierarchically follows information, then knowledge and finally on top of all wisdom. The following figure (Fig. 4.1) illustrates the relation between data, information, knowledge and wisdom.

Generally accepted definitions of these terms can be stated as follows:

Data: sets of distinguishable items of information about an event or a process which by themselves are not very useful unless converted into information. For example, the data can be related to numerical quantities or other properties derived from observations, experiments or calculations. Cost, speed, time, and capacity are quantified data.

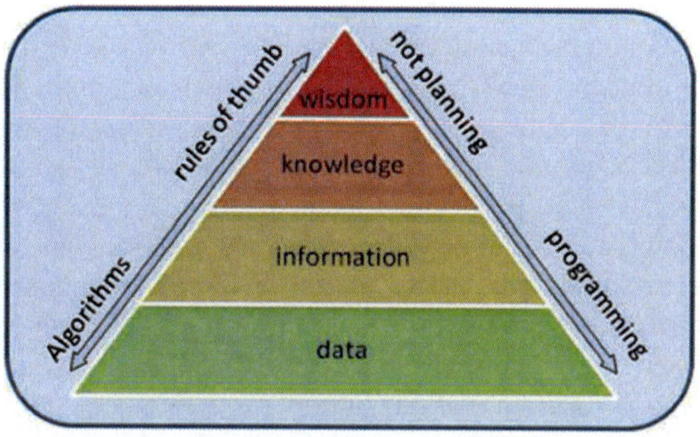

Fig. 4.1 The pyramid of wisdom [3]

Information: data that has been assigned relevance and purpose, that is data that have meaning and are organised to serve a particular purpose. For example, the information in a data collection may have the form of relevant explanation, interpretation and other attributes pertaining to a particular object, event or process.

The data can be converted into information mainly through five key processes [4]:

- Concentration—some data are summarised in a more concise form, while unnecessary details are being eliminated.
- Correlation—the purpose or reason for collecting data is known or understood in advance.
- Calculation—data are processed and aggregated in order to extract useful information.
- Classification—data are assigned to particular types or categories.
- Correction—errors are eliminated from data sets.

Knowledge: versatile combination of experiences, values, information about context or environment, insight, and reasonable intuition that provides an appropriate environment and framework for evaluating and incorporating new experiences and information. Knowledge derives from the human mind and is applied by it. In organisations, knowledge is often recorded not only in documents or records but also in the routine of the company, its procedures, practices and informal or formal rules that are already established.

Knowledge is based on information organised, summarised and synthesised to enhance its own understanding and awareness. Knowledge can represent a situation or the prospect of action and decision made by an individual, a group or an organisation. It can be modified during the learning process which causes changes in understanding, decisions and/or actions triggered by knowledge itself. A practical and symbolic definition describes knowledge as an "apple bite." To acquire knowledge, one must take a "bite" (information), chew it, digest it through the concept of process and, finally, put into practice what he/she has learnt. The usual questions about the data and information are related to "whom", "what", "where", "when", "how", and "why."

Wisdom: The ability to recognize the truth and make good judgments based on prior knowledge, experience and insight. Within an organisation, intellectual capital or organisational wisdom are the implementation of collective knowledge.

As an application of the above in public administration, we refer to an example taken from the Greek experience. The Ministry of Interior, among the programs implemented in the development of e-Government services, has included the program "Politeia[1]" (meaning "state") which aims, among other things, at enhancing the quality of services to citizens by providing telephone services. More specifically, with a call to a particular number, the citizen can request any of a number of supplied certificates, which are sent to him/her by post. If we examine the knowledge associated with the operation of this pilot program, we could mention the following:

---

[1] More information about the "Politeia" program can be retrieved from the official site of Ministry of the Interior of Greece and from the official documentation via the National Printing House of Greece (https://www.kep.unipi.gr/wp-content/uploads/2017/11/2880_2001.pdf).

- *Data*: The data may include measurable or qualitative elements, which are recorded on a daily basis, such as the distribution of telephone applications (measurable data) per category of certificate (qualitative data). The data recorded by employees serving citizens through telephone will make sense if placed in a specific structure and arranged through a special framework. The recording of comments or observations by both parties (employees and citizens) could in the future be transformed into valuable knowledge. If, for example, we consider the case of citizens who for various reasons can not easily respond to telephone calls, their difficulty to converse with employees may be due to insufficient knowledge of the Greek language or hearing problems, especially for older people, or even the inability to identify the citizen. Such problems should be recorded with the objective to identify solutions or adaptations of the overall system in the future.
- *Information*: When the employee data are properly structured and stored in a worksheet in order to present an overview that will concern specific characteristics, then relevant information has been produced. The corresponding information package will provide more information in the data entry process knowledge. For example, we could produce a worksheet where telephone calls are recorded by category of service, by gender of the citizen etc. as well as by specific events outside of the routine process.
- *Knowledge*: The careful interpretation of the worksheet that was handed by the employees to the person in charge of this task would help him/her to draw conclusions based on this process, which relies on knowledge held by the particular responsible. For example, comparison of worksheets on a weekly or monthly basis with the expected results of the project would require the person in charge to collect the relevant worksheets to consolidate and then put them into appropriate software that will help him/her to get the big picture. Thus, he/she will be able to judge what has not really worked in specific cases or to propose further implementation or rejection of the program.
- *Wisdom*: The wisdom in this example could be described as the responsible person's ability to improve his/her telephone service performance by reviewing the process or by changing the very way of evaluating the program based on experience gained from similar programs. The latter requires that problems and comments are recorded, as well as the way of processing by the person responsible for the project. Also, after a first period for providing the service, any interviews with the employees and recording of events and experiences that are not recorded on the worksheets, could lead to substantial changes of the service to better serve the actual needs of citizens.

4.1.2 Knowledge Creation

Knowledge creation always starts from an individual or a group of individuals who individually or as a group process and discuss new ideas, concepts, events, problems, etc. These issues may involve multiple cases such as innovative products or services, production processes, market developments, etc. [5]. The creation of knowledge can take place through research and development which requires innovation, experimentation, observation, etc. Knowledge creation usually begins with the formulation of an argument or idea, followed by individual or peer learning, to obtain information, assessment or assertion of the idea. Finally, organisational knowledge is constructed at higher levels [4]. It should be noted that in an organisation, the creation of knowledge is an endless process, practically without a beginning or an end. This process is sustained by continuous transformation.

The process of creation and the transformation of knowledge into an organisation is based on a simple framework that includes two dimensions. The first dimension shows that only individuals create knowledge while the second dimension relates to the interaction between tacit and explicit knowledge. These two dimensions are the basis for defining four processes of creation and transformation of knowledge, namely socialisation, externalisation, combination and internalisation. Specifically:

- During socialisation, tacit knowledge is transformed into tacit knowledge during discussions, communications, meetings, etc.
- During externalisation, tacit knowledge is transformed into explicit knowledge, recorded images, diagrams, documents, manuals, presentations, etc.
- During combination, explicit knowledge is transformed into another form of explicit knowledge by analysing and solving various complex problems arising constantly.
- During internalisation, explicit knowledge is transformed by people into tacit knowledge through reading, trial, repetition and many other instruments and statements, such as memory, impressions, etc.

The basic assumption of this well-known model [6] is that knowledge consists of tacit and explicit components. Specifically, four states are defined through which knowledge is transformed from one dynamic form (implicit) to another (explicit). Thus, through the process of socialisation we achieve sharing of tacit and explicit knowledge, while through the process of externalisation we convert tacit knowledge into explicit knowledge. Also, according to the model, explicit knowledge is combined with other explicit mechanisms through the set, and, through the process of internalisation, is converted into tacit knowledge. These transformations can be achieved only within organisations or communities characterised as learning ones.

The four different modes of transformation of knowledge create an endless helix of knowledge. This constant and dynamic process stems from the behaviour of the main generator of knowledge, that is humans. Let us consider the example of an employee in any department of the Government, who receives an application for citizenship. The employee initially refers to the relevant provisions (explicit knowledge) to check whether the application falls under the provisions of a certain law. When the law or the relevant provisions are ambiguous, then the employee goes back to any clarifying instructions and, if there aren't any, makes a request or a discussion (socialisation) either with other employees or in a public administration forum or with his/her superiors, to find a solution. If no solution is found from the discussion, the employee issues a clarifying instruction in order to both solve this problem and inform all employees concerned about the subject (externalisation). The implicit knowledge of the current state of public administration in Greece is diffused extremely slowly.

The conception of knowledge extends to all activities of an organisation from the structuring of information about customers and market to the collection of examples of best practice or "passive subjects" or the development of a counselling program. The tacit knowledge could include rumours, legends, stories, standards, beliefs, etc., while the explicit knowledge is being stored in books, documents, databases, websites, e-mails, etc. Thus, explicit knowledge is any systematic approach of conceiving, organising and perfecting information so as to facilitate finding information, learning and problem solving. The management of tacit knowledge is the process of capturing individual experiences and expertise within an organisation/institution and making them available to anyone in need of help.

In the case of public sector organisations, knowledge extends to all procedures to be followed, both internally between departments and externally, to provide public services to citizens or businesses. The tacit knowledge can be traced among employees with relevant experience in specific subjects and mainly informal situations, work culture or among informal groups created at the workplaces. Explicit knowledge is what characterises the overall operation of the public sector, as it is hardly recovered through the huge crowd provisions, legislation, circular letters, directives etc.

During the process of creating, conceiving and coding knowledge, new concepts or knowledge requirements are developed which generally require examination or penalty at a later stage in order to examine their sincerity, reliability and value. This means that new concepts have greater value than the existing ones. We should not forget, however, that the majority of companies and workers accumulate and capture knowledge unconsciously through various methods, practices and situations. Some of them are managed by company management and others are not, as they occur during daily work and practice (e.g. learning by doing, informal meetings, observation of others and following their sayings, so-called "passive lessons", etc.).

Generally, the acquisition of knowledge by individuals or groups can be defined as transport and transformation of valuable expertise from a source of knowledge, e.g. documents or employees with experience in a specific subject, to a setting of "storing" knowledge such as documents. Organisational memory includes all the elements of knowledge, from elements of tacit knowledge (based on the experience of the employees) to hard facts and data that can be stored in the archives of the organisation/institution. If knowledge is not stored in such concrete systems or if there is no immediate access to it through the effective involvement of people in intensive knowledge tasks, the organisation/institution is not able to handle the knowledge held by each of its members. The acquisition of organisational knowledge is the extension and articulation of individual knowledge into organizational level in order to internalise the knowledge base of the company [4].

4.1.3 *Knowledge Coding*

The sharing and use of existing knowledge within an organisation relies heavily on its codification. Knowledge can be spread widely and at a low cost particularly, when presented in a concrete, explicit form, e.g. in the form of an electronic or physical document. Knowledge must be encoded in order to be understood, maintained and improved as part of corporate or organisational memory [7]. In their daily practice either at work or in their personal lives, people use—usually without realising it—a kind of codification of knowledge in order to communicate with their environment and make their discussions more effective. Job or business jargon/slang, technical language of the programmer and even the communication of new messages via SMS or forums are such examples. However, the knowledge, skills, expertise and interest that an employee may have gained through his/her working years in a specific subject are always difficult and in many cases impossible to be consolidated in one document or in a database. In such cases, it is preferable to document existing linkages with knowledge sources and present them using knowledge maps or related guides such as the company phone book or a company directory.

The codification of explicit knowledge can be achieved through various techniques such as cognitive mapping, decision trees, knowledge taxonomies and task analysis. These are briefly analyzed next.

4.1.3.1 Cognitive Maps

Interviews with experts and professionals in specific subjects can lead to the transformation of the experience and knowledge from tacit to explicit form and its content can be represented by means of cognitive maps. A cognitive map is the representation of a "mental model" of knowledge of a person and a useful form of codified knowledge. The mapping of knowledge is based essentially on mapping concepts which in a knowledge map are represented by the nodes (circular areas), while the links between them (arrows) illustrate the connections between different concepts. The mapping enables experts to build knowledge and standards to reflect multiple perspectives or views on the content.

The mapping of knowledge and its expertise in an organisation is important in many ways. Easy access to knowledge maps facilitates the interconnection of the employees of an organisation in need of guidance or help by increasing their response rates, reducing repetition and improving efficiency by saving time and effort. Creating a knowledge map requires a special analysis of skills, experience, specialisation and their points of dispersion. Critical issues to ensure the validity and reliability of knowledge maps include information such as the "owner" of knowledge, degree of reliability, validity of the media, user rights, etc.

To illustrate the practical application of a knowledge map, we must first briefly introduce the twelve factors of topographic knowledge maps [8]. Then, we will discuss five important steps to be taken into account when applying a knowledge

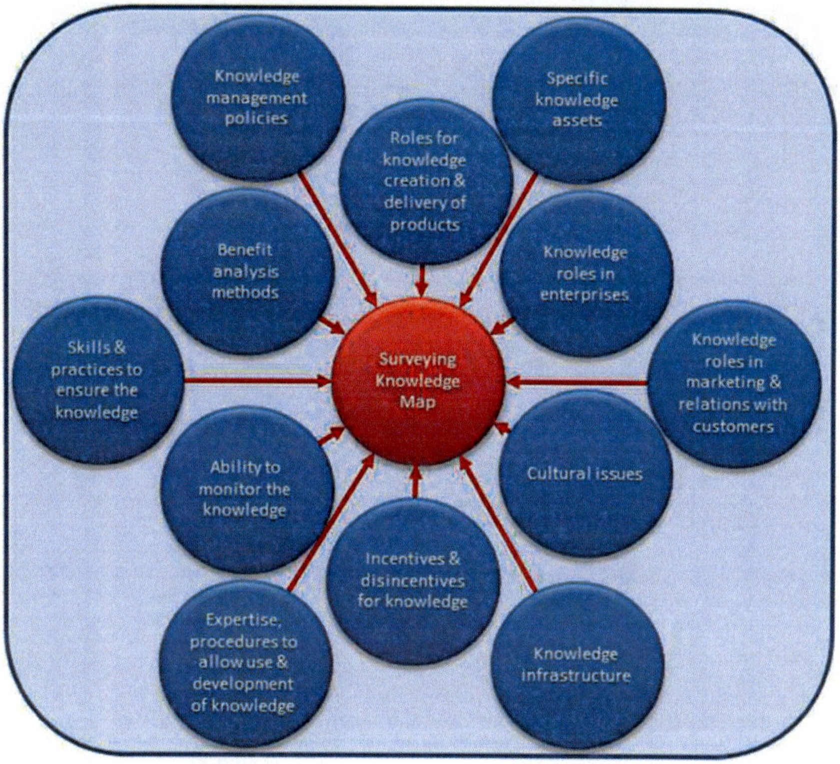

Fig. 4.2 Surveying knowledge map

map. The following figure (Fig. 4.2) shows the elements that can be integrated in a knowledge map.

In practice, it must be decided as an initial step, which of these items are important for the organisation/institution and which must be incorporated into the knowledge map. The topographic knowledge map generally provides short descriptions of key issues, programs, activities, practices, etc., concerning the knowledge. It also describes their situation in relation to competitors and existing markets and general developments within and outside the organisation/institution.

To build a knowledge map, we generally follow the next steps (Fig. 4.3):

Step 1: Building requirements and cognitive needs. It is recommended to test the strategy of the organisation and identify areas of knowledge required to develop, in order to achieve strategic objectives.

Step 2: Identification of knowledge required for specific tasks. We use methods such as:

- Interviews with very focused questions.
- Evaluate the work as it is carried out by the people involved.

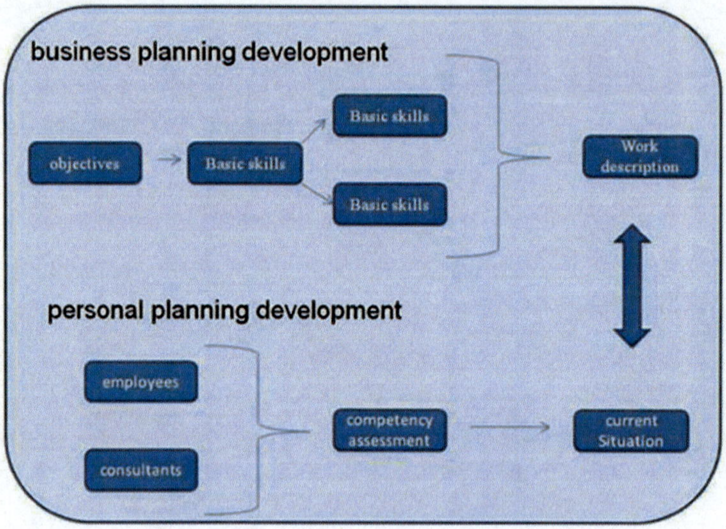

Fig. 4.3 Building a knowledge map

- Search for very critical objects (boundary objects).
- Search for logs related to the initiation in the business intranet.
- Investigation of common file types and documents of employees.
- Collection of political documents, organisational charts, documentation and procedures.
- Gathering information from external sources.

Step 3: Assessment of personal skills in relation to cognitive needs.

Step 4: Connecting knowledge map to education and staff development. In practice for comparing the desired corporate skills with the personal skills of the staff we usually use a two-way approach that starts from the corporate and personal level.

The following figure (Fig. 4.4) is an example of a knowledge map which could be taken into account by the political heads of a ministry to assess the current state of available experts in various fields of activities of the ministry. In this map, the space is divided into four sectors according to sectoral activity areas of the staff (administrators, technical staff, computer staff and project management staff). Employees may have specialised in issues within the ministry or in cooperating with other departments or have been engaged in projects commissioned by external customers in ministry. The analysis of this map can lead to definite conclusions, e.g. on technical or administrative subjects there are no external experts, etc.

Fig. 4.4 A staff allocation knowledge map

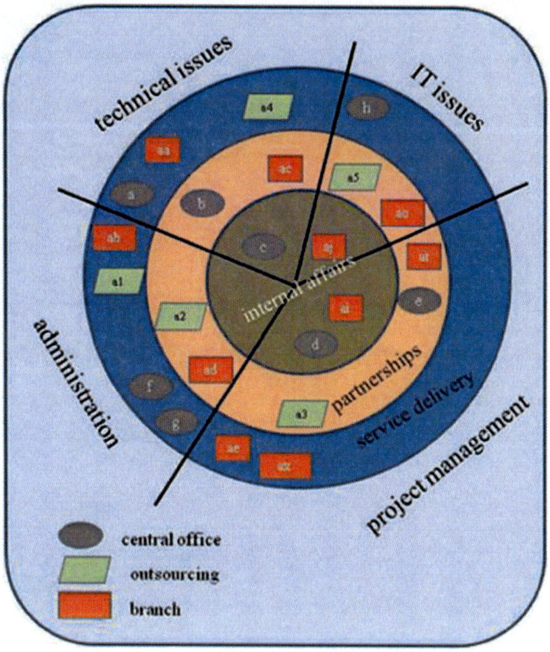

4.1.3.2 Decision Trees

They are usually presented with a flowchart (Fig. 4.5), where alternative routes deliver, using a visual way, the consequences or results that different decisions might have at each level that they can be taken. Its use for the representation of many rules is effective because the user can easily circumvent those rules which are not relevant to the cases on which he/she works, while he/she has oversight of all the rules and options.

4.1.3.3 Knowledge Taxonomies

As it has already been mentioned, concepts are the building blocks of knowledge and expertise. The taxonomies (Fig. 4.6) are essential classification systems that allow us to describe the concepts and their interdependencies—usually in a hierarchical fashion. The higher a concept is placed in the hierarchy, the more general or summarised it is. The lower it is placed, the more specific/special is the content of higher categories. This approach allows concepts that are lower in the hierarchy of taxonomy or are more specific to directly integrate the characteristics of the concepts that are higher in the hierarchy of parent concepts.

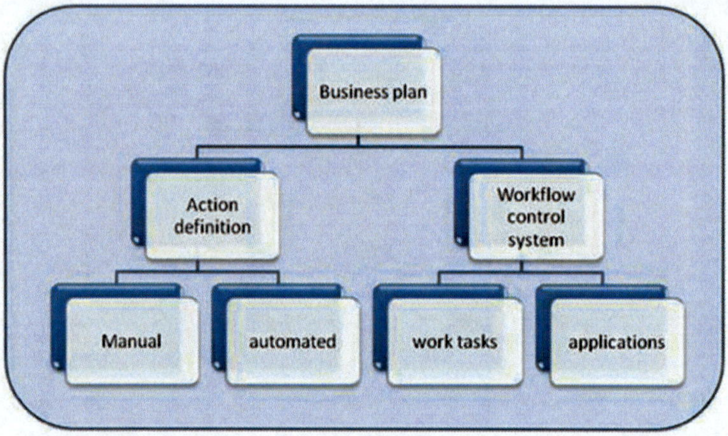

Fig. 4.5 Decision tree example

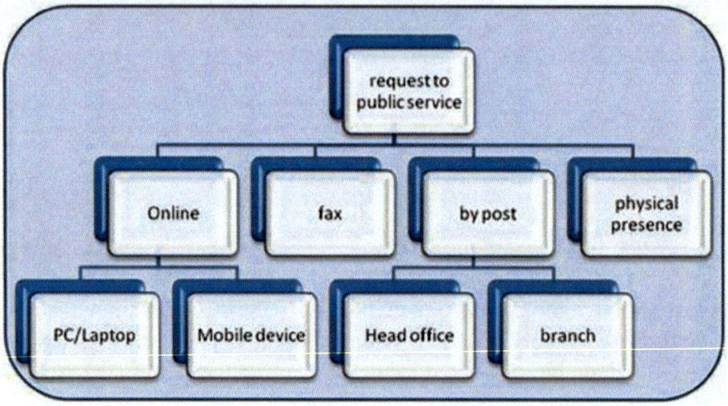

Fig. 4.6 The taxonomy of the request to a public service

4.2 Knowledge Management in Terms of Semantic Web—Critical Issues for Their Application in e-Government

4.2.1 The Concept "Knowledge Management"

According to a simple definition, knowledge management is "the process of converting information into useful knowledge, which is accessible when necessary" [9]. In other words, we could say that knowledge management in an organisation is a process of collecting the accumulated experience from its operation, and then disseminating and using it in those parts of the organisation where it will offer the maximum rewarding benefit. This knowledge is found in databases, or in documents, or even as intellect in the minds of the organisation's executives.

Knowledge management can be used by almost all government ministries and public organisations, although considered to have more visible results in relatively large companies and companies in technology-intensive or knowledge [10]. With knowledge management is possible to address problems of public administration such as:

- When there is a need for a specific expertise or skill and none of the existing staff seems to have such knowledge.
- When a solution to a problem requires some experience but the person who has this experience is missing.
- When the matching of the person (meaning its abilities) and the job does not work efficiently while optimal matching between position and specific, objective abilities.
- When an organisation decides to apply policies of staff training, but training needs, which reflect specific knowledge gaps are not known and have not been documented.
- When there is knowledge to certain individuals, however, either is not used or not effectively diffused the Organisation for a variety of reasons.

Although the application of knowledge management varies greatly between organisations in the public and private sector, generally the value of knowledge management can be found in the following areas:

- Decision making. With knowledge management can achieve better and faster decisions. This is because the experience of the organisation is recorded and this helps both to avoid mistakes and discover and use the best solutions to similar issues, saving time.
- Autonomy resources. The involved employees become able to access and use the knowledge of their peers while enhancing their responsibility and control of their performance.
- Learning. More rapid learning by compressing the length of the learning curve for any new object assigned to employees.

Although there are many business and organisational problems related to knowledge management, however oddly enough occur during the daily routine of organisations in both the private and public sectors:

- Employees have knowledge gained before entering the service or can be trained during their work (e.g. seminars, refresher courses, etc.) but are not allowed to use them for various reasons (e.g. political considerations, failure in matching job and person etc.).
- Employees during the exercise of their duties acquire knowledge but most do not transmit it to the other employees involved.

- Often there is the right person who can solve the problem, but few know how to locate him.
- Every organisation wants to have the best employees, but when they acquire them and train them properly, for various reasons those either change subject or even organisation.
- Employees are asked to solve problems and ask and seek the knowledge of their colleagues but in many cases they keep their knowledge to themselves.
- Employees must document their deliverables fully, but often do not have easy access to the relevant files.

4.2.2 Critical Issues for the Implementation of Knowledge Management in the Public Sector

Public bodies in order to have maximum results should align knowledge management with key strategic objectives. These targets are either known from the existing business plan of the entity or by conducting a SWOT analysis, from which we derive the corresponding priorities/objectives. This is crucial because in most cases, the results of knowledge management are not evident in terms of the costs or other quantitative characteristics and in order to ensure full commitment of management, the management knowledge team should present as soon as possible the first results.

A. *Diagnosis of the organisation's knowledge*

We must identify knowledge gaps, future requirements in knowledge and the specific knowledge characteristics of each institution. The well defined strategic objectives, providing a suitable framework for analysing the gaps in current knowledge and future needs of the body of knowledge. Knowledge gaps can arise by the absence of people (special expertise or expert) or lack of technology or both. Knowledge gaps should be filled by various techniques, such as internal training, hiring new staff, acquisition of expertise, production of new knowledge, etc. The potential knowledge gaps could be identified by problems that have been reported in the past, and after discussion with the heads of the organisations' units and the entire staff. This process is also likely to identify surplus knowledge in some units. This process can be done by using appropriate Diagnostics Knowledge Management Tools. These tools can provide the user with a detailed methodology on how to identify all areas (process identifier), which should be explored and further improved. In this way it is possible to identify some characteristics of the organisation regarding knowledge management and remove all the obstacles.

B. *Implementation of a policy changing culture*

In today's competitive work environment many employees do not feel safe in the job position and usually conceal their knowledge, which could be used to improve the overall efficiency of the department or the organisation. From another perspective, employees may hide critical knowledge, because they believe that if they share it with others they will lose their personal knowledge advantage. This is probably the biggest hurdle for any knowledge management activity. Employees must feel free to share ideas, make mistakes or ask other employees if they do not know something.

People in their workplace should not be afraid to discuss or read a book even while working, in order to gain new awareness. All these minor, but still very critical, daily activities can help an organisation to build a culture of knowledge management. It is equally important that the hierarchically highest executives communicate to the lower ones the strategic goals of the organisation in order that the necessity of any changes will be clearly known.

C. *Using appropriate technologies*

For Knowledge Management, the role of innovative ICT systems and the Internet next generation is catalytic. However, it is useful to note that their own infrastructure and applications are not sufficient for optimal results. Examples of some of the technologies that assist in knowledge management.

D. *Tools for the identification and organising of knowledge*

These include the following:

- Search applications based on a key word or an expression, such applications are provided so far by various private providers, but they do not focus on the semantic analysis of terms. However, for the modern needs of the public sector it is essential to use search techniques that provide solutions to the problem of semantic correlation between search terms.
- Tools such as intelligent agents could offer solutions to the issue of filtering the information requested or applications depending on the needs of the employees.
- Pattern recognition techniques and text comparison techniques will facilitate the clustering, categorization and classification of public government documents based on similarities in the terminology they use.
- Data warehouses, in the sense of linked databases on all public sector bodies.
- Smart tools tracking such information or document-based question answering undirected. Such tools can be used for data mining.
- Application of a knowledge mapping tool. The knowledge mapping is a process that aims to link knowledge to describe the people who possess it. In most cases it is difficult to map the knowledge itself and is easier and logical to create links between a brief description of knowledge and its holders. However, the knowledge in an organisation exists in many forms and refers to various subject areas (administration, production, technical issues, etc.). Therefore, the knowledge in order to be described and fully understood by all, has to be expressed in a familiar and easy language (codified knowledge). There are many types of coding and the most appropriate should include knowledge of the specific characteristics of the company.

E. *Collaboration tools*

Herein are meant those technologies that support continuous communication and sharing of knowledge, ideas and solutions from all the officials concerned. In particular characteristic functions must be:

- The real-time communication via messages between agents are logged into the system each time
- The asynchronous communication, to the logic of e-mail
- The ability to form communities with common objects, employment and knowledge level.

F. *Analysis and visualisation tools*

Herein means tools that are able to interpret the input information but also to transform input data into detailed graphics. In this area, the virtual environment will allow users a more efficient collaboration between remote partners compared to traditional techniques.

G. *Decision support systems*

Herein means systems interacting through electronic means, in order to assist users in problem solving and decision making. Such systems take into account the input data and use specific models to replace the administrative crisis. The development of intelligent decision support systems that learn from the usage history of each user and the organisation would have significant advantages over existing systems that have been developed.

4.2.3 Knowledge Management Procedures

Knowledge management focuses on capturing the expertise of organisations/institutions and individuals, as well as the allocation and application of this expertise to maximise benefits and profits derived from the assets of knowledge. Knowledge management helps to transfer the appropriate knowledge to the right people at the right time, giving them the power to make the right decisions. Still, involves the identification and analysis of the existing and the required knowledge, and the subsequent planning and control of operations required for the development of knowledge assets and the fulfilment of the objectives of the organisation/institution.

In order for the knowledge to be transformed into a valuable corporate asset, knowledge, experience and expertise should be standardised, distributed, divided and implemented [11]. Knowledge Management is an integral part of any strategy that utilises the expertise to create a sustainable competitive advantage in today's business environment. Many researchers formulated templates for Knowledge Management

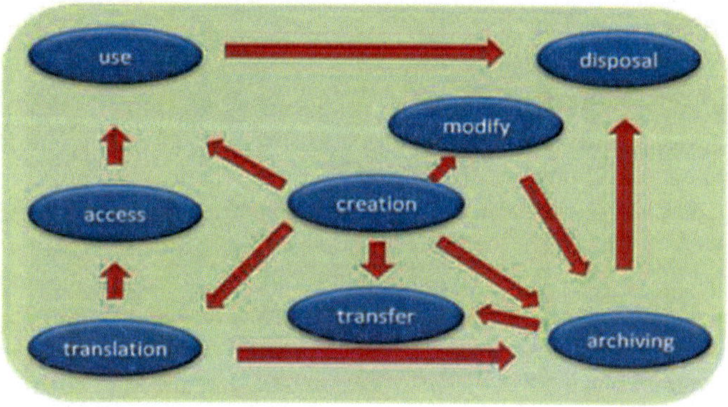

Fig. 4.7 Processes of knowledge management [12]

comprising from 2 to 8 or more different processes (Fig. 4.7) among which the following [12]:

- Creation or acquisition. Knowledge either created or collected by knowledge workers.
- Modify. Knowledge modified, adapted or combined to meet immediate and future needs.
- Use. This knowledge is used for a particular useful purpose.
- Archiving. The knowledge is stored in a specific format (format) which will be maintained in perpetuity and will be accessible and useful for future use by knowledge workers of the organisation/institution (encoding).
- Transfer. Knowledge can be transferred or communicated from one person or place to another.
- Translation/redefinition of purpose. This knowledge is translated from its original form to a new form which is most suitable to achieve a new goal.
- Access user. Knowledge is accessible to knowledge workers depending on the position of responsibility held by the organisation.
- Disposal. It is important to recognize which information/knowledge one should keep and which one to delete.

We note that in the context of knowledge management practice, these processes can be grouped into more general processes that are more appropriate for business processes implemented by the organisation/institution.

4.2.4 Knowledge Management Systems

With the concept of "system" knowledge management refers to a collection of technological infrastructures which are connected with each other and with their environment with specific interfaces [8, 13]. People, content, technology (software and hardware) and their interfaces are the interdependent and inseparable components of such a system (Fig. 4.8). People who possess knowledge (explicit or implicit) offer content-knowledge, and the transfer and diffusion of the whole organism is through technology. This combination provides effective and efficient management of an organisation's core knowledge.

From a technological point of view, an interesting architecture of such a system is shown in the following table (Table 4.1) of the 7-level subsystems.

A. *The User Interface Layer*

The first layer of the proposed architecture is actually a unique subsystem-level of interaction between user and machine. It can have the form of a browser or a machine wiki. The aim of this module is to support the flow of both the explicit and the tacit knowledge and information to and from the man and the machine. This transfer can be made more efficient and effective by using a personalised system that will use appropriate teaching aids, such as graphs, tables, charts and text. Thus, the explicit and tacit knowledge will be presented in a consistent, navigable, friendly and consistent manner regardless of the source of the turnout.

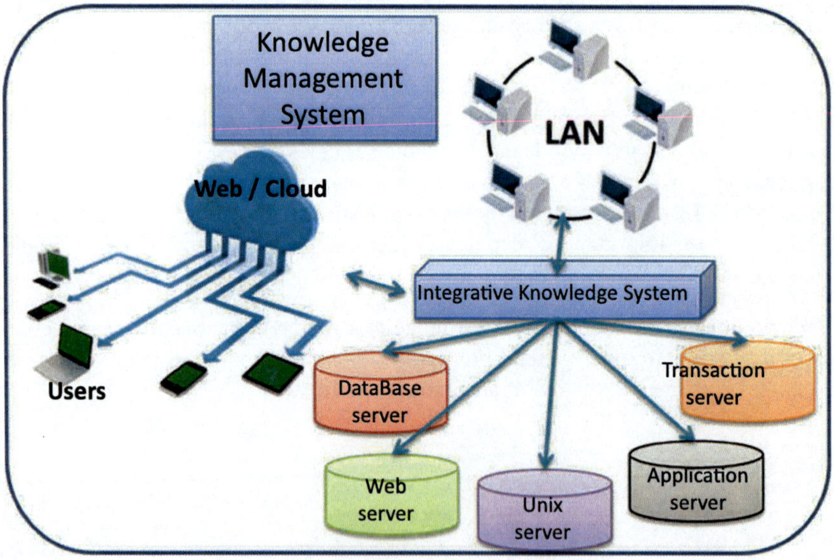

Fig. 4.8 The structure of a knowledge management system [8, 13]

Table 4.1 The 7-level subsystems

User interface layer	• browsers
Authorized access layer	• support : authentication, identification, security, firewall, tunneling
Collaborative intelligence & filtering layer	• intelligent agent • Personalization • quest • content
Knowledge enabling application layer	• Knowledge sharing environment • Process automation • Imaging tools
Transport layer	• LAN's • WAN's • Intranet – Extranet - Internet
Middleware layer	• Communication support
Repositories layer	• Collaboration software • Data bases • Warehouses – data mining

Since this layer acts as an interface between the user and the system, the question of design is extremely important and various features must be taken into account like:

- Consistency. As with any other software we must use menus, buttons and icons with the same appearance and functionality.
- Relevance. The information provided must be strictly relevant to the user's requirements.
- Optical clarity to the extent of saving space in the output means of information.
- Optical clarity to the extent of saving space in the output means of information.
- Easy navigation between pages and related files.
- Maximum usability. The usability of the various menus and icons, the speed of access to files and websites are among the basic users required.

At this level, we must address the issue of managing tacit knowledge. The tacit knowledge embodied in the intellect of experts or employees engaged for a long time to a specific object, it is difficult to codify and share with existing systems. It is therefore a major challenge to develop a subsystem that supports the automated recording and mapping of tacit knowledge of government officials.

B. *Authorised access layer*

Here is where the protection of the system and controlled access to knowledge found in computer systems of the body are ensured. It is extremely important because such systems enable remote access via Internet or via intranet, so they are particularly vulnerable to malicious attacks. A modern access layer focuses on safety, on the use of protocols such as passwords, on certification to ensure that they are authorised users and software tools such as firewalls that prevent the exit of certain sensitive personal data or the entrance of information that can destroy the records of the organisation.

C. *Collaborative intelligence and filtering layer*

The third layer provides in a personalised way the stored information. Designed to reduce the seek time of information and the relevance to the actual knowledge searched in conjunction with the user profile. This feature is provided via a search engine such as a browser or client/server network of the company in conjunction with the first level of the architecture. In this field, intelligent agents can give meaningful solutions. Intelligent agents are active tools that can understand, analyse (learn from past experiences) and take appropriate action to help solve a problem.

D. *Knowledge-enabling application layer*

Most of the applications developed at this level, aim at knowledge management and learning by providing users better ways to perform their work. May include discussion databases, automation tools for often repetitive tasks, phonebook decision support, tools, visualisation tools or other specialised applications. The ultimate goal is to provide users with knowledge about specific tasks that they perform.

E. *Transport layer*

This level includes local area networks—LANs to wide area networks—WANs, the intranet and extranet as well as the Internet. The basic technical characteristics which must be considered are the interface speeds, bandwidth (bandwidth), the tools for managing web traffic/load.

F. *Middleware layer*

Here there is the interoperability of different systems. During the system design we must take into account the way the system AIs communicate with existing systems and applications that have been developed on different platforms or run on different operating systems.

G. *The Repositories layer*

This is the lowest layer of the architecture and DK represents the physical layer in which the points of data storage are located. Includes data warehouses, legacy applications, operational databases and custom applications to manage network traffic/congestion and ensure the integrity of the architecture AIs. Each storage location has the required structure for the type of knowledge that is stored.

4.3 Semantic Tools for Knowledge Management in the Domain of Public Administration

4.3.1 The RDF Data Model

The Resource Description Framework—RDF as it was mentioned before is a W3C standard [14] for simple description of metadata on the Internet although there are many who consider it as an ontological language. This model enables you to implement graphs of data and to share it between users and machines. For this reason various programming languages such as JAVA and Python, support through appropriate libraries, function and procedures the RDF data. Also many software applications have been developed for the use of such data and the production of this knowledge. A detailed technical description of RDF technology can be found at the official W3C website [15].

4.3.1.1 Semantics of the RDF Model

A key advantage of this technology is the ability to render semantics to the concepts we handle when browsing the Internet automatically so that it can be understood by both humans and machines. The description of this semantics should be done in a specific and standardised way, so that it becomes explicit and accessible by the machine. It will also be possible to automatically draw conclusions through reasoners, as well as to provide logical consistency assessment services in RDF documents. The formalism that can be used to describe semantics is based on predicate logic. For example in each data triple (s, p, o) the arguments "s" and "o" are connected through the property/predicate "p" while implying the statement (p, RDF: type, RDF: property) that the resource/predicate "p" is a "property" that connects the resources/arguments "s" and "o". This way of representing knowledge is compatible with the syntax used in natural languages and is therefore easily understood by humans. However, since the main goal is the understanding of this knowledge by the machines, specific conditions covered by the technological solutions of the existing World Wide Web are necessary, and include:

- A system of identifiers that define the subject, predicate and object of a statement in a unique way, eliminating the possibility of confusion between similar identifiers on the World Wide Web. The technology that supports this requirement is the Uniform Resource Identifier—URI.

- Computer-understandable and computer-editable language for representing and exchanging these statements between machines, such as technology provided through XML.

4.3.1.2 Basic Concepts

The basic concept of the representation of knowledge through the RDF is the "resource". In this sense, we mean any entity on the World Wide Web such as a website, a part or a group of web pages, files, multimedia files, literals or even physical things that are not available online, for example a book, a person or a building. The description of the resources of the Internet with the RDF focuses primarily on efficiency of metadata such as title, author's name, creation date, etc. The general idea is that each resource has one or more predicates/properties which have specific values or they may be themselves resources from the Internet. To summarise, the RDF information is given as suggestions in the form of triples subject—predicate—object. In this triple, the subject is a resource on the Internet which is identified by a specific unique URI. The predicate/property attributed to the subject is specifically described in another resource of the Internet (and therefore is itself a resource). The object can either be a literal or is itself another resource. A set of RDF triples can be perceived as a graph. In this graph (Fig. 4.9) the objects and subjects play the role of nodes and predicates/properties play the role of the connecting edges.

There can be three kinds of nodes in an RDF graph: IRIs, literals, and blank nodes [16].

Since description using graphs is not an easy practice, especially in complex concepts—statements, it is preferable to use triples which can be in the form of text. The syntax used in the RDF representation of the data, is based on the XML standard referred to as RDF/XML syntax. For reasons of generality the RDF uses URI references to identify the entities that are in the position of the subject, the property and the object.

For example if the resource "XXXX" is described by the URI: <http://ministryofinterior/path1/decisions#XXXX>, the resource "YYYY" by the URI: <http://ministryofinterior/path1/persons#YYYY> and the property "hasSigner" described by the URI: <http://ministryofinterior/properties#hassigner> then the graph of the Fig. 4.10 described by the triple:

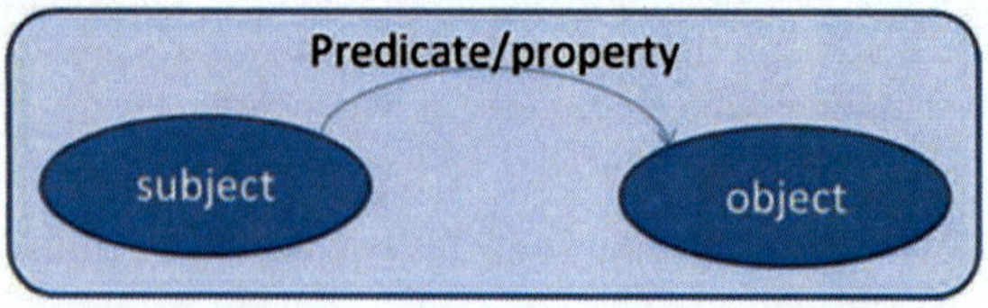

Fig. 4.9 The basic subject—predicate—object RDF triple

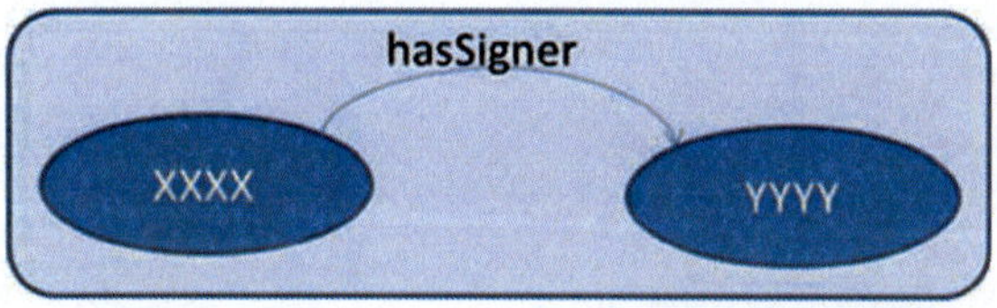

Fig. 4.10 The RDF graph for the representation of the statement: "the YYYY has signer the XXXX"

```
(<http://ministryofinterior/path1/decisions#XXXX>
<http://ministryofinterior/properties#hassigner>
 http://ministryofinterior/path1/persons#YYYY).
```

In this case the resource `<http://ministryofinterior/properties#hassigner>` that connects the two adjacent resources is taken as a property and is referenced with the triple (`<http://ministryofinterior/properties#hassigner>, RDF:type, RDF:property`).

Any statement that corresponds to a real situation and that can be more complex can be transformed and represented by a simple triple of the "subject—predicate—object" form or by a combination of many simple triples of the same form. For example in the model of public administration that we have developed in this study, we consider the case/statement: "XXXX decision is signatory to YYYY who is the Minister of the Interior". The RDF graph that describes the above statement is shown in figure below (Fig. 4.11).

4.3.1.3 Verbal and Blank Nodes

Two key elements of RDF are Literals and Anonymous or Blank Nodes. Verbs can substitute the object in an RDF triple and it is essentially a set of characters, i.e. text.

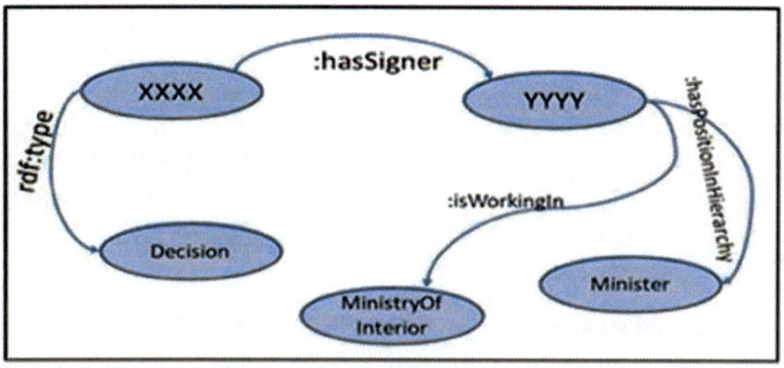

Fig. 4.11 An RDF graph describing a statement in e-Government domain

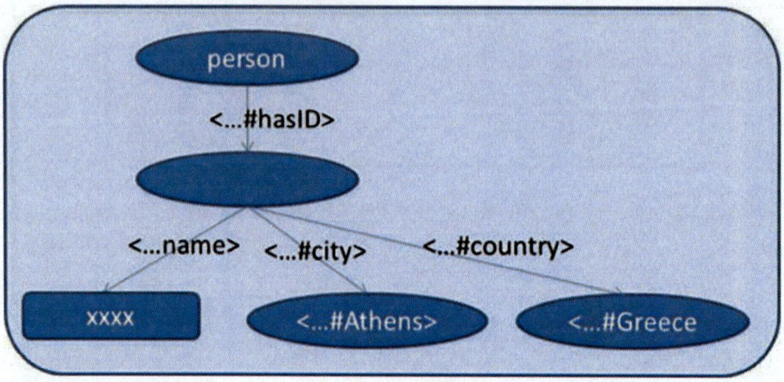

Fig. 4.12 An example about the "blank node" for the concept of "person"

Unlike URIs, which correspond to real-world objects, verbal values are plain text data entered in this form into the graph. For example, words can be used to associate a person with his name or a document with the register number etc.

Blank nodes are nodes in the graph that do not have a name. These nodes were not labelled either because they are unknown or because they do not exist. A key use of empty nodes is in representing cumulative concepts, such as the address of an individual or a building. In such a case, if we want to create statements by separating the individual elements of the cumulative concept—such as the identity of a person which includes elements such as: surname, first name, address, place of birth, etc.—a blank representation could be created that is, an empty node which in turn is connected to all the individual elements. One such example is given in the following figure (Fig. 4.12). The empty ellipse corresponds to an empty node that expresses the identity of a specific person.

Blank node identifiers are local identifiers that are used in some concrete RDF syntaxes or RDF store implementations. They are always locally scoped to the file or RDF store, and are not persistent or portable identifiers for blank nodes. Blank node identifiers are not part of the RDF abstract syntax, but are entirely dependent on the concrete syntax or implementation. The syntactic restrictions on blank node identifiers, if any, therefore also depend on the concrete RDF syntax or implementation. Implementations that handle blank node identifiers in concrete syntaxes need to be careful not to create the same blank node from multiple occurrences of the same blank node identifier except in situations where this is supported by the syntax [16].

4.3.1.4 Reification

Very often the need arises to express statements through RDF for other statements. This is the case if, for example, we want to record specific information about a statement or concept, such as the person who made it, the time at which it was made, the statement being issued, or when we generally want to comment on a

Fig. 4.13 Reification example

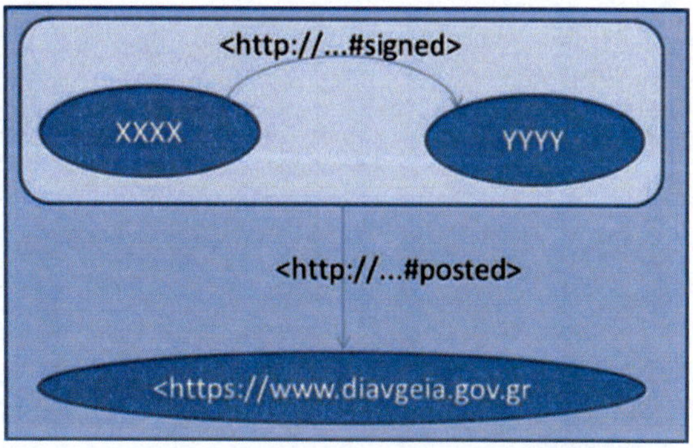

statement or concept. To make statements about other statements we need to build a model of the original statement by expressing it as a resource to which we can then attach properties. This process is called "reification". RDF provides a built-in vocabulary specifically for describing statements for RDF statements, which is referred to as substantiation vocabulary. It consists of the type "`RDF: Statement`" and the properties "`RDF: subject`", "`RDF: predicate`" and "`RDF: object`". Note that the original statement and its substantiation are two completely different entities and neither implies the other.

For example, through the following statement: "The website of the "Diavgeia" program mentions the decision YYYY issued by XXXXX", a statement is made that is attributed to a specific entity. Schematically, in the form of a graph, it is rendered as follows (Fig. 4.13).

4.3.1.5 Valuation of the RDF Model

The representation of RDF statements, however, involves certain limitations that for some may be considered a disadvantage. For example, the duality[2] of the properties used implies the need to transform the more complex properties in combination with other simpler properties. For example in the statement "XXXX has signed the documents AAAA and BBBB" the property "has signed" connects three arguments, so in order to represent it with RDF graph this should be done as follows: "XXXX has signed the document AAAA and XXXXX has signed the document BBBB" (Fig. 4.14). However, this solution increases the complexity of the graphs and consequently the difficulty in understanding them.

Another issue is related to the method of property declarations as a special case of resources. Although this method offers great flexibility it is an unusual practice and can confuse designers. Also, the reification mechanism introduces a high degree of complexity that is rather unnatural and unnecessary for a basic language like RDF.

[2] According the RDF model, the relationships connecting entities are considered binary, according to the "entityA-property-entityB" scheme.

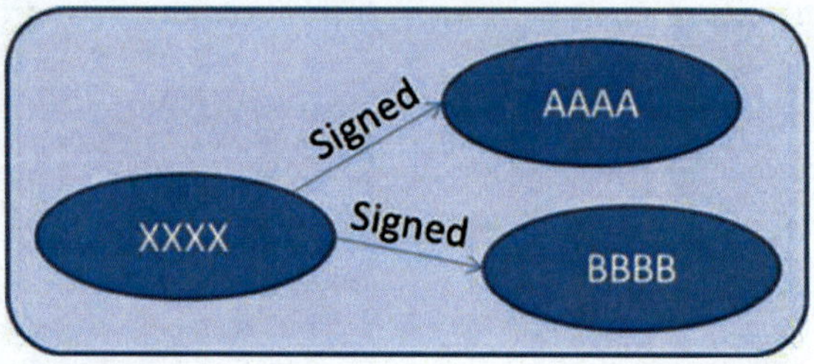

Fig. 4.14 RDF graph describing that "XXXX has signed the documents AAAA and BBBB"

However, it should be noted that RDF, which has already become a W3C standard as we mentioned, is not intended to be the language in which Semantic Web documents are written, but as the basis for higher-level representations. Thus, our interest in this work is focused, among other, on the development of an algorithm for the automatic production of data in RDF format from various data formats in a structured format, such as data stored in relational databases, which is analysed in the next chapter of this work.

4.3.2 RDF Schema Specification Language

The RDF data model does not offer the possibility of restrictions on reporting relationships that exist between the properties and resources. For example in RDF there is no statement to define that "signatory of a decision can only be a minister" or that "the MinistryOfInterior is an individual of the class Ministry". So apart from the RDF descriptions of resources, the RDF data model is also supported by a schema definition language (sets of classes and properties), the RDF Schema Specification Language (RDFS). With the help of RDFS, it is possible to identify mechanisms for determining classes of resources and the limitation of possible relationships between classes using appropriate correlations. An RDF schema consists of statements, classes, attributes and relationships between classes. Similar resources are grouped under the same heading [16].

Based on the above, we can distinguish three different levels of abstraction in the data model RDFS:

- In the lower level there are the same resources (documents, websites, persons, or whatever).
- The next level of abstraction is the data level, which is the description of information resources using specific vocabularies, which are described in the last level, the shape level.
- The shape level is the abstraction level where RDF shapes are developed to facilitate semantic resource description. At this level, the classes r present abstract entities and collectively refer to sets of similar objects.

For the graphical display of RDFS descriptions and figures a directed graph model is used, with labels both to the edges and nodes which can easily combine many different vocabularies and be expanded simply by adding more edges. The nodes represent objects (resources or classes) and edges represent relationships between nodes (properties). As nodes we also represent Literal types, i.e. strings and other basic data types as defined in the standard XML Schema. The nodes are represented as gaps and individual values as parallelograms. The edges can be of three types: Performance attributes, creating instances and employees. The performance attribute edges represent attributes of nodes and relations between them while the subsumption edges used to indicate a shape layer that a node (class) or property, are a subclass of a broader semantic node or property respectively. Finally, the instances creating edges form the link between RDF standards and RDFS, allowing the creation of instances of a class and assigning types to information resources described.

For better sharing of resources RDF descriptions and schemas on the web the model RDF "borrows" the same syntax as XML. This leads to the RDF/XML syntax of RDF schemas and RDF description of resources shown in the graph representation.

4.3.2.1 Basic Principles

As mentioned above, while RDF allows the user to define the required vocabulary to describe a web resource there is a specific deficit in defining the semantics of the resource. But this is covered by the extension of RDF to the RDF Schema which is essentially an ontological language.

A. *Classes*

In order to describe a field through the RDF Schema we use the object classes in a way similar to an object-oriented programming language. Thus a class defines a type of objects which are resources at the RDF level. In the example we gave above, we could consider that the "Ministry of Interior" is an instance of the "Ministry" class which will obviously have other instances. Also the "Ministry of Interior" is a resource at the level of RDF which is mentioned uniquely through its own URI—for example from "http://www.ypes.gr". The relationship between classes and instances is defined using the "RDF: type" statement. Thus this example is implemented as follows using the RDF Schema: <owl:NamedIndividual RDF:about="&pa; MinistryOfInterior"> <RDF:type RDF:resource="&pa;Ministry"/> .

The RDF Schema makes it possible to define hierarchical relationships between classes. Thus a class C′ will be a subclass of a class C if every instance of class C′ is also an instance of class C. This subclass relationship defines a hierarchy of classes, which need not be strict. That is, a class can have many superclasses. This happens when every instance of a class is at the same time an instance of each of its superclasses. For example (Fig. 4.15) the class "Information Services" is a subclass of the classes "G2C services", "G2G services" and "G2E services", since

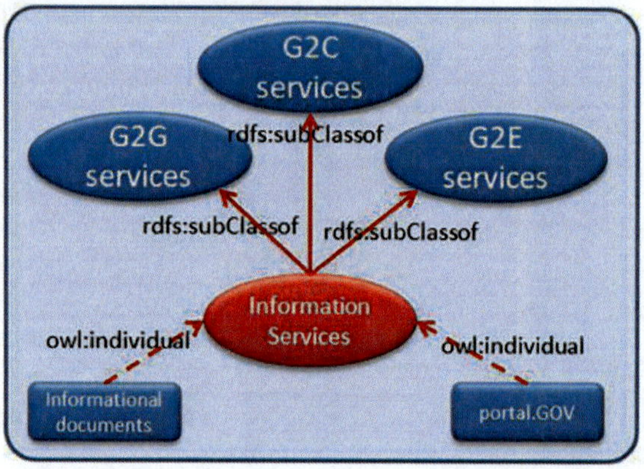

Fig. 4.15 RDF graph describing about the "Information Services"

the information services can be provided as services to both citizens and enterprises and services to the public sector itself.

It is also possible for an individual to belong to two or more different classes. For example, as shown in the Fig. 4.16, the person "XXXX" is at the same time a member/instance of the class "`Citizen`" and the class "`Civil servant`" since in practice a civil servant is at the same time administered as a single citizen.

In general, the hierarchical relationship can be represented graphically as in the figure below (Fig. 4.17). The acne from a lower class to a higher class means that the first class is a subclass of the second. The fact that the hierarchy is not strict means that such a graphic representation does not have to be a tree.

By stating in the RDF Schema that one class is a subclass of another, it is implied that each instance of the class is also an instance of its superclass. The importance of this is that it is presumed automatically and we do not need to state again for this class that every property we have declared for its superclass applies to itself as well as to every instance of it as it inherits every property of the class to which it belongs.

B. *Properties*

Using RDF Schema we can define properties of classes but also properties on properties since in RDF each property is an instance of the class "RDF: property". A key

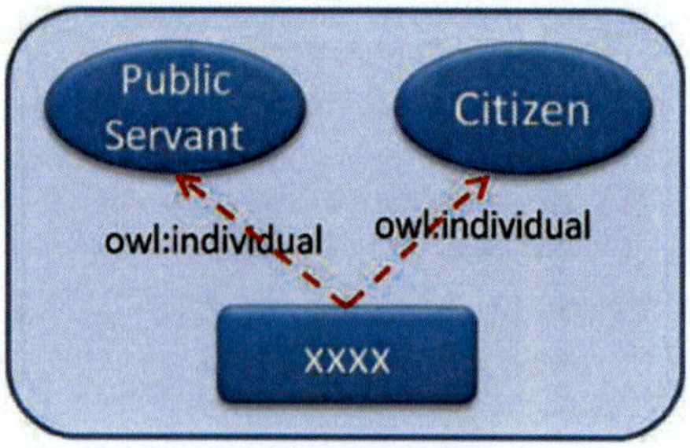

Fig. 4.16 RDF graph describing the "XXXX" individual

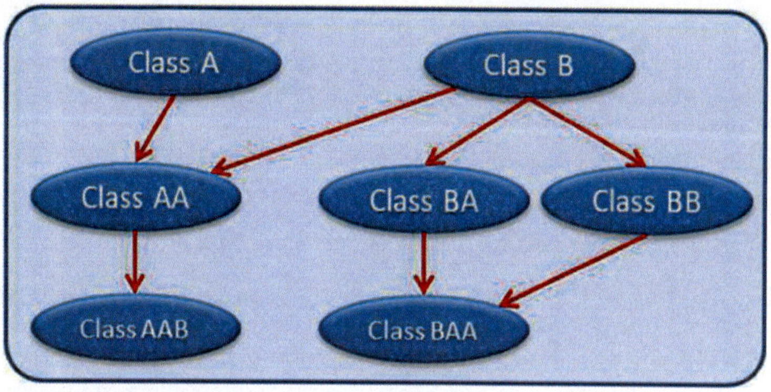

Fig. 4.17 Hierarchical relationship among RDF classes

feature of the RDF Scheme is that in addition to declaring properties and classes, we can set constraints such as defining the value field of a property or its set of values to which specific classes belong. In RDF Schema, unlike an object-oriented programming language, properties are universally defined as classes, not just class properties. It is possible to define new properties so that existing classes can acquire these properties. This is a very powerful mechanism of RDF Schema as it allows us to use classes defined in either another format or by another user and customize their properties, but it is also a special logic of RDF Schema that is not followed in object-oriented programming languages.

The mechanism of hierarchy also applies to properties, i.e. it is possible for one property to have sub-properties or to be a sub-property of another. This means that if a property holds for a subject and an object, and that property is a sub-property of a second property, then it is automatically assumed that the subject and the object are also associated with the second property.

4.3.2.2 RDF Schema Basic Elements

The RDF Schema provides specific *classes* [17] such as:

- `RDFs:Resource`. All things described by RDF are called resources, and are instances of the class `RDFs:Resource`. This is the class of everything. All other classes are subclasses of this class. `RDFs:Resource` is an instance of `RDFs:Class`.
- `RDFs:Class`. This is the class of resources that are RDF classes and it is also an instance of `RDFs:Class`.
- `RDFs:Literal`. This is the class of literal values such as strings and integers. Property values such as textual strings are examples of RDF literals. `RDFs:Literal` is an instance of `RDFs:Class` and a subclass of `RDFs:Resource`.

- RDFs:Datatype. This is the class of datatypes. All instances of this class correspond to the RDF model of a data type described in the RDF Concepts specification. RDFs:Datatype is both an instance of and a subclass of RDFs:Class. Each instance of RDFs:Datatype is a subclass of RDFs:Literal.
- RDF:Property. This is the class of RDF properties and an instance of RDFs:Class.

The RDF schema provides specific *properties/predicates* as well:

- RDFs:range. This is an instance of RDF:Property that is used to state that the values of a property are instances of one or more classes. For example the triple "P RDFs:range C" states that P is an instance of the class RDF:Property, that C is an instance of the class RDFs:Class and that the resources denoted by the objects of triples whose predicate is P are instances of the class C. Where P has more than one RDFs:range property, then the resources denoted by the objects of triples with predicate P are instances of all the classes stated by the RDFs:range properties.
  The RDFs:range property can be applied to itself. The RDFs:range of RDFs:range is the class RDFs:Class. This states that any resource that is the value of an RDFs:range property is an instance of RDFs:Class.
  The RDFs:range property is applied to properties. This can be represented in RDF using the RDFs:domain property. The RDFs:domain of RDFs:range is the class RDF:Property. This states that any resource with an RDFs:range property is an instance of RDF:Property.
- RDFs:domain. This is an instance of RDF:Property that is used to state that any resource that has a given property is an instance of one or more classes. For example the triple "P RDFs:domain C" states that P is an instance of the class RDF:Property, that C is a instance of the class RDFs:Class and that the resources denoted by the subjects of triples whose predicate is P are instances of the class C. Where a property P has more than one RDFs:domain property, then the resources denoted by subjects of triples with predicate P are instances of all the classes stated by the RDFs:domain properties.
  The RDFs:domain property may be applied to itself. The RDFs:domain of RDFs:domain is the class RDF:Property. This states that any resource with an RDFs:domain property is an instance of RDF:Property.
  The RDFs:range of RDFs:domain is the class RDFs:Class. This states that any resource that is the value of an RDFs:domain property is an instance of RDFs:Class.
- RDF:type. This is an instance of RDF:Property that is used to state that a resource is an instance of a class. For example the triple "R RDF:type C" states that C is an instance of RDFs:Class and R is an instance of C.
  The RDFs:domain of RDF:type is RDFs:Resource. The RDFs:range of RDF:type is RDFs:Class.
- RDFs:subClassOf. This is an instance of RDF:Property that is used to state that all the instances of one class are instances of another. For example the triple "C1 RDFs:subClassOf C2" states that C1 is an

instance of `RDFs:Class`, C2 is an instance of `RDFs:Class` and C1 is a subclass of C2. The `RDFs:subClassOf` property is transitive. The `RDFs:domain` of `RDFs:subClassOf` is `RDFs:Class`. The `RDFs:range` of `RDFs:subClassOf` is `RDFs:Class`.

- `RDFs:subPropertyOf`. This property is an instance of `RDF:Property` that is used to state that all resources related by one property are also related by another. For example the triple "`P1 RDFs:subPropertyOf P2`" states that P1 is an instance of `RDF:Property`, P2 is an instance of `RDF:Property` and P1 is a subproperty of P2. The `RDFs:subPropertyOf` property is transitive. The `RDFs:domain` of `RDFs:subPropertyOf` is `RDF:Property`. The `RDFs:range` of `RDFs:subPropertyOf` is `RDF:Property`.

4.3.2.3 Semantics of the RDF Schema

The standardisation offered by the RDF Schema in combination with the standardisation of the RDF has a direct application in the semantic rendering of the statements submitted on the Internet. At the same time, this standardisation serves the mechanisms of automatic inference through special inference tools (reasoners).

Of particular interest are the properties supported by the RDF Schema, as they give the semantics of RDF triples in practice. As a typical case we mention that in RDFS `RDFs:subClassOf` is a property which states that if class c′ is a subclass of class c then every instance of c′ is also an instance of class c. That is, if the triples `(c' RDFs: subClassOf c)` and `(i RDFs: type c')` are valid, then the automatic inference mechanism will output the triple `(i RDFs: type c)`. Transferring this reasoning to a real example, we refer to the statement: "The Ministry of Interior is an instance of the class Ministries which in turn are a subclass of the class "Public Administration". The obvious conclusion from the triples `(MinistryOfInterior RDFs: type ministry)` and `(ministry RDFs: subClassOf PublicAdministration)` is the triple `(Ministry-OfInterior RDFs: type publicAdministration)`.

An important feature of `RDFs:subClassOf` is that it is a transition property. That is, if the triples `(c RDFs:subClassOf c′)` and `(c′ RDFs:subClassOf d)` apply then the triple `(c RDFs: subClassOf d)` is also assumed. As a practical example, we mention the statement that "Ministry is part of Public Administration and the Public Administration is part of the State", so it is logically assumed that "Ministry is part of the State". This statement can be coded using only two RDF triples: `(ministry RDFs:subClassOf publicAdministra-tion)` and `(publicAdministration RDFs:subClasssOf state)`, as the third triple `(ministry RDFs:subClassOf state)` is immediately assumed.

Likewise for RDFs: subPropertyOf. If the property p′ is a sub-property of the property p and the triple (s, p′, o) is valid, then the triple (s, p, o) is also valid. For example, from the statement "XXXX has a mother, YYYY" and the fact that "mother" is a subdivision of the status of "parent", then it is automatically assumed that "XXXX has a parent YYYY". This example is particularly important if we consider the complexity of the queries on traditional databases that keep data from Citizen Registers in order to extract the information described above.

The RDF Schema provides an equally important mechanism in order to limit the scope of application properties resulting in savings in resources and search time as well as logic control. Here we talk about the properties RDFs: domain and RDFs: range. If the RDFs: domain of a "p" property is "d" and the RDFs: range is "r", then for every triple (s, p, o), "s" belongs to class "d" and "o" belongs to "r". The result of the definition of domain and range.

From the above we can deduce the following:

- The domain of the range property is the Property class.
- The range of the range property is the Class class.
- The domain of the domain property is the Property class.
- The range of the domain property is the Class class.

The result from the use of these properties is twofold. On the one hand it can conclude that a particular instance belongs to a particular class and on the other hand it can be used to detect logical errors. For example from the triples (`AAAA RDFs: type b`), (`p RDFs: range a`) and (`BBBB, p, AAAA`) it appears that on the one hand the range of the property "p" is the class "a" and on the other hand that the instance "AAAA" is connected to "BBBB" via the "p" property. Of these three, it is presumed that the "AAAA" instance also belongs to class "b". This conclusion could also be an indication of a logical error or attribute to the "AAAA" instance the property that belongs to two classes "a" and "b".

4.3.2.4 Valuation of the RDF Schema

Both the RDF Schema and the RDF have specific capabilities for encoding statements, although this seems to be based on the logic of the different levels on which the Semantic Web is built. Thus, in RDF the properties connect two parts, i.e. they are characterised as binary properties while the RDF Schema is limited to class hierarchy and property hierarchy relations, to the domain definition and property value constraints and to instant class relations.

In the issues that create relevant malfunctions when building the statements of an RDF document we mention the following.

- Range of properties. By default the properties are "public" throughout the document so it is possible to link instances from any of the classes. In case we define `RDFs:domain` or `RDFs:range` in a property this applies to all classes. However, sometimes it would be desirable for the same property to have a different price range for different classes. For example, if we define the property "`hasTopic`", we want it to have different `RDFs:range` for instances of the "`decision`" class and different for the instances of the "`book`" class.
- There are cases mainly for reasons of control of logical errors to be able to express the fact that two classes are foreign to each other e.g. the "`decision`" and

"`circular`" classes, as it is not acceptable for one instance of one class to belong to the other. This could happen if the `RDFs:range` and `RDFs:domain` are incorrectly declared.
- Ability to define operations between classes such as merging, intersecting and complementing classes.
- Ability to set the number of different values that a property can get or even the constraint on the elements of a set. There are cases where it is desirable for a property to take a single value or to be able to count the different values it has taken.
- Ability to declare relationships between properties, such as that one property is the inverse of another or that it has a unique object or subject or that it is transient.

All of the above are issues that can be addressed with the help of technologies at a higher design level such as the OWL ontology language discussed below.

4.3.3 The URI and URI's Use

In order to implement the RDF representation mentioned above, where the basic concept is the "resource" we use as unique identifiers the Universal Resource Identifiers (URIs). The known Universal Resource Locators (URLs) or the strings that we use to identify each site on the Internet is a subset of URIs. The URIs generalise logic of URLs in the sense that anything that can be described and recovered can be identified in a manner analogous to websites. The URI that is used in an RDF statement is characterised as a URI reference (URIRef). A URIref comprises a URI and an optional fragment identifier. For example the `URIRef:http://www.example.org/index.html#doc1` consists of the URI `<http://www.example.org/index.html>` and the fragment identifier `<doc1>` which distinguish from the URI of the use the # symbol. It is used for the first part of the URI to be entrusted to a relative prefix which is then accessible. In this way each person can identify the resources they wish to publish electronically uniquely worldwide. So for example, the Ministry of Interior can use for one of its resources the URI: `<http://www.ministryofinterior.org#path1/doc1>`, which is a component of a unique URL assigned to the institution `<http://www.ministryofinterior.org#>` followed by the corresponding internal path `</Path1/doc1>` with which each resource is identified in the internal system feature. With the RDF statement: "`PREFIX mint: <http://www.ministryofinterior.org#>`", the URI could be accessible in the form of "`mint: Path1/doc1`". Because the URI are used in the manner mentioned as unique identifiers of resources worldwide they are considered as strong identifiers.

4.3.4 Web Ontology Language—OWL

OWL was created to satisfy the need for a web ontology language and is one of the W3C's recommendations on implementing Semantic Web technologies. OWL is designed to be able to process the content of information. The purpose of OWL [18] is to provide a standard format that is compatible with the Internet architecture and the Semantic Web. The standardisation of ontologies in OWL language will make the data on the Web more machine processable and reusable applications. Thus, scalability, convertibility and interoperability possess high priority in language design. Compared with existing technologies, OWL exceeds the known and widely used languages XML, XML Schema, RDF and RDF Schema as it supports greater clarification of the Web content than machines, providing additional vocabulary along with a formal semantics.

The OWL adds larger vocabulary for the description of properties and classes such as relations between classes (e.g. disjointness), the number of elements in a set (cardinality), logical combinations of classes (union, intersection), equality, richer typing of properties, characteristics of properties (e.g. symmetry, uniqueness, transitivity), and enumerated classes (enumeration).

An advantage of OWL ontologies is the availability of tools which may draw conclusions about these (reasoners). These tools produce general support related to stated rules or events that follow specific morphology and vocabulary. OWL is developed as an extension of RDF vocabulary. It derives from the web ontology language DAML+OIL and it is written in XML format that can easily be independent of the operating system and language implementation of a computer. An OWL ontology includes descriptions of classes, properties and their instances. Taking into account such an ontology, the formal semantics OWL specifies how to produce the logical conclusions, i.e. facts not literally present in the ontology, but implied by the semantics. These implications can rely on a single document or multiple distributed documents that have been combined using defined OWL mechanisms.

As the Semantic Web is distributed, the OWL should allow it to gather information from distributed sources. This is done partly by allowing ontologies to be associated, including clearly the import of information from other ontologies. The practical applications of OWL, include Web portals, which can be used to create the classification rules, in order to improve the search, the Multimedia Collections, which can be used to enable content-based searches of media and Web Services, where it can be used for the discovery and composition of Web Services as well as for rights management and access control.

Table 4.2 shows the correspondence between the abstract syntax of OWL class descriptions and descriptions of concepts of descriptive logic. As we can see, this match derives the semantics of the structural elements of the OWL language. A similar correlation can be set in the case of class axioms of OWL language. These axioms may be assigned to axioms of affiliation and equivalency between concepts of a Description Logic. The correlations are shown in Table 4.3 and in most cases are quite obvious. The only case that is worth commenting on is the one of foreign

Table 4.2 Correspondence between the abstract syntax of OWL class descriptions and descriptions of concepts of descriptive logic

OWL syntax	DL	Example	Semantic annotation
IntersectionOf	$C \cap D$	CivilServant ∩ CitizenAttributes	All the Civil Servants that have specific attributes
UnionOf	$C \cup D$	CivilServant ∪ Professor	Anyone that is civil servant or a professor
Atomic megation	$\neg C$	¬CivilServant	Anyone than is not a Civil Servant
hasValue(o)	$\forall R.\{0\}$	∀hasSigner.CivilServant	The signer of a document must have value CivilServant
allValuesFrom(C)	$\forall R.C$	∀CivilServant.{isWorkingIn some State}	Civil servant is anyone that is working in some unit of State

Table 4.3 Class axioms in OWL

OWL syntax	DL
Class(A partial C1,…Cn)	$A \subseteq C1 \cap \ldots \cap Cn$
Class(A complete C1,…Cn)	$A \equiv C1 \cap \ldots \cap Cn$
EnumeratedClass(A O1,…On)	$A \equiv \{O1\} \cup \ldots \cup \{On\}$
SubClassOf(C1, C2)	$C1 \subseteq C2$
EquivalentClasses(C1,…Cn)	$C1 \equiv \ldots \equiv Cn$
DisjointClasses(C1,…Cn)	$Ci \cap Cj \subseteq \perp$

classes. From the set theory we know that two sets are disjoint if their intersection is the empty set. If transferred from the field of sets to the abstract space of concepts then we can define two concepts as foreign if the intersection of them falls into the empty concept. It is pertinent to note that there is an alternative statement that two concepts are disjoint. More specifically the term C is foreign to the meaning D if-f $C \sqsubseteq \neg D$.

Table 4.4 shows the correlation between the properties of the OWL axioms and axioms of roles in a descriptive sense. Once again we observe that OWL does not offer anything more than the expressive ability that gives us a Description Logic System even where this is not immediately obvious, as in the case of functional roles explained above. From this table however we observe something interesting. The axioms of definition of range and domain of a property, which we met in language RDF-S, may themselves be encoded by using Descriptive Logic axioms. More specifically a form axiom definition of the range of ObjectProperty (R domain (C)) corresponds to the office of entry for $\exists R.\top \sqsubseteq C$, while an axiom definition of the form of range ObjectProperty (R range (C)) corresponding to the entry office, $\top \sqsubseteq \forall R.C$. Intuitively the first axiom tells us: "If an object is connected through the relation R to something, then this object is of type C". So this axiom tells us that all individuals who are in the first position of a pair (a, b) which belongs to the relation R is type C. The second axiom tells us that "for all objects of the world, whenever they are

Table 4.4 Property axioms in OWL

OWL syntax	DL
SubPropertyOf(R1,R2)	$R1 \subseteq R2$
EquivalentProperties(R1,R2)	$R1 \equiv R2$
ObjectProperty (R domain (C))	$\exists R.\top \sqsubseteq C$
ObjectProperty (R range (C))	$\top \sqsubseteq \forall R.C$

connected to another object via the relationship R then the object is of type C". So all individuals that are in the second position of a pair (a, b) which belongs to the link.

4.3.5 Reasoning Tools

The acquisition of knowledge from a database or from a base of knowledge is often associated with the submission of relevant questions through appropriate interfaces. The answer to a question in Semantic Web, is not just about data control (such as databases), but it requires the implementation of complex considerations that cover a huge crowd of classes and instances. Classes and instances are linked together under declared rules and constraints that describe specific events and roles. For this reason, the combination of declared facts and concepts in an ontology are particularly important so that the correctness or not of another event occurs as a conclusion or so that a new unexpected fact is produced. This process is described as reasoning and is one of the main differences between databases and knowledge bases of the Semantic Web. Of particular interest is the sub-process of reasoning concerning the production-addition of new facts in ontology referred also as inferencing.

The reasoning depends on the way of representing the events in the ontology. This representation is formulated by using formal languages based on various, such as predicate logic and descriptive logic. The last, are used mainly for supporting the Semantic Web and based on Open World Assumption. In descriptive logic, we use the concepts, objects and people in order to define the nomenclature relating to the field of interest. A typical architecture of a system is given below (Fig. 4.18) based on descriptive logic.

The knowledge base is the combination of statements of two ingredients: (a) Terminology Box—TBox comprising the vocabulary of the field of interest through the concepts and definitions of properties and (b) Assertion Box–Abox comprising all axioms/allegations on individuals included in TBox.

As a vocabulary we mean all the concepts (which usually group together individuals) and the roles referring to binary relations between individuals. The concept and roles description languages for the knowledge base, are based on a basic language, Attributive Language, that provides key manufacturers with common syntax rules.

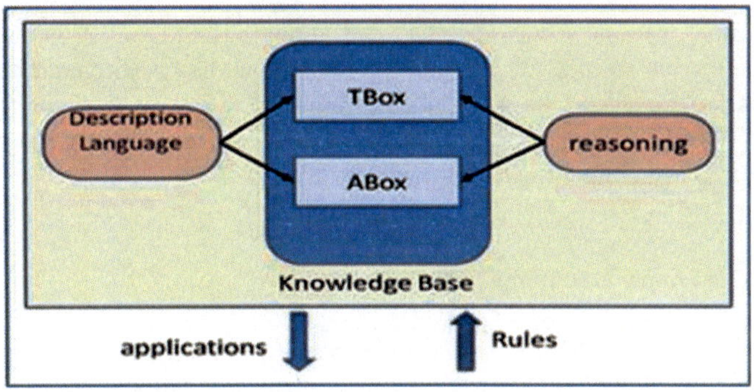

Fig. 4.18 A typical architecture of a reasoning system

Special software has been developed under this logic (for example OWL2 Query Tab of Protégé presented below) which supports the operation of ontologies management applications by using reasoners such as Pellet, Fact++ and HermiT used in Protégé. The reasoning (reasoning) provided by the software are a particularly useful service and mainly include the following categories of controls: (a) knowledge base consistency checking, (b) concepts classification and (c) instance checking, with the sense of the calculation of classes, which include instances of the database.

4.3.5.1 SWRL Rules

The SWRL (Semantic Web Rule Language) is a language of the Semantic Web standards and is based on a combination of OWL DL and OWL Lite sublanguage of OWL with Unary/Binary Datalog RuleML sublanguages of the Rule Markup Language. This proposal extends the set of OWL axioms by introducing Horn-like rules [19]. In this way it allows a combination of Horn-like rules with an OWL database. At the same time, it provides a high-level abstract syntax for Horn-like rules, which are expressed in terms of the OWL language (classes, properties, individuals). The rules are in the form of contributory fault between an antecedent and a consequent part. The antecedent part is the main body of a rule and it is known as body, while the consequent part is the consequence of the rule and is known as head. The point at which a rule is intended to be interpreted as follows: whenever applicable conditions laid down in the antecedent of a rule, you should apply the conditions laid down in the consequent part of the rule. The rules are stored as part of the ontology and also it is possible that they interact with reasoners. While the provision of reasoning tools that support SWRL is constantly growing, we can distinguish Bossam, R2ML, Hoolet, Pellet, KAON2, RacerPro and SWRLTab. It is known to carry OWL inference capabilities through the features of OWL properties, such as inversion, symmetry and composition of indirect relationships through a combination of direct relations (transitive property). The SWRL carries inference capabilities through the rules. To avoid repetition, necessity between OWL extrapolations and SWRL extrapolations would be desirable so that rules engines could follow the OWL characterizations. This means that the OWL features should be translated to SWRL equivalents. For this reason, in SWRL it is perfectly possible to define rules for attributes that represent symmetry, inversion and transitive possibility.

4.3.5.2 The Query Language SQWRL

The library of Protégé-OWL contains a number of built-ins that extend the SWRL in SQWRL. The SQWRL (Semantic Query-Enhanced Web Rule Language) is a language based on the SWRL standard for performing queries in OWL ontologies. It has similar methods to SQL to retrieve knowledge from OWL. The determination of SQWRL is done using a library of SWRL built-ins that structure an efficient query language over the SWRL. The built-ins contained in this library are determined by the SQWRL Ontology. The common prefix is sqwrl. The Jess Rule Engine is necessary for the performance of SQWRL questions. The SQWRL questions being built on SWRL can be used for retrieving knowledge which has been deduced from SWRL rules. Moreover, SQWRL questions are free to cooperate with other built-in libraries. This free choice and use of built-ins to the questions, provides a continuous expansion means of the expressive power of the query language.

4.4 Modelling and Extraction of Knowledge in the Field of e-Government—Our Proposal as "The e-Government Ontology"

4.4.1 The e-Government Ontology Motivation

The term e-Government as presented in detail in a previous chapter is used worldwide by all economically developed countries. However, both the existing administrative structures and the services provided in this context vary from country to country. Consequently, the issue of non-use of common terminology arises which is observed not only between states but also within the services and citizens of the countries themselves. This implies the emerging question of the semantic interoperability of systems which extends beyond and independently of functional interoperability. What is observed mainly in the case of Greece is the use of different vocabularies for the rendering or description of the concepts used in the moving documents even between public services resulting in inefficiency and waste of resources in the back-office of the public administration. As a typical example we mention the case of the concept of "decision" that is widely used in the daily lives of employees in the Greek Public Administration. This means an administrative act signed by the competent body and subject to individual or collective arrangements. However, a similar administrative act exists in the security forces, such as the Greek Police, where it has the title and character of the "Order". These two types of documents, while having corresponding issuance procedures and yielding corresponding results, are treated as two distinct concepts. Another example is related to the issuance of certificates, attestations, excerpts and other administrative documents in order for the competent administrative bodies to certify various statements concerning citizens and businesses. These documents are usually issued on the basis of laws, circulars and instructions that are generally a complex institutional framework and result in frequent misunderstandings of citizens and malfunctions in the procedures followed in the back-office services of public administration. This phenomenon makes it difficult even for the electronic search of the relevant services since the lack of interoperability mainly concerns the information systems. The situation is further complicated when there is a need for transnational cooperation or exchange of documents between the respective public services due to the use of different languages that require the translation of terminology into different languages.

The need to record the administrative concepts and procedures that use (with various variations) in their daily activities both services and citizens in Greece and then the development and maintenance of a common unified vocabulary gave us the motivation for this work. For this reason in this Book, we develop and document the "e-Government ontology". In the next step, this ontology is published in the form of open data on a corresponding server that we have developed to be public knowledge and to be reused where possible.

4.4.2 The Ontology Development in Protégé 4.3

The aim of this study was to develop the "e-Government ontology" so that it describes the actual image recorded today in the Greek public administration and not a theoretical model of knowledge in this field. This ontology models the overall governance environment, structures and functions of public administration and the relationships between them. So the concepts and properties that we defined correspond to reality while many of the instances are images of the actual structure of public administration in Greece. However, many of the instances are virtual data in order to check the logic and consistency of the ontology with the help of the appropriate tools listed below.

In this study we used the 4.3 version of Protégé, as one of the most updated, oriented to OWL ontologies. As a big advantage of the 4.x. versions of Protégé, we highlighted the friendliness towards the creation of the properties and their limitations on the classes and instances. This is mainly due to the use of the OWL ontology language which implements the semantic rendering of our ontology concepts at a higher design level than the RDF Schema. On the other hand, the concealment of SWRL and the replacement of the corresponding tab that the versions 3.x offered with the rules tab, restricts the possibilities for developing complex rules on classes or instances. At the same time, the visualisation tools provided through Protégé, such as OWLViz for the graphical representation of classes and their subclasses at various levels and Ontograf for the graphical representation of classes, subclasses and instances help both in the supervision of the ontology. As well as in controlling the logic and consistency of the ontology. In particular, the Ontograf, which also depicts the properties of the instances, perfectly renders the format of the RDF graphs we talked about above.

4.4.2.1 Defining Classes

The latest version of our ontology describes the interaction environment of public administration as a system under the tripolar model comprising three main pillars: the state, citizens and businesses. In our ontology, this is indicated by the definition of the class `Trader` which has as subclasses the: `Citizen, Enterprize, State`, and an additional two (excluding basic pillars): `NGO (Non Governmental Organizations)` and `EuropeanUnion`. The case of subclasses of `State` interests us particularly, since each country has its own state structure but also internally in each country there are different views as to what is considered as state. Here, the class `State` depicts the Greek reality as it emerges from relevant legal texts e.g. the `Constitution` of our country. By this logic structure created shown in the figure below (Fig. 4.19).

We also considered citizens as a member of the system to the same class. Fundamental also is the character of the profession which gives us a way to distinguish citizens. So the `civil servants, judicial officials, professionals, pensioners, politicians, private employees, teachers, students, the unemployed,` are represented through their respective subclasses of Class `Citizen`.

To describe the properties of the citizens we defined the class `CitizenAttributes` with subclasses the `Gender, IllegalActivity, Married`. Class `Gender` has two disjoint individuals `(man, woman)`. The class `IllegalActivity` has additional subclasses according to the type specified by the statements in the General Class Axioms, shown in the figure below (Fig. 4.20).

This model supports the so-called "*separation of powers*" of the Constitution into `executive, judicial and legislative`. These concepts are provided by `StateAuthorities` class and subclasses `Executive, Judicial` and `StateLegislation` respectively as shown in the figure below (Fig. 4.21). At the same basic level with the class `Trader` and `StateAuthorities`, we defined the class `Services` to describe the categorization of the services provided by the state to citizens (subclass `G2C`), to businesses (subclass `G2E)` and to the state itself (`G2G`).

From another perspective, the services and procedures also cover the so-called front-Office and back-Office of Administration. The visual is covered by the class `Process` and respective subclasses that appear in the figure below (Fig. 4.22).

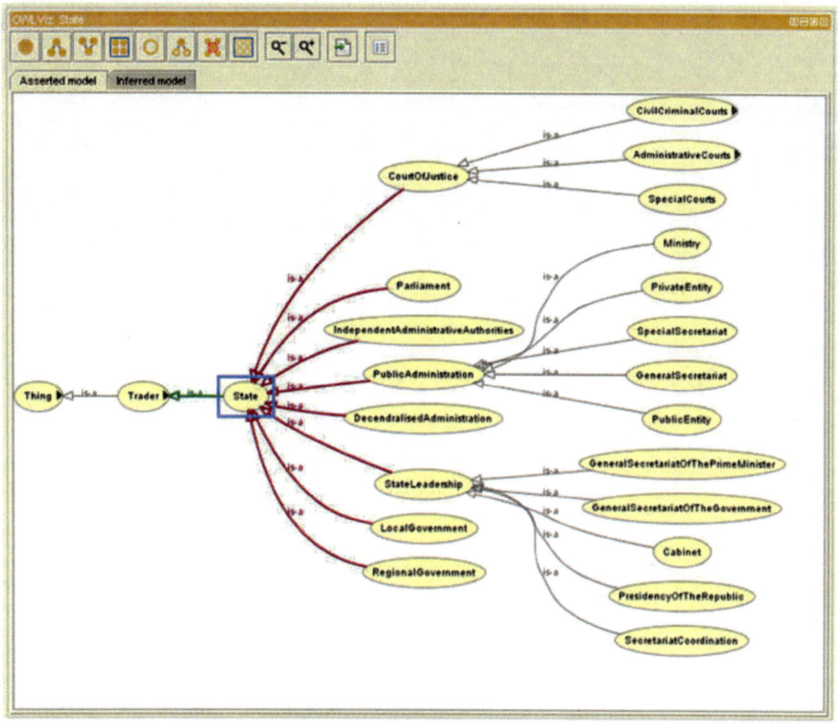

Fig. 4.19 The state class and subclasses in OWLViz tool of Protégé

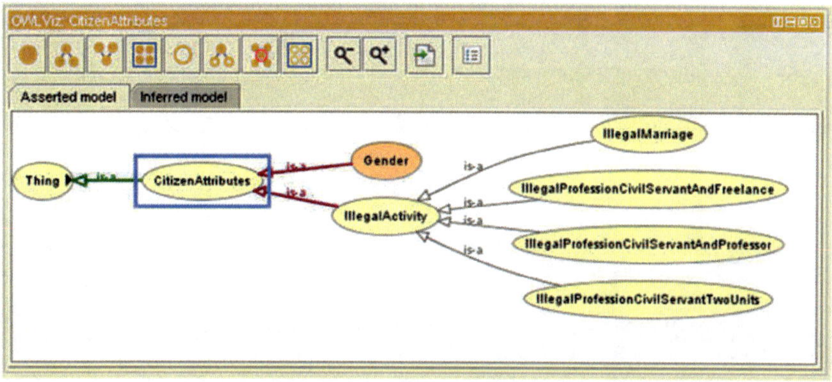

Fig. 4.20 The CitizenAttributes class and subclasses in OWLViz tool of Protégé

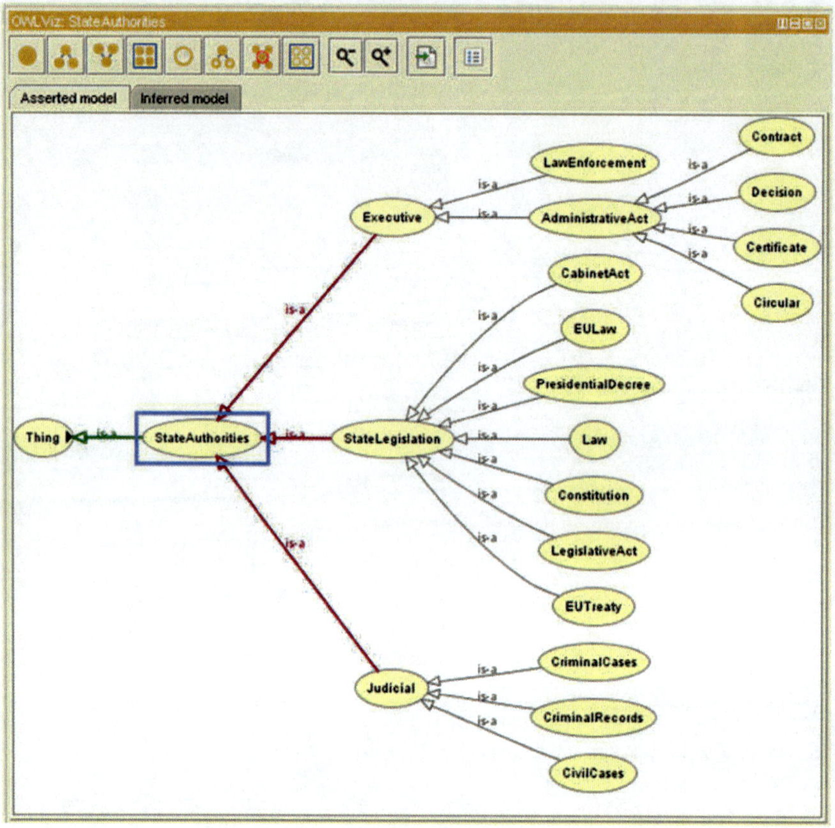

Fig. 4.21 The StateAuthorities class and subclasses in OWLViz tool of Protégé

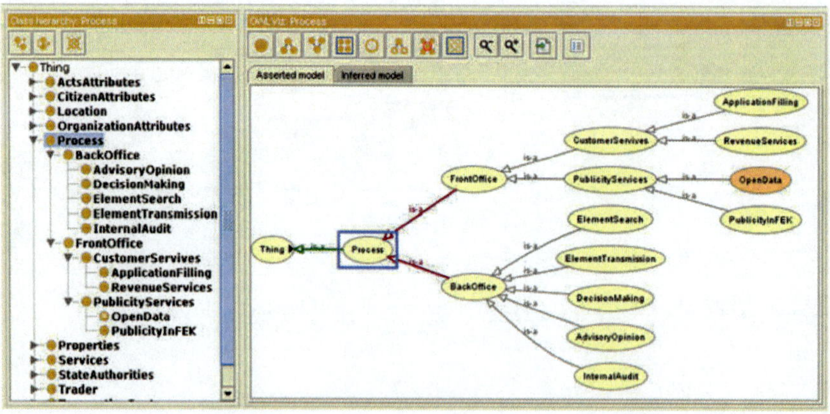

Fig. 4.22 The process class and subclasses in class hierarchy view and OWLViz tool of Protégé

In this ontology, we chose to describe the properties of organisms as a class `OrganizationAttibutes` to include two main characteristics of agencies as subclasses: the `bureaucracy` (in the sense of hierarchical organisation) and `organizational structure` (in the sense of the units which constitute organisations).

For a description of the operations of the state, we defined the class `ActsAttributes` with subclasses `Gradation, Kind, Priority, Topic` for describing their essential characteristics. In parallel, we have provided a description of the transaction points to the Public Administration with the class `TransactionSpot`. Transaction points can be understood either by their physical nature (`AgentOffice`) or their electronic form (`DiavgeiaSpot, WebSite`). The class `Location` and the subclasses `City, Prefecture, Region` implement the requirement that transaction points are located within a specific spot.

4.4.2.2 Defining Properties

The Protégé enables definition of object properties and datatype properties. The object properties have the role of correlations among individuals and practically implement the triplets RDF. The object properties that we stated in our ontology, are shown in the Fig. 4.23 below. We note as particularly important the `belongsTo` which is transitional and the inverse of the `comprises` which we use to attribute semantically that "`a City belongs to a Prefecture`" or "`a Prefecture comprises a Region`". Also for the corresponding requirement that "`a Prefecture belongs to a Region`" or "`a Region comprises a Prefecture`". Also important is the `hasPositionInHierarchy` which has the union of the domain classes `CivilServant` and `PrivateSectorEmployee` and range class `Bureaucracy.` This property is used to describe the requirement that "`an Employee has a specific position in the bureaucracy of the organization`". We also note the `hasSigner` (the inverse of the `isSignerOf`) and `hasIntroduser` (the inverse of the `isIntroduserOf`) properties. These properties have as a subject the acts adopted by the executive authority (`domain = Executive`) and as an object the civil servants (`range = CivilServant`). With these qualities we implement the description of the requirement that "`an Executive act has signer a civil servant`" or "`an Executive act has introducer a civil servant`". By the property `hasIssuingAuthority` we describe the requirement that every transaction (of the `Executive` or of the `Legislative authority`) has a competent authority which issued it. This is achieved with the statement as a domain of the association of classes `Executive` and `StateLegislation`.

The datatype properties we have set are shown in Fig. 4.23. The properties of this kind are used to connect the logic of RDF of a certain individual with specific literal. For example, every legislative act or some of the administrative acts shall be published in the daily newspaper of the Government and take a unique number (FEK). This requirement is indicated by the use of the property `hasFEK` which

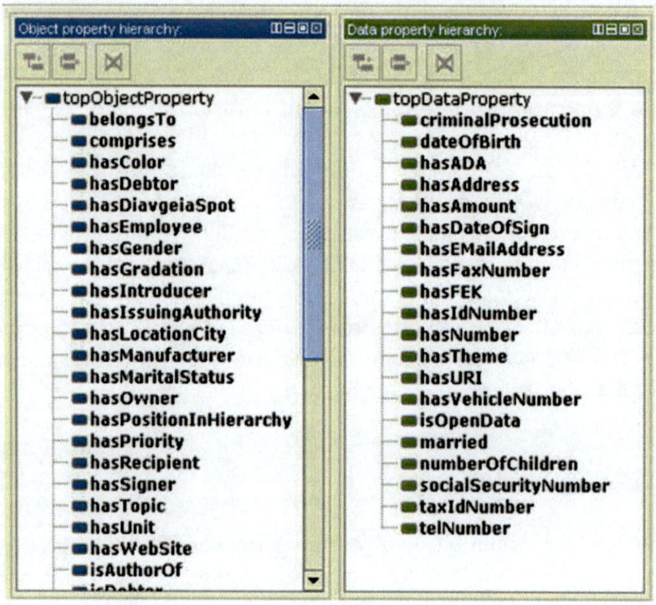

Fig. 4.23 The ontology properties in hierarchy view of Protégé

has as a domain the union of `AdministrativeAct`, `CabinetAct`, `Decision`, `Law`, `LegislativeAct`, `PresidentialDecree` and as a range `type:string`. Also, according to the "Diavgeia program", specific categories of administrative acts and instruments are published on the Internet, making the so-called open public data. Any document notified on the specific website (www.diavgeia.gov.gr) is characterised by a unique number ADA. This requirement is described in the property `hasADA` which has the same domain and range with the capacity `hasFEK`. Also by using the property `isOpenData` which is `type:Boolean`, we describe the fact that "`a particular kind of the administrative acts or Laws are Open Data`". This has been achieved by the definition of the general class axiom: "`hasADA some string SubclassOf isOpenData value true`".

4.4.2.3 The Use of RDF, RDFS, OWL and SWRL Through a Case Study

The RDFS statements involving domain-range are particularly important. This is because the declaration of a class as a domain of a property implements the part of the RDF triplet concerning the subject. In other words, the individual having `RDF:type` the domain of a property is the subject and the particular property is the predicate of a RDF triplet. Correspondingly, the individual having `RDF:type` the range of the property can be the object of the triplet.

For example the property *hasSigner* we have stated to have the domain and range Class *Executive* Class *CivilServant* as shown in the following code section RDF/XML.

```
<owl:ObjectProperty RDF:about="&pa;hasSigner">
  <RDFs:range RDF:resource="&pa;CivilServant"/>
   <RDFs:domain RDF:resource="&pa;Executive"/>
          </owl:ObjectProperty>
```

As shown, the `hasSigner` resource is `owl:ObjectProperty` while `CivilServant` and `Executive` resources are `RDFs:range` and `RDF:domain` respectively. In this way we implement the triplet that performs the semantic requirement that "`all acts of the executive authority has as signatory a civil servant.`"

We should note at this point that the domain and range do not function as limitations, such as the domain of a mathematical function, but function as a classifier for specific individuals. For instance with the statement:

```
<owl:NamedIndividual RDF:about="&pa;Ind_Test1">
     <hasSigner RDF:resource="&pa;Citizen5"/>
           </owl:NamedIndividual>
```

is implied that `Int_Test1` the subject is associated with the object `Citizen5` with the property `hasSigner`. We have stated that "`Ind_Test1 has Signer Citizen1`". As shown in the figure below (Fig. 4.24) using the reasoner's protégé, it is concluded that the `Int_Test1` is type `Executive`, which has not been declared and we do not know if it is a correct conclusion for the classification concerned.

Also important is the function of the features of the properties. Such features are "functional", "transitive", "symmetric" etc. Their use requires attention and understanding of their operation because although they are very useful, they can easily lead to logic errors.

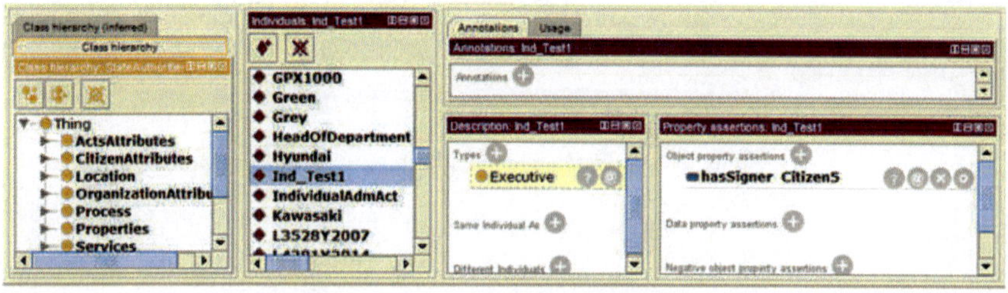

Fig. 4.24 The reasoner inference about the classification of Ind_Test1 individual

For example, when a property is classified as functional, this means it can get at most one value. That means that the subject in this property can be associated with at most one object. This feature does not work limited as to the subject, but conclusively. For example, if you declare that:

```
<owl:NamedIndividual RDF:about="&pa;Citizen1">
<hasPositionInHierarchy RDF:resource="&pa;Introducer"/>
<hasPositionInHierarchy RDF:resource="&pa;Rapporteur"/>
        </owl:NamedIndividual>
```

i.e. if we declare that "`Citizen1 has Position In Hierarchy`" both `Introducer` and `Rapporteur`, (with the use of functional property `hasPositionInHierarchy`), then the reasoner of Protégé concludes that individual `Introducer` and `Rapporteur` is identical, as shown in Fig. 4.25.

For a higher level of expressivity we can use OWL. Relation between classes can be formally modelled based on description logics (mathematical theory). Because OWL relies heavily on the reasoner, it is possible to express complex constructs such as chained properties for instance or restriction between classes. OWL serves to build ontologies or schema on the top of RDF datasets. As OWL can be serialised as RDF/XML, it is theoretically possible to query it via SPARQL, yet it is much more intuitive to query an OWL ontology with a DL query (which is usually a standard OWL class expression). An example of OWL constructs serialised in RDF/XML is given throughout the object Property `hasPositionInHierarchy`.

The `hasPositionInHierarchy` property connects the `Citizen` classes and `PrivateSectorEmployee` with the `Bureaucracy` which has as domain the union of the first two and the third range from those classes. As shown in the following part of the RDF/XML code, the statements of the domain and range follow the

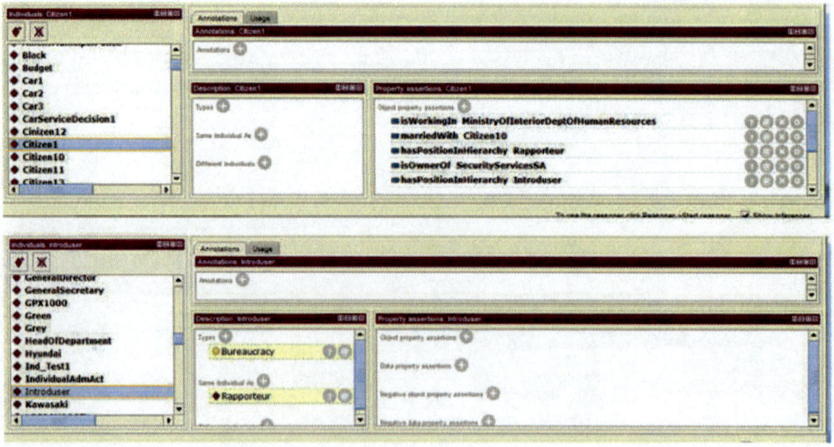

Fig. 4.25 Classification of Citizen1

standard RDFS and the concept of classes and their union are attributed with the use of OWL.

```
<owl:ObjectProperty RDF:about="&pa;hasPositionInHierarchy">
      <RDF:type RDF:resource="&owl;FunctionalProperty"/>
         <RDFs:range RDF:resource="&pa;Bureaucracy"/>
                        <RDFs:domain>
                           <owl:Class>
            <owl:unionOf RDF:parseType="Collection">
           <RDF:Description RDF:about="&pa;CivilServant"/>
                            <RDF:Description
       RDF:about="&pa;PrivateSectorEmployee"/>
                           </owl:unionOf>
                          </owl:Class>
                       </RDFs:domain>
                  </owl:ObjectProperty>
```

A more complex requirement is presented in the following example. One of the requirements of the Civil Servants Code specifies among others that: "1. Civil servant is any citizen who works in a public body. 2. Each civil servant holds a position in the bureaucracy of the organization, including: Rapporteur, Head of Department, Director, General Director. 3. A civil servant cannot possess two positions in two public bodies."

The requirement 1. is described in the declarations:

```
<owl:Class RDF:about="&pa;CivilServant">
             <owl:equivalentClass>
                 <owl:Restriction>
    <owl:onProperty RDF:resource="&pa;isWorkingIn"/>
     <owl:someValuesFrom RDF:resource="&pa;State"/>
                </owl:Restriction>
            </owl:equivalentClass>
  <RDFs:subClassOf RDF:resource="&pa;Citizen"/>
               </owl:Class>
```

The statement describes the bureaucracy of an organisation that is necessary for requirement 2. is enumeration. This is done by the OWL statement <owl: equivalentClass> together with the statement of the members made with the statement owl:oneof RDF: parseType='Collection' *and RDF* statements of the form RDF:Description RDF:about=NamedIntividual to give the respective individual.

```
<owl:Class RDF:about="&pa;Bureaucracy">
    <owl:equivalentClass>
        <owl:Class>
    <owl:oneOf RDF:parseType="Collection">
    <RDF:Description RDF:about="&pa;Director"/>
        <RDF:Description
RDF:about="&pa;GeneralDirector"/>
    <RDF:Description RDF:about="&pa;Rapporteur"/>
        <RDF:Description
RDF:about="&pa;GeneralSecretary"/>
        <RDF:Description
RDF:about="&pa;HeadOfDepartment"/>
    <RDF:Description RDF:about="&pa;Minister"/>
        </owl:oneOf>
        </owl:Class>
    </owl:equivalentClass>
<RDFs:subClassOf RDF:resource="&pa;OrganizationAttributes"/>
</owl:Class>
```

The requirement 3. is described with the class definition `IllegalProfessionCivilServantTwoUnits`, with the condition that an individual in order to belong to the class must meet the status `isWorkingIn` at least twice, as seen in the following code fragment:

```
<owl:Class>
    <RDFs:subClassOf
RDF:resource="&pa;IllegalProfessionCivilServantTwoUnits"/>
    <owl:intersectionOf RDF:parseType="Collection">
        <RDF:Description RDF:about="&pa;Citizen"/>
            <owl:Restriction>
    <owl:onProperty RDF:resource="&pa;isWorkingIn"/>
        <owl:onClass RDF:resource="&pa;State"/>
            <owl:minQualifiedCardinality
RDF:datatype="&xsd;nonNegativeInteger">2</owl:minQualifiedCardinality
>
            </owl:Restriction>
        </owl:intersectionOf>
</owl:Class>
```

4.5 Knowledge Acquisition from "The e-Government Ontology"

As mentioned above, the knowledge held by experts in each field can be described based on a common vocabulary with the help of ontologies. Many times, however, the knowledge derived from Data mining tools and the knowledge of experts contained in an ontology is often written in different knowledge representation models or in different shapes. So the first step is the representation of knowledge in a common format using translation mechanisms from different formats to a common one. In the next step we confront issues of resolving any disputes and then processing them.

4.5.1 SPARQL

Knowledge in Semantic Web is in the form of triplets RDF, to which specific languages are focused. Their main object is the recovery of results upon submission of tailored questions in the form of triplets. Such languages are RQL and SPARQL. With these languages ontological knowledge that has been represented in RDF can easily be questioned, cut, transformed or concentrated. This knowledge often stems from a Web Service and is updated either from a specialist, or it is the result of a Data mining process. In our work we are interested in SPARQL language, which we can use in different ontology management tools such as Protégé.

The SPARQL (SPARQL Protocol and RDF Query Language) is the query language for RDF documents which has become a prototype for the W3C in January 2008 [15]. The SPARQL is based on previous query languages for RDF, like RDQL and SeRQL with which they form a family of query languages. Their use depends on the applications developed and run every time. One of their common features is that they comprehend the RDF data as simple triples without some shape or other ontological information unless it is explicitly set to the RDF document. Also, these languages are all SQL-like, i.e. mimic the SQL syntax styles. The basic building block of SPARQL is the triple pattern and is an individual (elementary) question. A standard triple is actually an RDF triple (subject—predicate—object) which may contain a variable in one or more of the three positions. A prerequisite for the question to give a result, in other words for the variable to get some price is the existence of pairing with some RDF triple of the document upon which the question is put. A typical example is the question of the form:

```
SELECT ?subject ?predicate ?object
   WHERE {?subject ?predicate ?object}
```

which returns all RDF triples contained in the document.

Remarks by the implementation of SPARQL query.

The Protégé environment allows the submission of SPARQL queries. In this environment we submitted specific questions shown in Fig. 4.26.

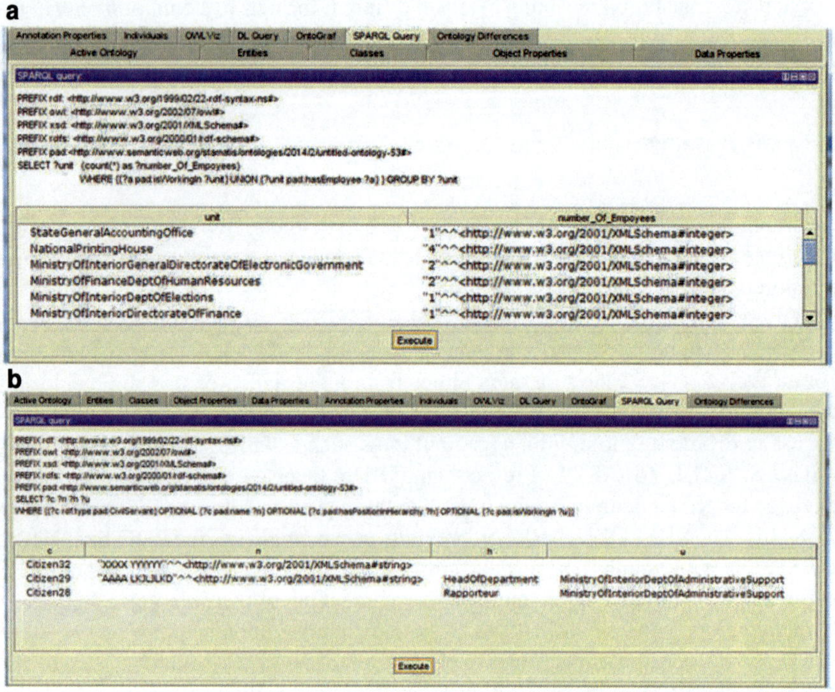

Fig. 4.26 **a** and **b** Queries and their results in SPARQL Query Tab of Protégé

a. Number of employees per organisation unit:

```
PREFIX
pad:<http://www.semanticweb.org/stamatis/ontologies/2014/2/untitled-
ontology-53#>
SELECT ?unit   (count(*) as ?number_Of_Empoyees)
WHERE {{?a pad:isWorkingIn ?unit} UNION {?unit pad:hasEmployee
?a} } GROUP BY ?unit
```

b. The data about employees in the units of organisations:

```
PREFIX
pad:<http://www.semanticweb.org/stamatis/ontologies/2014/2/untitled-
ontology-53#>
SELECT ?c ?n ?h ?u
WHERE {{?c RDF:type pad:CivilServant} OPTIONAL {?c pad:name ?n}
OPTIONAL {?c pad:hasPositionInHierarchy ?h} OPTIONAL {?c
pad:isWorkingIn ?u}}
```

From these questions, we note the following points.

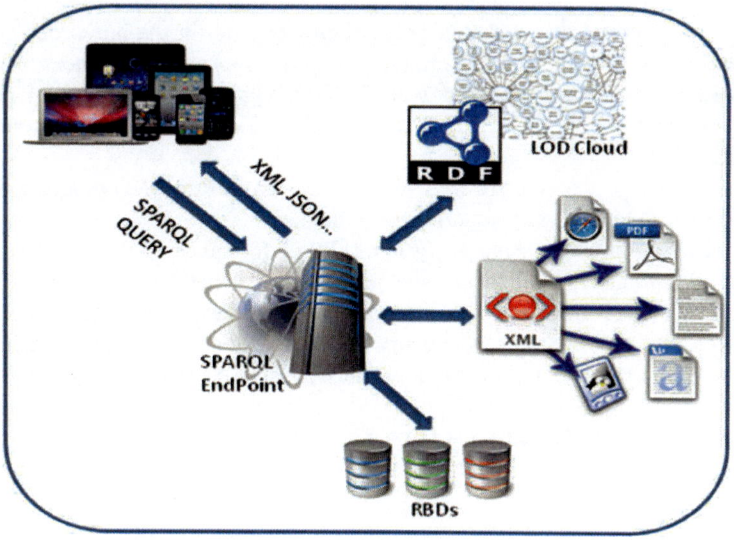

Fig. 4.27 The SparqL Endpoint architecture

- We used the SPARQL Query tab of the Protege, searching for information from the knowledge base created through the ontology, in the local environment of our computer. We are particularly interested in this tool, because SPARQL provides the ability to retrieve information by querying knowledge bases mainly oriented to RDF triples, but without excluding relational databases and datasets with semantics derived from spreadsheets. Full utilisation of these questions in the Internet environment is achieved through SPARQL EndPoints on the general architecture of Fig. 4.27.
- A SPARQL EndPoint is nothing but a Web Service through which a server can accept a SPARQL query from the client, then perform the appropriate search depending on the data set that is addressed by the client and finally return the results through the protocol HTTP [15]. However there are still problems to be solved such as the availability, maintenance and consistent performance of endpoints.
- Particular care must be taken when modelling knowledge to be as compatible as possible with RDF triples.
- When defining concepts, initial declarations of classes and their properties are important in order to avoid inconsistencies and logical errors which do not appear in the original design. Although protégé provides the ability to control the accuracy and consistency of the ontology through the reasoner, this is not useful in the performance of queries. This happens because queries are executed on the RDF triples created by the definition of ontology and not in subsequent relationships.
- When implementing a SPARQL EndPoint we must predict the recovery and presentation to the end user, of the properties associated with the concepts of ontologies. Although the basic concepts are usually known in advance to the end

users, their properties and their relationships depend on the initial design and modelling of knowledge in accordance with the visual of the designer.

- The results of queries submitted to the SPARQL Query Tab of Protégé, originally depend on declared properties rather than those deduced from the reasoners of Protégé.
- A disadvantage we note for SPARQL Query Tab of Protégé is that it does not provide the ability to store queries for future use.
- There is no comprehensive analysis on the performance of the relevant questions. Empirically, we can say the performance depends on the interrogated database in the form of storing information, the complexity of the query itself, the optimisations introduced by the query engine and other environmental factors.
- SPARQL does not support the semantics of RDF Schema. For example, the query: "`SELECT ?s WHERE {?s RDF: type pad: Citizen}`", where prefix pad regards the namespace of our ontology and Citizen argument the corresponding class, will return only instances of Class `Citizen` and not instances of the subclass, as shown in Fig. 4.28.
- In its current version it is not possible to modify an RDF document, i.e. there are no structures of the respective INSERT, UPDATE, or DELETE SQL. The SPARQL is currently a pure query authoring language for knowledge acquisition.
- The SPARQL specification defines the query results based on the RDF implication system. Nevertheless, in the specification there is a general, parameterized definition matching model graph, which can be extended to any implication system. Alternatively, the specification acknowledges that inquiries may be made to a virtual graph which has no explicit definition. Using this feature with suitable software, it is possible to produce a graph based on the rules of RDFS implication which are simple SPARQL questions.

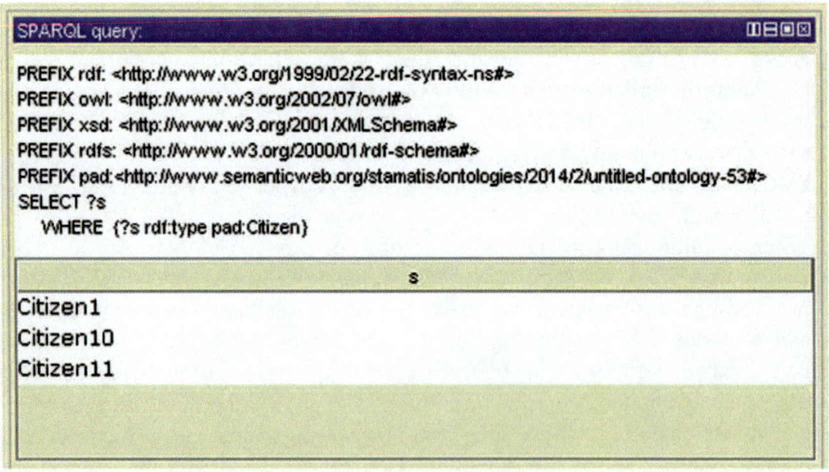

Fig. 4.28 Members of Citizen class as a response from SPARQL

- As regards the OWL-DL ontology, the OWL-DL axioms do not always lead to a unique graph. Therefore, to answer SPARQL queries to an OWL-DL ontology based on the implication rules of OWL-DL, the implementation of the general customised default template matching securities mentioned is required, suitable for implication based on the rules of the OWL-DL. Such an extension of SPARQL is the language SPARQL-DL.

4.5.2 SPARQL-DL in OWL2 Query Tab of Protégé

SPARQL-DL was designed as an extension of SPARQL, since, as already mentioned, was not supported by the discovery of knowledge from OWL-DL documents [20]. It appears as a query language capable of retrieving information but also as a separate query submission tool on OWL-DL documents [21]. SPARQL-DL combines TBox/RBox/ABox queries and templates used by SPARQL to ensure interoperability from the Semantic Web.

In our research we installed OWL2 Query Tab as a plugin in Protégé 4.3 and we used it (as well as the DL Query tool and SPARQL tool) for querying the Knowledge base that we have developed. It is a conjunctive query and meta-query engine (for SPARQL-DL and negation as failure), and visualisation plug-in. It facilitates creation of queries using SPARQL or intuitive graph-based syntax, and evaluates them using any OWL API-compliant reasoner.

It also provides the possibility of submitting additional atom queries related to the hierarchy of classes as: `directSubClassOf`, `strictSubClassOf`, `directSubPropertyOf`, `strictSubPropertyOf`, `directType`. These questions do not represent ontology data but are implemented through the existing questions of SPARQL.

4.5.3 DL Query Tool of Protégé

The DL Query tab provides a powerful and easy-to-use feature for searching a classified ontology. It comes with the standard distribution of Protégé Desktop (versions 4.x and above), both as a tab and also as a view widget that can be positioned into any other tab. The query language (class expression) supported by the plugin is based on the Manchester OWL syntax, a user-friendly syntax for OWL DL that is fundamentally based on collecting all information about a particular class, property, or individual into a single construct, called a frame [22].

With this tool, we are able to submit questions which are based on inference of RDFs shapes. As seen above, the questions of SPARQL retrieve information from the RDF statements in the relevant documents without using inferences. This weakness is covered with the DL Query of the Protégé. The search results correlate triples on the basis of logical conclusions generated by the chain (transitional, reverse, etc.)

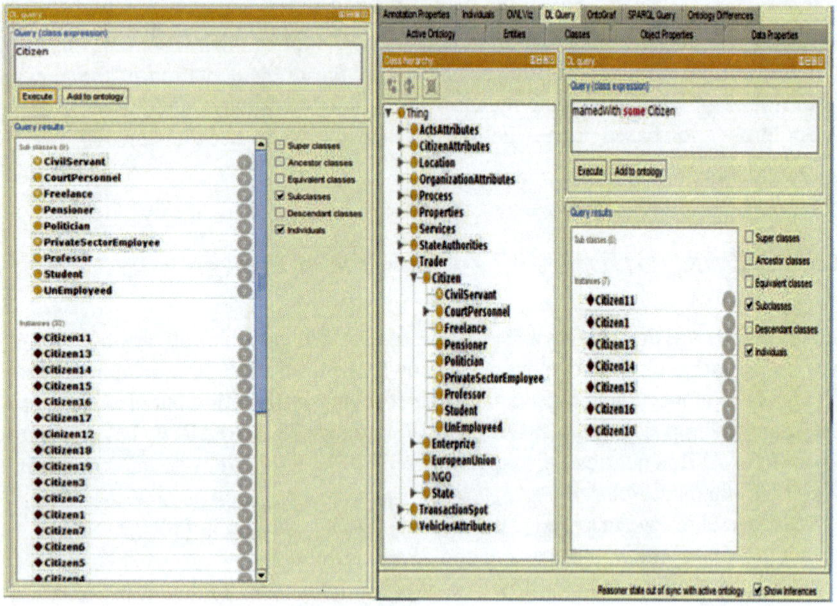

Fig. 4.29 The DL query results in Protégé

properties and relations between classes and subclasses. For example, it supports the logic that an instance of a class P is at the same time instance of superclass Q. With mathematical symbols would you describe this logic as follows:

$$(w \in P) \cap (P\mathrm{C}Q) \rightarrow w \in Q$$

In Fig. 4.29 the reasoners show instances of Citizen Class following the relevant question. We note here that the corresponding SPARQL query had no results found.

This tool allows the exporting of the results as general axioms in the form of classes. For example the question "`marriedWith some Citizen`" we can define the results in the form of general axioms as shown in Fig. 4.29.

4.6 Evaluation of Ontology

4.6.1 Categorization of the Ontology

Generally, an ontology includes a vocabulary of terms and some form of specifications for their importance. Concerning the degree of formality of the representation of an ontology can be:

- Informal, expressed in a natural language.
- Semi-informal, formulated in a limited and structured subset of a natural language.
- Semi-formal, formulated in an artificial and strictly certain language.
- Rigorously formal:definitions of terms with semantics, theorems and proofs of properties such as soundness and completeness.

The ontology developed in this work, is an OWL ontology, i.e. formulated in a certain artificial language. In the Protégé environment we have defined a set of rules, axioms and restrictions on classes and properties according to which any stored instances are linked. The concepts and properties that we defined follow the semantics based on the actual situation so as listed in the official legal sources. So we could classify our ontology in semi-formal ontologies with several elements from strictly formal ontologies.

A different categorization of ontologies is as follows:

- Knowledge representation ontologies: entities provide representation without specifying what this represents.
- General or common ontologies: aim to capture general knowledge around the world, providing basic concepts such as time, space, events etc.
- Top-level ontologies: provide general concepts under which all terms are associated with existing ontologies.
- Metadata ontologies: provide a vocabulary to describe the information content, which is available electronically.
- Domain ontologies: represent knowledge about a particular field, e.g. medical, administration etc.

Our ontology is at the intersection of general ontologies and of the domain of definition ontologies. This is because, in our ontology we have defined various concepts of our everyday life, such as the concept of citizenship, the concept of the state, etc. That is the nature of general ontologies. It also has basic characteristics of domain of definition ontologies, since our main goal was the description of the scope of public administration and public open data. Finally, our ontology, combines features of business ontologies, information ontologies, as shown below in Fig. 4.30.

4.6.2 Basic Principles of Design

When designing the ontology of authentic instruments and public administration, we have considered basic principles [18] of ontologies design, among which we mention the following:

- Clarity and objectivity. The ontology is intended to describe a particular subject field. This field should be attributed clearly and if possible objectively. Our ontology was developed based on existing legislation for the separation of powers, the Greek legislation which defines the criteria of open data and concepts set out in the provisions of administrative law.

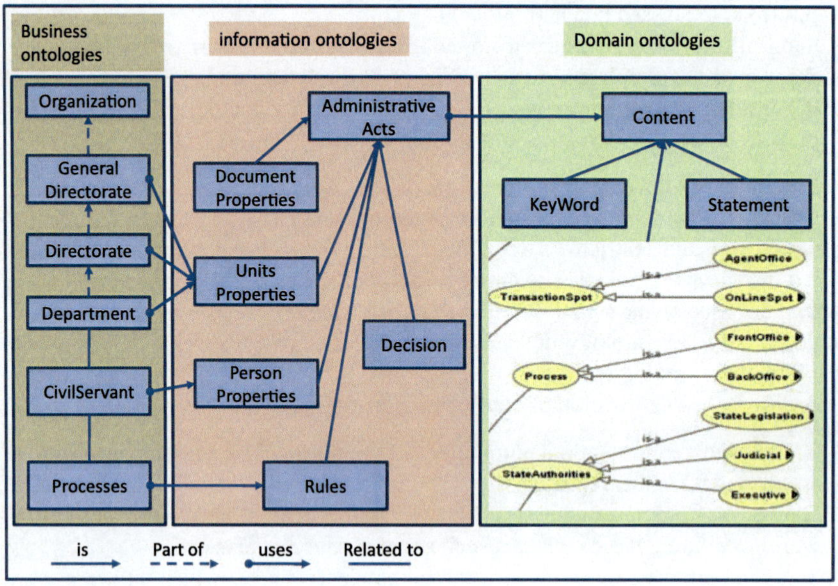

Fig. 4.30 Categorization of our ontology

- Consequence: the concepts and axioms declared do not contradict or conflict with the axioms or conclusions deduced. This principle was tested extensively using reasoners available in Protégé and analysed in detail in the preceding paragraph.
- Scalability—reuse of concepts: the design of the ontology should provide scalability and expertise at all levels, i.e. the classes, properties and instances. This point is important, given that the needs required to cover the ontology may change compared to the original design and that it may be necessary to add a new concept or relationship between concepts. For example, in our ontology in order to describe the concept of "*responsible position Head of Department in the Ministry*" instead of considering a class with all the Heads of Department in ministries (e.g. *HeadOfDtpMinistry*) as individual, we chose to define the following:

  - class Ministry with individual all ministries
  - property hasPositionInHierarchy, as a predicate for specific individual having the type Citizen
  - The individual HeadOfDepartment, type: Bureaucracy.
    With these definitions we achieve two objectives: (a) achievement of the wanted description and (b) the possibility of reusing the concepts to describe others.

- Minimal Ontological commitment: The axioms declared should be as generalised as possible. The possibility of extending or application to other fields of interest, shall be given without a clear reference to the specific area covered by the ontology. For example, the statement of the rule that "`whoever works`

`in a department ->also works at the corresponding direction`" was implemented by the rule:
isWorking (? x,? y), isUnit (? y,? z) -> isWorking (? x,? z),
so that it can also be applied in the case where we want to describe that "`anyone who works in an direction -> work and the corresponding general direction`".

- Minimise coding: because the general idea is to reuse the ontology and systems with different approaches, the coding should be as simple as possible and not to affect the information we want to model.
- Semantic analysis of conditions. For example, the concept of Citizen is often associated with the property of professional status. That is, we consider a civil servant or a retired or a student etc. as a citizen of a country. So we considered that kinds of professions are subclasses of the Citizen.
- Analysis is aiming to the reuse of concepts.

4.6.3 Methodology of the Ontology Development

In the bibliography several models of ontologies' development are recorded, depending on the scope and needs to be covered. Among these models we mention the model of Uschold-King [23] and the model of Gruninger-Fox [24]. The development of the ontology in the context of this work follows the basic steps of the model of Uschold-King enriched with elements of the methodology of Gruninger-Fox [24]. In more detail, the steps we followed in the development of the ontology were:

Step 1: Identification of the purpose for which the ontology was built in conjunction with the capturing of motivating scenarios

The purpose usually arises from the need to provide solutions to specific problems that either have not been resolved or the solutions found do not correspond to reality. In our example, the sector of public administration with the associated public open data is a field that involves many different aspects and concerns such a huge number of concepts and relationships. As mentioned above, the Semantic Web is expected to provide solutions in this area.

Step 2: Ontology Building

a. Standardisation of informal competency questions (formulation of competency questions) is considered to be important. These questions are based on motivating scenarios and form a sort of requirement to be met by the ontology. These questions must be answered with the terminology developed in the ontology.
b. Conception of ontology. At this stage we make the identification of the most important concepts and relationships between them and then we make their registration.
c. Definition of axioms. The axioms define any restrictions of the concepts combined with the atypical sufficiency questions.

d. Coding. It is done in the Protégé environment with the help of Description logic.
e. Integration of existing ontologies. We examined already published ontologies in the field of public administration with a view to incorporating any concepts or relationships. No models that align with our model were found.

Step 3: Assessment of the ontology

Besides texts of the documentation of the ontology, the evaluation record results from the use of a reasoner featured by Protégé. Their use enables us to ensure any issues of inconsistency restrictions and general axioms.

Step 4: Maintenance of the ontology

In the final stage, which may be the first of any revision of the ontology, we considered any omissions or the need to adapt to new requirements.

4.7 Semantic Modelling in the Domain of Official Statistics

4.7.1 The Official Statistics Domain

Businesses and public sector organisations have been facing a particular economic juncture in a changing international business environment in recent years, which makes it difficult to further develop them. Making the right decisions is a key success factor for both the private and public sectors. Executives charged with this responsibility have to take into account the huge amount of information in the field before they make decisions that will lead the public body to a successful path. This information is often linked to the statistics based assessment.

Regarding the functioning of the public sector, decision-making by the political leadership is often based on the suggestions of the public administration officials, who have to analyse and appreciate a huge amount of information derived from an equally large volume of primary data. In support of this function, executives are often required to have quality and reliable statistical analyses on large databases in their possession. Data and/or information is often required to be retained and managed by various operators whose systems are not interoperable. The current situation is characterised by a really large number of statistics produced by public sector bodies, which belong to a large number of categories while being produced and processed with software tools on a case-by-case basis. More efficient data processing and better use of statistical products require the adoption of innovative methods and tools associated with the emerging concept of Semantic Web, which is the smart extension of the existing web.

4.7.1.1 Motivation

Coordination of the services of the Ministry of Interior in order to support the program of certification of the produced statistics by Hellenic Statistical Authority—ELSTAT [25] highlighted specific management issues within the organisation within the responsibilities of the Statistical Head [26].

Among the issues that triggered this work are the following.

- Statistics production services are characterised by heterogeneity in terms of administrative structure and on the processing methodology and data analysis
- a large number of regulatory provisions on the production of statistical products which often produce conflicting results
- non-interoperable software applications for processing and managing statistical products
- difficulties in understanding the terms and procedures for certifying statistics
- non-standardized procedures both during production and dissemination of statistical products
- possible overlaps in the production and diffusion of statistics.

Semantic Web technologies have very positive results to offer in addressing these issues and are mainly related to the modelling of knowledge and the intelligent management of statistics. In this work, we focus on the modelling of knowledge in the field of official statistics of the Greek Statistical System through the development of a corresponding OWL ontology. We also highlight the possibility of intelligent search on the data stored in the knowledge base we developed along with ontology.

4.7.1.2 Recording and Presenting Basic Concepts of the ELSS

In accordance with the statistical law [26] and the provisions of the institutional and legal framework, as detailed on the official website of the ELSTAT, the ELSS bodies, which produce and disseminate statistical products, have the obligation to certify as official statistics by ELSTAT. An important parameter of the success of this effort is the understanding of the terms and procedures as well as the implementation of concrete steps/actions by the stakeholders. In detail, the definitions and necessary steps have been exhaustively depicted in the relevant circulars and instructions of ELSTAT [27].

As indicative terms and procedures we mention:

- The main components of the ELSS are the "National Authorities of ELSS" and the "ELSS bodies". In detail the composition of the two entities refers to the relevant decisions of the President of ELSTAT [27].
- Each statistical product produced by the ELSS institutions after being certified by ELSTAT is characterised as "Official Statistics", as provided for in the relevant provisions of the institutional and legal framework.

- Each statistical product is characterised by a unique code that follows it throughout its lifecycle, irrespective of the operator, the subject and the domain that belongs to it as well as the producer that produces it. These data are recorded in the corresponding tables of ELSTAT.
- Producer and production manager is considered to be the service belonging to Ministries or Independent Authorities or other bodies of the Greek Public Administration.
- Each entity in the system can be considered either as a producer or as a user of statistics on a case by case basis. In the meaning of the producer, it is understood that the body is independent of the organisational unit that has the substantive competence, whereas the user can be understood as either an entity or individuals/citizens.

4.7.2 Developing an Ontology for the Modelling of Knowledge in the Field of Official Statistics of the ELS

Information retrieval with innovative—intelligent methods and the automated management of the knowledge produced by them have as a prerequisite the appropriate modelling of information and knowledge. An important factor in this direction is the annotation of data, using semantic terms and methods of the Semantic Web. Following the primary step in the study of the ELSS environment, which was the recording of the basic concepts, correlations and procedures of the field, we developed the ELSS ontology in order to create a knowledge base. The development of the ontology and the feeding of the base was done using Protégé 5.1.0, a modern open source tool [28].

The methodology we have followed includes the following stages:

- Initially, we developed all the necessary entities that correspond one another to the ELSS concepts we developed above.
- Then we created the necessary instances to complete the set of entities and create the knowledge base.
- Then, the most important step of the process is the semantic annotation entities and instances. Properties and relations between entities play an essential role in rendering semantics to the concepts and entities of the database and is the key difference from a relational database. Their appropriate use also supports the efficient production of RDF triplets that essentially support the open and interconnected distribution of our information.
- We also defined, where appropriate, restrictions on the domain and range of classes as well as specific rules for extracting conclusions on the Protégé SWRL tool.

In particular, we present the most basic of the classes we have defined and their respective subclasses in correspondence with their real concepts and associations as discussed above.

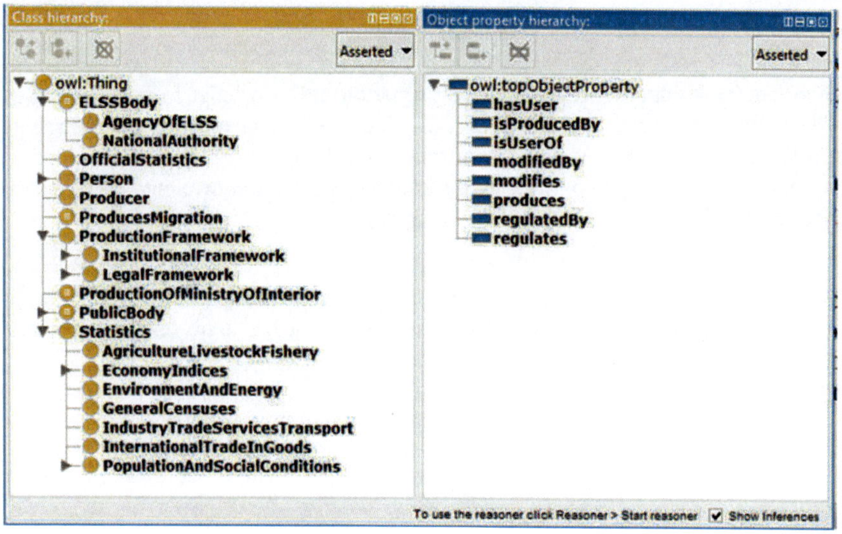

Fig. 4.31 Class and Object property hierarchy

4.7.2.1 Definition of Classes/Subclasses

As mentioned above, the ELSS entities are in full correspondence with the classes of ontology as shown in Fig. 4.31.

- We have used the "`Person`" class to attribute the roles of staff and citizens involved in the production and use of statistics respectively, while the "Producer" class corresponds to the bodies that produce and disseminate statistics.
- The "Statistics" class corresponds as the basic taxonomy of the statistics produced, based on the classification set by ELSTAT [27]. Individuals of the class and its subclasses are the statistics produced by the ELSS agencies.
- The "`OfficialStatistics`" class implements the "official statistics" entity
- The core components of the ELSS are implemented with the "`AgencyOfELSS`" and "`NationalAuthority`" classes.

4.7.2.2 Definition of Properties

Particularly important are the properties we have defined and which essentially materialise the semantic annotation of the entities (Fig. 4.31). From the properties we have developed, we consider as important, the "produces" property, used to implement the basic function of the ELSS in the production of statistics. Correspondingly, the inverse "isProducedBy" property is used to mark the relevant statistics produced by the operators.

4.7.2.3 Definition of Rules and Restrictions

Important for automated production of information and knowledge from our ontology is the definition of rules through the SWRL language and the corresponding Protégé tool as well as property statements through the domain and range definitions as appropriate. Another way of limiting is the statement of the equivalent class we have used for the class "`ELSSBody`". For example:

- The "`produces(? x,? y)-> Producer(? x)`" rule results in automatic updating of the Producer class when using the reasoner provided by the application by the respective individuals, implementing the logic that "Producer is someone who produces a statistical product".
- The "`Statistics(? x), isVerified(? x, true) -> Official-Statistics(? x)`" rule results in the stamping of a statistic as "`official statistic`" by including it in the corresponding class, when the property "`isVerified`" gets the value "`true`".

4.7.3 Results

With OWL ontologies, we have the ability to declare the relationship between the classes, by defining appropriate properties in the logic of the representation of RDF triples. This way and with the help of the reasoners provided by the application, concrete conclusions are made on a case-by-case basis.

Protégé [28] provides two main ontology query tools: the DL Query and the SPARQL environment. The first tool is based on the function of the reasoners also provided as protégé plug-ins, and uses the reasoning that OWL supports on the statement of declared properties. The use of this tool produces information that is the product of a conclusion, without directly arising from the statements of classes, members and their properties. On the contrary, the SPARQL environment operates on RDF data logic, and runs through all of the declared RDF triplets without applying any other kind of conclusion. In this work we present some typical examples of using DLQuery.

Figure 4.32a shows the results from the question "`produces value Migration`" on "`which body produces the statistics called Migration`", while the Fig. 4.32b shows the results from the query "`isProducedBy value MinistryOfInterior`" which statistics are produced by the Ministry of the Interior.

Figure 4.33 shows the results of the reasoner in our attempt to state that the "Migration" statistic is produced in addition to the Ministry of Migration Policy and the Ministry of the Interior. In this case, the inconsistency of the database is recorded, registration is not accepted and the database maintains its integrity and consistency.

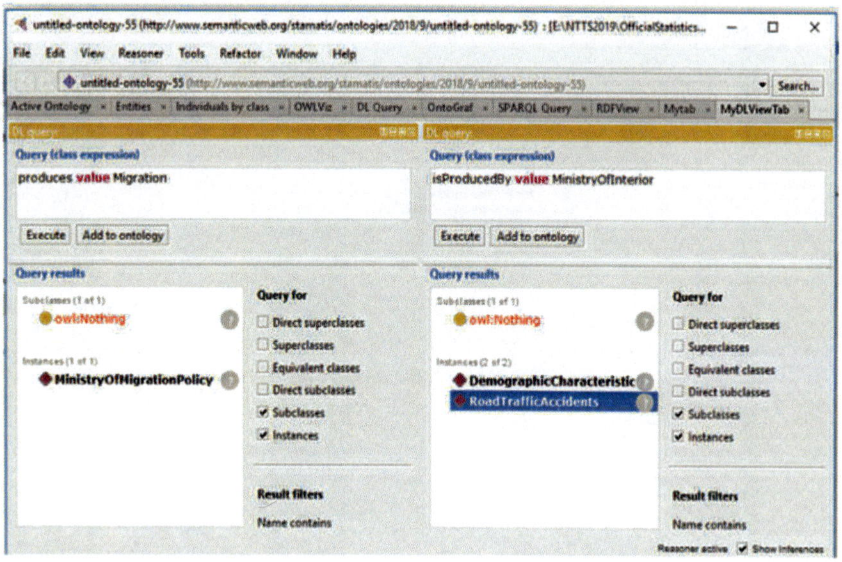

Fig. 4.32 **a**, **b** Results from DLQuery

4.7.4 Assessment and Evaluation of Ontology

The base we developed was tested using the reasoners tools provided by Protégé and the ontology is checked for its consistency and integrity. The ontology and the corresponding basis we developed are useful in many ways, including (a) the possibility of semantic search of information based on relations between classes and between individuals, (b) as a training material for those involved in the development and dissemination of statistical public bodies; (c) the visualisation of a complex system of concepts, entities and associations, (d) future use of the database as an input file for the production of linked statistical data. This was made possible by modelling the relative knowledge of the object in the form of an OWL-ontology.

In terms of ontology assessment, we have to mention the following advantages of the proposed modelling.

- Simplicity of class statements, instances, and relationships between them.
- Ability to directly enrich the base without requiring a change of the existing structure from the beginning.
- Consistency with the actual ELSS data.
- Possibility to query based on reasonable conclusions.

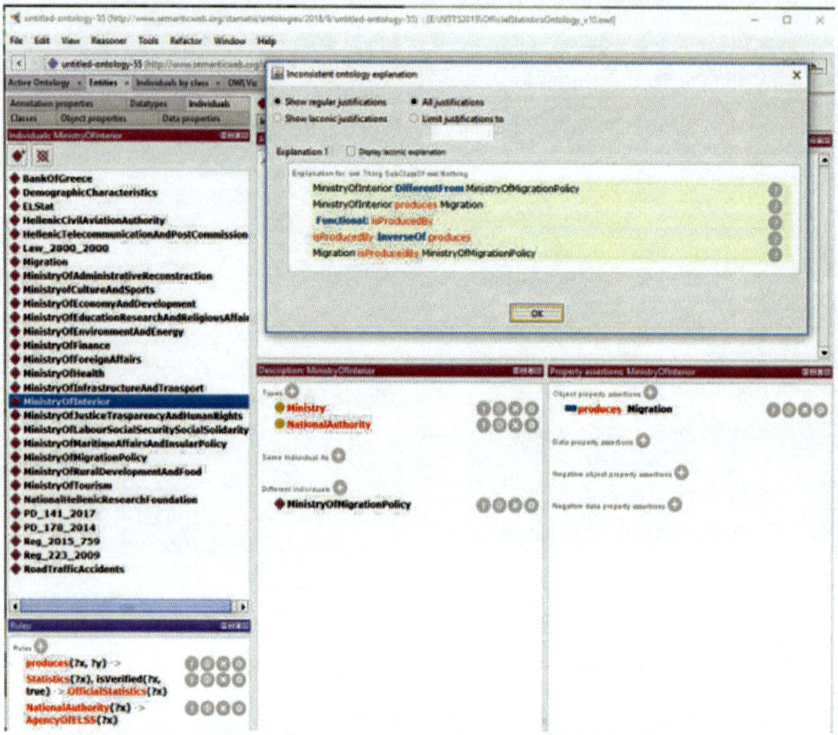

Fig. 4.33 Results from DLQuery

4.8 Knowledge Representation in the Internal Audit Field

4.8.1 Introduction and Motivation

Businesses and public sector organisations face in recent years a very complex and competitive environment which hinders their operation and further development. This environment is characterised by volatile financial figures and the existence of specific risks arising either by increasing global competition or other economic or inherent operational factors. There are, however, risks associated with errors or non-normal situations related to the business or the organisation itself. In the area of dealing with these risks, we consider that the development of the auditing science and in particular the internal audit is of great importance. The control no longer concerns only the financial figures of a company or organisation but also the compliance with the policies espoused, the legality of the acts of employees and finally the prevention of corruption phenomena.

The concept of internal audit concerns the monitoring and evaluation of the overall operation of enterprises and organisations "from the inside" and is an essential tool of the authorities in their attempt to shield their services from voluntary or involuntary acts, detrimental acts, and behaviours which have a negative impact on the organisation's interests. For this purpose the administrations of organisations usually adopt an Internal Control System as a set of operational actions for the purpose of their effective, efficient, and safe operation. By the Internal Control System specific procedures and methods of operation or production of goods and services are set out and specific safeguards for their compliance are provided. In the public sector this is already being addressed systematically through the institutional framework—legislation, circulars, and directives—defining the function of public sector entities. In the private sector this is the responsibility of each administration and varies according to the size and scope of business.

The compliance with the rules set by the Internal Control System, and any inconsistencies or failures are subject to relevant controls of internal audit units. These services are typically staffed and operate within the organization itself and aim to assure the administration about the adequacy and consistency of the Internal Control Systems that have been adopted. Their operation concerns all the entity's functions, and covers both administrative matters as well as productive and financial matters. In recent years, the ubiquitous use of information systems and the Internet, has led the Internal Audit to extend to the field of Information Technology Governance, which is of particular interest.

Motivation for this work has been the activation of Internal Audit Units in the public sector entities in Greece, about three years ago although the Greek government had institutionally planned to operate them in the public sector since 2006 [29]. Related reports regarding the listed companies have existed in Greece since the beginning of 2002 [30]. Internal audit is for the Greek Public Administration a new operating area in which special attention should be paid to the administrative and functional integration and the further consolidation and development. Especially for the authors of this paper, this is an area of concern as the internal control is associated with the processes and the information systems concerning the application of e-Government.

In this work an attempt is made to map the environment of internal audit in terms of Semantic Webby using relevant tools, like Protégé. This was considered appropriate since the audit environment and the procedures followed in the internal audit units are particularly complicated and combine various cognitive objects. The study of internal audit as an object is summarised in the second part, as an essential structural part of this work. Then, by using a special tool we implemented specific ontology in order to record the terminology and the corresponding concepts and entities that characterise the field of internal audit. The reasoning and the result of this work is presented in the third part. In the fourth part we present the results of semantic search with the help of relevant tools and in the fifth part we evaluate the ontology developed by using relevant tools and in the sixth part we give the overall conclusions of our work.

4.8.2 *Presentation of the Audit Field*

4.8.2.1 Generally About Audit

Auditing as a branch of economics and management is both science and technique. It is characterised as science because through the investigation, it constantly generates new knowledge and as a technique because along with the scientific knowledge generated, it meets the needs of audit thus achieving its mission.

More specifically by the term auditing we mean the systematic process of gathering and evaluating audit evidence from an independent, capable, and appropriate person. In greater detail, the content of the auditing may be categorised into the following three themes:

a. The audit object, i.e. what is audited, why it is necessary to carry out an audit and what objectives are pursued and achieved through audit.
b. The subject of the audit (auditors) i.e. which persons perform auditing and what should be their qualifications, their skills and the necessary experience.
c. The auditing procedures (methodology and technique of audit), i.e. how the audit is carried out.

4.8.2.2 Audit Categories

The controls are divided, according to various criteria, in the following main categories [31].

A.1. Internal audits are the audits organised by the body itself and carried out by qualified and properly trained executives, internal auditors, who are employees of the organisation.

A.2. External audits are the audits performed by external auditors who have no employment relationship or employee status with the body that they are invited to audit.

A.3. Mixed audits are organised and coordinated by the Internal Audit Units and carried out in collaboration with external auditors.

B.1. Special audits are those who are exploring a specific field or object, such as procurement, revenues, payments, inventory management, quality assurance procedures of all the Agency's activities etc.

B.2. General audits are those extending across the management for a specified period.

C.1. Preventive audits are those which aim to prevent the risk and are carried out by the same organisational units involved every time.

C.2. Repressive audits are those carried out subsequently. They aim at testing the implementation of procedures and the confirmation that the objectives were achieved in full in an effective and efficient manner. They also aim at revealing and suppressing errors and omissions, irregularities, theft and mismanagement of funds and resources.

D.1. Permanent or ongoing audits are those conducted on an ongoing basis and take place during the execution of the financial act, transaction or productive operation.
D.2. Regular or periodic audits are those carried out at regular periods.
D.3. Temporary or occasional audits are those carried out in exceptional cases either by the internal audit service initiative or by order of the Administration or following specific complaints. These are audits carried out after assessing that there has been suspected waste, abuse, theft or corruption and generally wherever failures identified in the functions and immediate determination of their size is required in order to take administrative measures.

4.8.2.3 Concept and Internal Audit Content in Public Administration

Internal audit should not be confused with the Internal Control System. Usually, an internal control system adopted by the management body in order (a) to protect the company's assets (b) to secure and check the accuracy and reliability of its accounting data (c) to improve the effectiveness and efficiency of its operations and (d) the compliance and implementation of its operational policies. According to [32], an internal control system has specific elements—components that make it strong [33]. (a) Audit environment as the foundation for all other components, setting the stigma and the general spirit of the enterprise. The factors that affect it are the integrity, the moral values, the abilities and skills of the organisation's people, etc. (b) Risk assessment. A prerequisite for risk assessment is establishing objectives at different levels of the organisation that are linked to each other and are internally coherent. The risk assessment lies in identification and risk analysis regarding the achievement of goals and objectives of the organisation and determining how they should be managed. (c) Auditing activities are all the policies and procedures that ensure the compliance with the instructions of the organisation's administration, (d) Information and communication. That is, all the tools to manage and disseminate information which assist in making business decisions, (e) Monitoring. That is, the quality assessment procedures of system performance.

On the other hand, the internal audit is an independent, objective, guaranteeing and consulting activity designed to add value to the organisation and improve its functions [34]. It also contributes to achieving the organisation's objectives through a systematic method of valuation and improving the efficiency of processes, risk management, internal control and corporate governance.

The internal control can and should control all Directorates, Departments, services, functions, activities as well as policies, procedures, regulations, and applied practices that constitute the internal control system of the institution. It is a tool of the Administration, which enlightens, advises and guides aimed at overall improvement. It intervenes to examine a process or an activity of the organisation and deposits (a) a diagnosis confirming more or less the good operation, (b) a prognosis to alert those responsible for the central administration and (c) a treatment that aims to ensuring the organisation's interests, the reliability of information, the effectiveness of operations and the competitiveness.

The primary purpose of internal audit is to assist all executives of the organisation (at all levels of administration) to effectively perform their duties and in this way to provide estimates, recommendations, opinions and information relating to the activities in question. Its object is to examine and evaluate the adequacy and implementation of various functional systems of an organisation and their control systems, as well as the examination of the quality of the organisation's actions within the framework of undertaken responsibilities. In addition, auditors should protect the essential values of public administration as they serve all citizens.

The auditing procedures as formulated by the Institute of Internal Auditors [34], adapted to the nature and the public sector scope are:

- Oversight. The auditors help the directors to supervise whether the public sector services do what they ought to do, if they use the available funds for the intended purposes and if they comply with the laws and regulations.
- Detect. Detect aims to identify improper, inadequate, illegal, fraudulent or abusive acts already done and to obtain the information that will lead to decisions related to criminal prosecutions, disciplinary actions, or other corrective actions.
- Deterrence. Deterrence is designed to identify and limit the circumstances that allow corruption.
- Insight. The controllers provide insight to help managers to determine which programs and policies work and which do not.
- Foresight. The auditors help organisations have foresight identifying future trends and focusing their attention on the potential challenges before they grow into crises.

4.8.2.4 Audit Types

The types of Internal Audit are:

- Production Audits.
- Financial Audits, which comprise the auditing procedures relating to the security of the business's wealth and assets and aim to verify the accuracy, honesty and reliability of financial statements and data and identify its legality.
- Operational audits, which include the framework and the assessment procedures on the compliance with the policies and the company's operating procedures. The objective of operational audits is the evaluation and assessment of the existing system structure and the proposal to develop new systems and improve existing ones in areas that have been discredited.
- Management audits, which include the organisational framework and the procedures related to decision making, compliance with them and their evaluation. In contrast to the financial and operational audits, they have broader objectives which consist of examining and evaluating, based on objective and scientific methods, the overall administrative efficiency of the business/body.

- Information technology and information systems audits. They aim to establish the extent to which reliability, confidentiality and integrity of information available is ensured. These lie in the effectiveness of information systems, the proper use of resources, the strengthening of infrastructure and information security, the correctness, completeness and accuracy of transactions and the updating of information files.

4.8.3 Modelling of Knowledge Within the Auditing Sector

The knowledge of the organisation's structure and the risks of which it is threatened, the internal controls designation and proper functioning, the auditing procedure, reports issued to the management and continuous monitoring of the organisation's work. All the above are the main work areas of an Internal Audit Unit. This consideration, however, must be continuous, since things often change, especially safeguards, the person authorised to apply them, the organisation risks and the organisation's objectives. A prerequisite for achieving the above is the retrieval of information in innovative ways and then the effective management of knowledge generated by them. It is therefore an important factor of success for the internal audit unit, the modelling of the overall environment in a semantic approach. As a primary step in the study of audits' environment and in order to create a knowledge base, we considered appropriate, to present in this work, the recording of key concepts and entities of the audit as well as the properties and the relationships that characterise them. The properties and relationships between concepts and entities are essential for the performance of semantics in concepts and individuals in this database and constitute the key difference from a relational database. This database becomes useful in many ways, including (a) the ability to directly semantic search for information based on the relationships between classes and between individuals, (b) as an educational material for those involved with the audits, (c) visualisation of a complex system of concepts, entities and associations, (d) future use of the base as an input file to generate associated data. This was made possible by the modelling of the relevant knowledge of the auditing field in the form of an OWL ontology. For this purpose, we used the application Protégé 5.1.0 [35]. The ontology we present is equipped with all the basic concepts, with some corresponding virtual individuals and the properties that interconnect (Fig. 4.34). Also we defined, where necessary, restrictions on the domain and range of classes and specific inference rules in SWRT tool of Protégé.

In particular, we present the most essential classes we defined and their respective subclasses in correspondence with the actual concepts and their relationships, as discussed above.

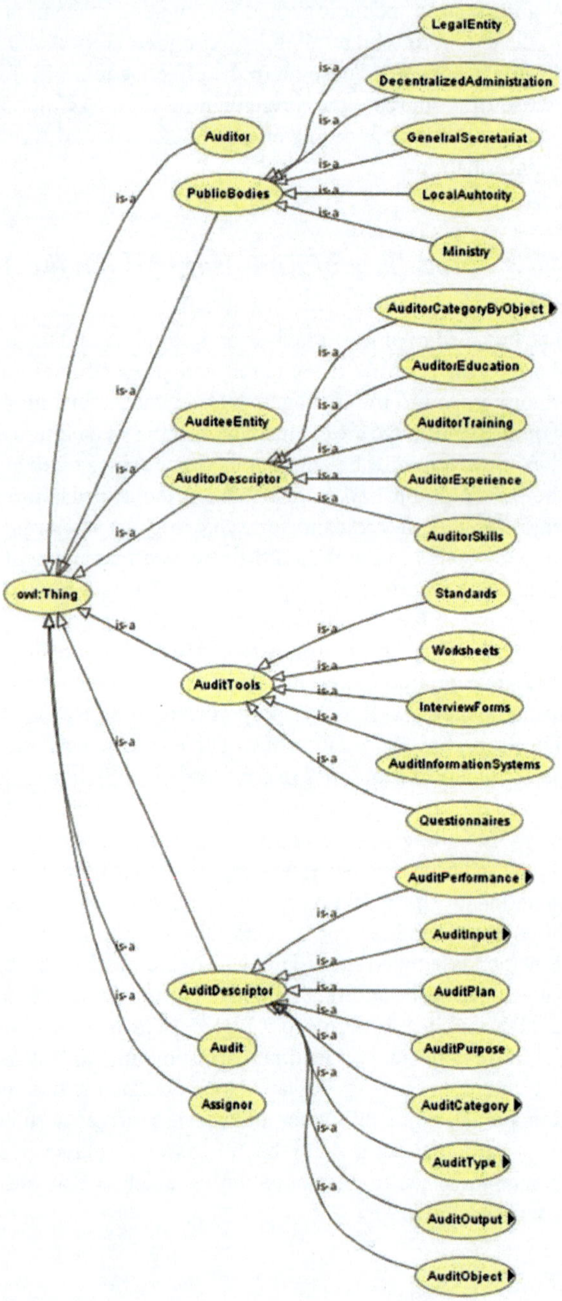

Fig. 4.34 The ontology view from the OWL-Viz Tab of Protégé

4.8.3.1 Definition of Classes

- Classes "Audit" and "Auditors" resulting from the basic entities of the field of interest: audits and auditors respectively while the physical entities constitute the corresponding individuals of the related classes.
- The class "AuditDescriptor" was set to correspond to the characteristics that describe the audits. Here we have defined subclasses of "AuditCategory" and "AudiType" to classify the corresponding instances of audits in the appropriate categories and genres. Also we have set the remaining subclasses shown in Fig. 4.35 by formatting the assumption that "any audit—as an individual—belongs to at least one category ("AuditCategory"-class) and type ("AuditType"-class), has a specific object ("AuditObject"-class), while based on a specific plan ("AuditPlan"-class) and function ("AuditPerformance"-class) it produces effects ("AuditOutput"-class), fulfilling the purpose ("AuditPurpose"-class) specified ("hasAssignor"-object Property) by the assignor ("Assignor"-class)".
- Similarly we have set the class "AuditorDescriptor" in order to format the characteristics of each auditor (Fig. 4.36): "each auditor ("Auditor"-class) is classified into a specific category ("AuditorCategoryByObject"-class) related to his object, has specific educational characteristics ("AuditorEducation"-class), special training ("AuditorTraining"-class), experience ("AuditorExperience"-class) and skills ("AuditorSkills"-class)".
- An audit is an individual of the Audit class and simultaneously one of the classes that describe audit. For this reason the "Audit" and "AuditDescriptor" classes are not disjoint to each other. For example, the control of the procedure for obtaining a citizenship decision.
- "AuditRelatedToCitizenship" is an individual of "Audit" class, "Regular" class and "InternalAudit" class. Similarly, each auditor (physical entity), is an individual of the class "Auditor" and simultaneously the "AuditorCategoryByObject" class as a subclass of "AuditorDescriptor".
- As each audit is mandated by the responsible body, we define as base the "Assignor" class which has as individuals all the assignors who commands the carrying out of the audits. Also as in each audit specific tools are used, such as Information Systems, Interview Forms, Questionnaires and Worksheets, we defined the corresponding classes with individuals and the instances of each case.
- As each audit refers to one auditee entity and that entity belongs in the public body we respectively defined the classes "AuditeeEntity" and "PublicBodies". Each auditee entity constitutes an individual of the "AuditeeEntitee" class while the particular public sector organisations are represented as subclasses of the "PublicBodies" class.

Fig. 4.35 The ontology class hierarchy in Protégé

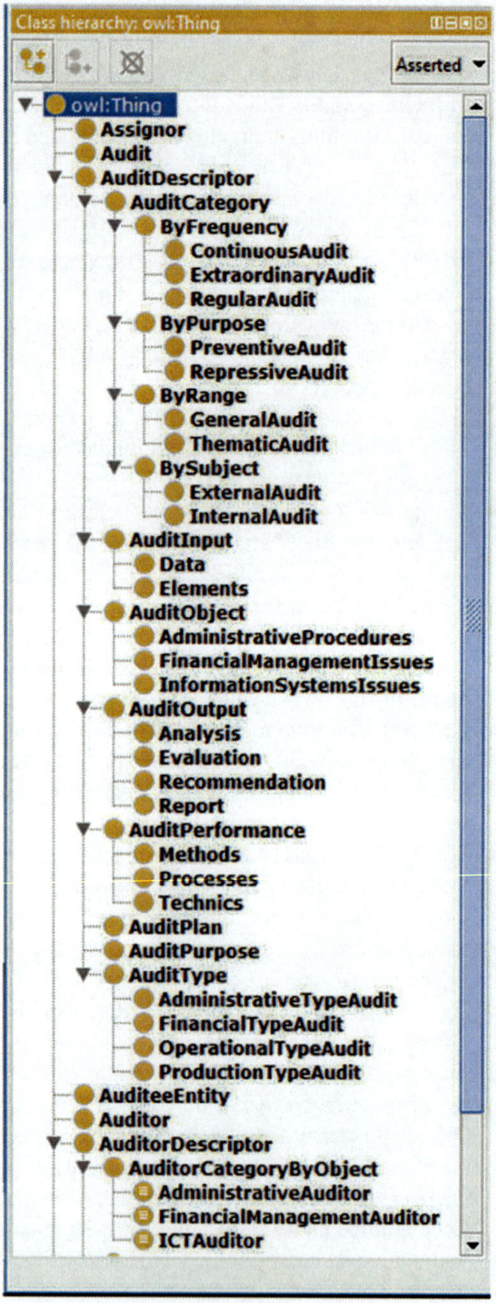

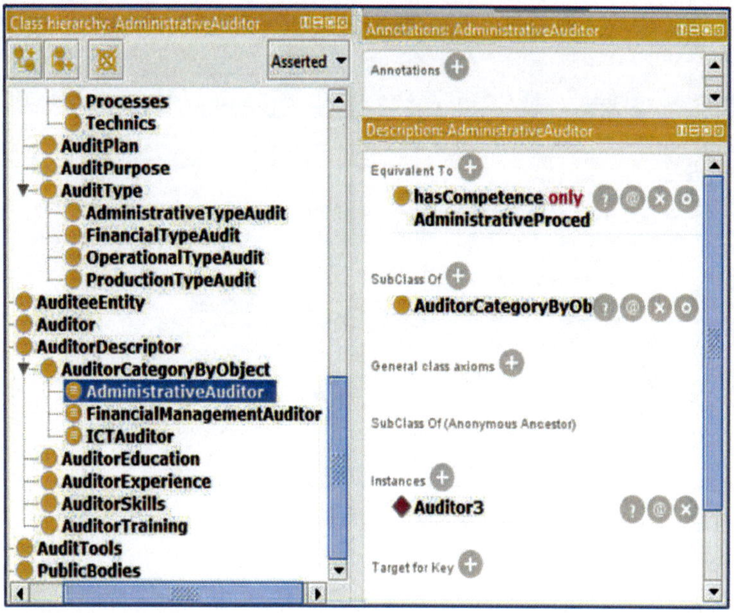

Fig. 4.36 Part of class and subclasses in Protégé

4.8.3.2 Definition of Properties

The properties defined by the user are of particular interest as they appear in the corresponding tab (Fig. 4.37).

- In particular, we mention the properties "`hasAuditor`" and "`hasObject`" that connect the audits with the auditors and their objects. These properties

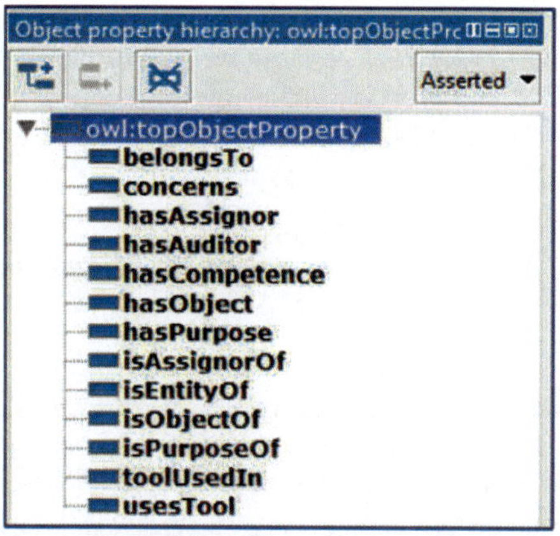

Fig. 4.37 The object properties in Protégé

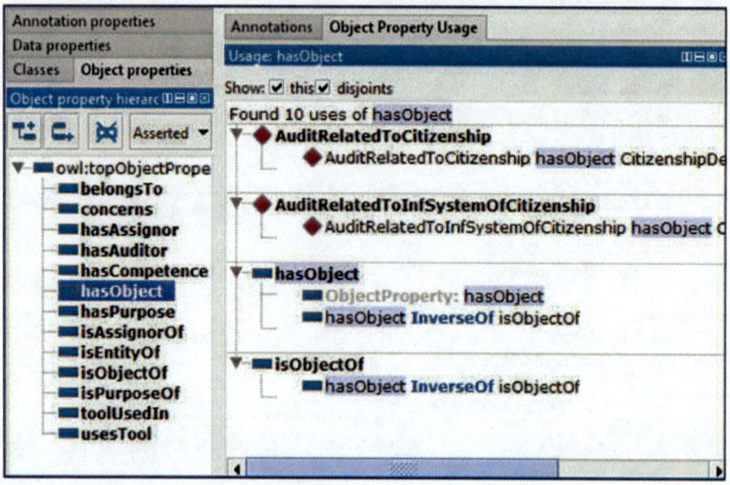

Fig. 4.38 The *hasObject* property and its use

combined with the respective classes, perform semantically the knowledge that "each audit ("`Audit`"-*class*) carried ("`hasAuditor`"-object property) by an auditor ("`Auditor`"-class) and has a specific object ("`hasObject`"-object property)". Their use appears in the corresponding tab of Protégé (Fig. 4.38). Relative is also the property has Assignor, connecting the audits with their assignors, in order to perform the knowledge that "each audit ("`Audit`"-class) has ("`hasAssignor`"-Object Property) an assignor ("`Assignor`"-class).

- Also interesting is the property "`hasCompetence`" which connects the "`Auditor`" class to the object controlled. Its use is shown in the corresponding tab of Protégé. It is used as an example to identify as "`ICTAuditor`" class any individual whose duties include matters relating to information systems (Fig. 4.39).
- Useful property when searching for information in the field of audits is "`usesTool`", linking the audits and tools used to them. An example of the use of this property in the ontology ISO_27007 [36] particular tool is shown in figure (Fig. 4.40).

4.8.4 Examples of the Application of Restrictions, Rules and Queries in Ontology

With our OWL ontologies we are given the opportunity to state the relationship between classes, with the definition of suitable properties within the logic of the data representation in RDF. For example we declared the property "`hasObject`" and we set it up as a domain class "`Audit`" and range class "`AuditObject`". This way we

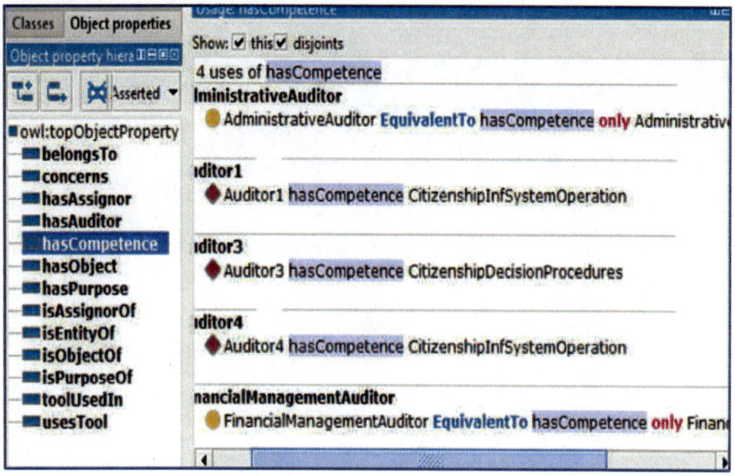

Fig. 4.39 The *hasCompetence* property and its use

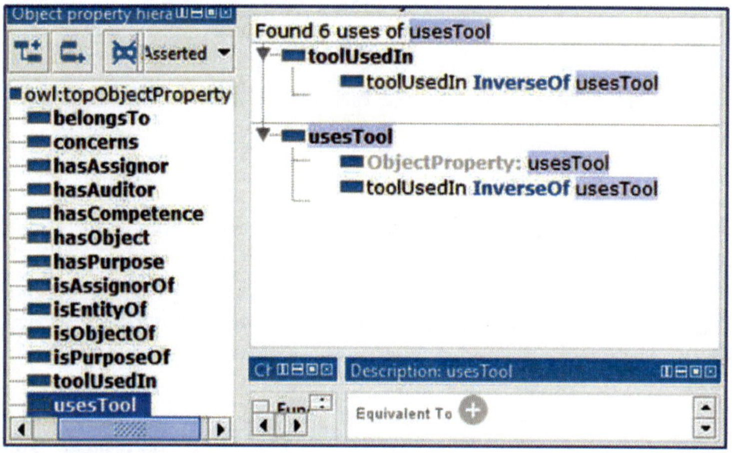

Fig. 4.40 The *usesTool* property and its use

declare the reasoning that if the individual "AuditRelatedToCitizenship" is associated with the individual "CitizenshipDecisionProcedures" then these belong to Classes "Audit" and "AuditObject" respectively. In this way and by means of the reasoners provided by the application specific conclusions are produced as appropriate (Fig. 4.41).

In order to state specific restrictions on the members of classes of the ontology, we used the concept of equivalent class. This was achieved with the statement of logical expressions in OWL, which use properties and classes appointed by the user. For example in "ICTAuditor" class we stated as an equivalent class the one derived by_

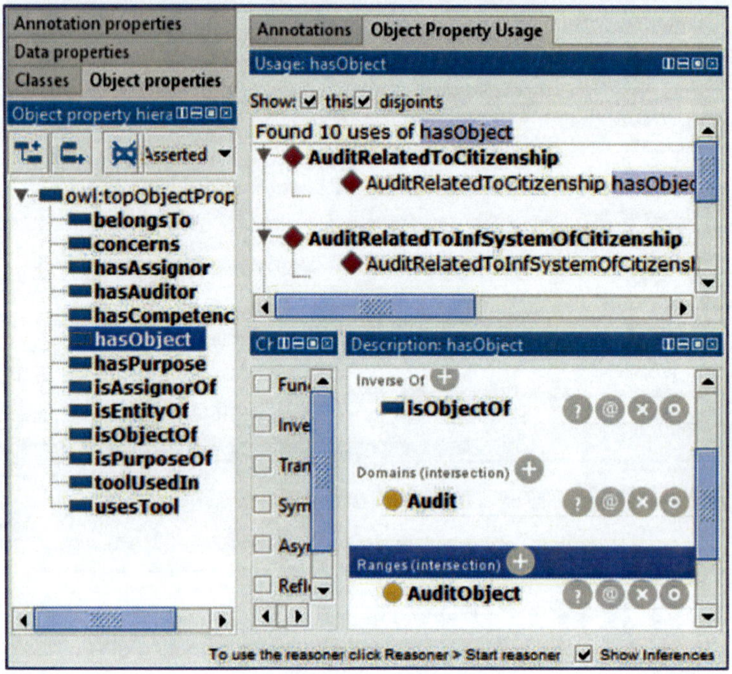

Fig. 4.41 The *hasObject* property and its domain and range statement

the logical expression "hasCompetence only InformationSystemsIssues" (Fig. 4.42). This declaration serves the need to reduce the members of the "`ICTAuditor`" class only to the members that satisfy the condition that "the information systems auditors have responsibilities only to information systems issues". If no such restriction is satisfied the systems of the reasoners of the application display a message for data inconsistency.

Another way to impose certain restrictions or to state facts which lead to the conclusions on the initial data is the SWRL tool offered by the Protégé. This tool uses the same language to designate the desired events which will form the basis of the inference in DL Query tool Protégé. For example in Fig. 4.43, we have stated that "whoever is auditor of information systems, then belongs to the class of auditors".

The Protégé provides two basic query tools for the ontology: the DL Query and the SPARQL environment. The first tool is based on the function of the reasoners also provided as plug-ins of the Protégé and uses inference supported by OWL on the stated properties. By using this tool information is produced that constitutes an inference product, without directly resulting from the statement of classes, members and their properties. For example Fig. 4.44 shows the effect of the relevant question we posed to the ontology about the members of the "`Auditor`" class. Initially member "`Auditor1`", was not declared as a member of a class. By applying the inference of the reasoner and DL Query we see that the member "`Auditor1`" conclusively is a

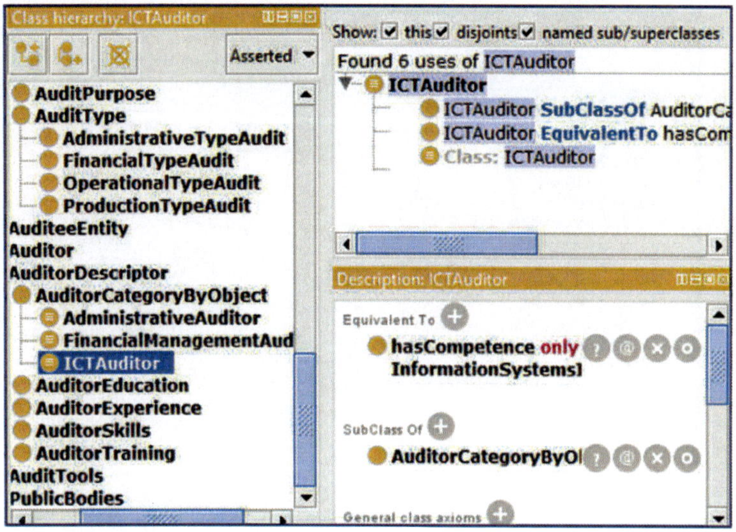

Fig. 4.42 The *ICTAuditor* equivalent class statement

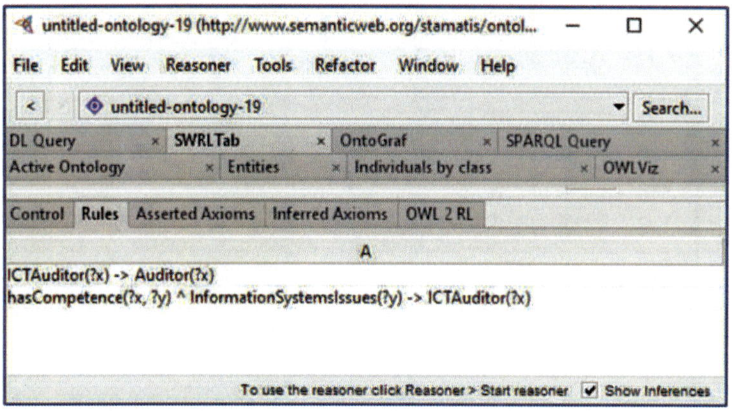

Fig. 4.43 Rules declaration in the SWRL tab in Protégé

member of the class "Auditor". In contrast, the SPARQL environment operates on the logic of the RDF data, and runs all the stated RDF triples without applying any other kind of inference. A typical result is shown in Fig. 4.45, where the "Auditor1" is not forming part of the results of the relevant search.

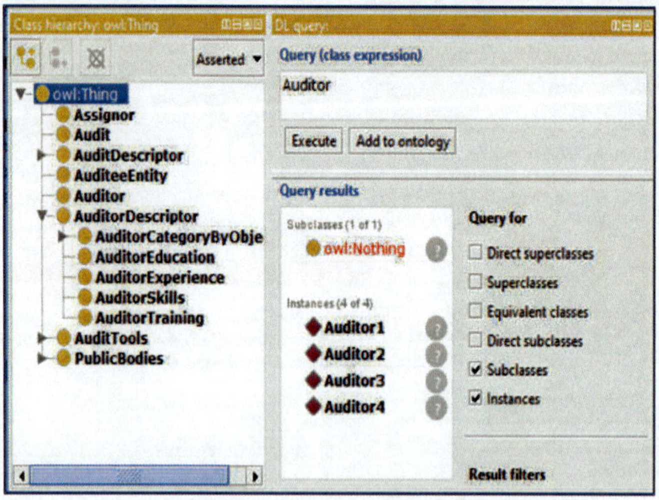

Fig. 4.44 The effect by the relevant query about the Auditor class

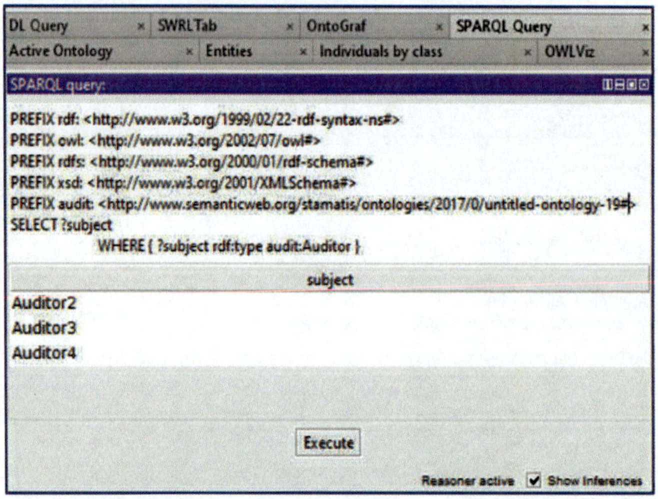

Fig. 4.45 The response of SPARQL query in Protégé

4.8.5 *Evaluation—Assessment of Ontology*

In order to examine the level of consistency of the ontology under the actual circumstances of the audit, we recovered the official text of the ISO/IEC 27007: 2011 [36]. By this standard, we cover issues related to the planning of audits, the object, the frequency of audits, the procedures to be followed and the qualifications of suitable auditors in the field of IT security. We also cover design and auditing content issues, auditing activities and compliance issues. All model's elements listed, are already included in the ontology developed as basic classes. Note that if judged necessary to add new concepts or properties, this does not affect the operation of the original version. We conclude that the ontology could be integrated into productive operations to be fed with real data.

Moreover, the Protégé offers through reasoners control mechanisms for the consistency and precision of ontology. Any problems are encountered during the running of the DL Query independently or by calling the specific reasoners. As shown above when running relevant DL questions, there were no inconsistencies in the ontology.

Regarding the valuation of the ontology we can mention the following advantages of the proposed modelling.

- Simplicity of starting classes, instances and relations between them.
- Possibility of direct enrichment of the base without the requirement to change the existing structure from the outset.
- Consistency with truly international standards applying on audits.
- Ability to query the basis of logical inferences.

On the downsides we note the manual query process. However this issue can be addressed by appropriate programming techniques applied on the data that can be extracted from the ontology, which is not the subject of this work.

4.8.6 Conclusions

Due to constant changes in economic, entrepreneurial, or regulatory conditions, it has become urgent for the auditing bodies to adopt and use appropriate mechanisms and tools to identify and counter the risks associated with the expected changes. At the same time there is an enlargement of the audit environment, with the addition of an additional audit field, that is related to the governance of information systems and electronic communications. As a consequence, we have a multiplication of the operational activities of audit services—internal or external on one hand and the need to modernise and/or fully review information systems to support their work on the other. At the same time, the work of dissemination of information related to audits, in the higher hierarchical levels of organisations, where the decisions are made, must be supported.

Existing information systems used by the auditing bodies are now facing their needs as appropriate and are based on traditional technologies of relational databases and decision support systems. The field described belongs to those cases that can be supported more successfully than the technologies of the Semantic Web, as it is characterized by the large volume of data which is still increasing, the various categories and types of information that change rapidly and must be shared with different rights to users. These features cannot be addressed in terms of resource economy with traditional programming techniques and tools. Moreover, this area is not characterised by strong commercial demand leading to a reduced supply of software applications. This was one of the motivations to investigate the possibility of software development based on Semantic Web technologies and associated data.

In order to meet the operational needs of the audit mechanisms of the public sector, the systematic study of the audit area is essential to develop a "smart" integrated information system that meets the real needs of the auditors. The issues that emerge in this study are the following.

- Diagnosis and identification of operational needs. To serve the work of audit mechanisms, we must take into account the differences of the audit environment between the various organisations, but also the similarities of the audit procedures followed by all internal audit units or other control mechanisms. In any case it is necessary to have common information search and management tools and resources, automated planning and control, measurement tools and risk assessment, task management tools, etc.
- Determination of the information system purpose. A common requirement of all control mechanisms is mainly to provide information. Specifically: (a) the production, management and intelligent search of information materials relative to audits, such as operating manuals and legislation, (b) auditing equipment like questionnaire forms, audit reports, check-lists (c) analyses of issues, findings and points of attention per audit category, as resulting from the corresponding reports.
- Access to the information system data. Key features are the large volume of data and information that can constitute sources and/or audit results and the fact that some information may be public and others of special classification. For this reason it is necessary to use "smart" technical data and search interface. Such techniques are offered through the technologies of Semantic Web and linked data.
- Maintenance—management. The maintenance and management of a centralised system must be the subject of a central public service that will ensure the stability, confidentiality and integrity of information and to ensure proper operation.

References

1. Nonaka, I. (2007). The knowledge creating company. *Harvard Business Review.*
2. Wang, J., & Wang, H. (2008). *The research of e-Government governance mode under knowledge management*. IEEE. https://doi.org/10.1109/WiCom.2008.2220
3. Bender, S., & Fish, A. (2000). Transfer of knowledge and the retention of expertise: The continuing need for global assignments. *Journal of Knowledge Management, 4*(2), 125–137.
4. Chen, Y.-L., Hong, Z.-W., & Chuang, C.-H. (2012). A knowledge-based system for extracting text-lines from mixed and overlapping text/graphics compound document images. *Expert Systems with Applications, 39*(1), 494–507 (2011, Elsevier).
5. Paralic, J., & Sabol, T. (2001). *Implementation of e-Government using knowledge-based system* (pp. 364–368). IEEE. https://doi.org/10.1109/DEXA.2001.953088
6. Nonaka, & Takeichi. (1995). *The knowledge creating company.* Oxford University Press.

7. Wimmer, M. A. (2006). Implementing a knowledge portal for eGovernment based on semantic modelling: the e-Government intelligent portal. In *Proceedings of the 39th Annual Hawaii International Conference on System Sciences (HICSS'06)*. IEEE.
8. Dang, Y., Wang, Y., & Wang, H. (2011). The design and implement of e-Government information system based on knowledge management. In *International Conference on Mechatronic Science, Electric Engineering and Computer (MEC)*. IEEE.
9. Varna, V. Use of ontologies for organizational knowledge management and knowledge management systems. In *Integrated Series in Information Systems Book Series (ISIS)* (Vol. 14, Chap. 2, pp. 21–47).
10. Sarantis, D., Charalabidis, Y., & Askounis, D. (2010). A goal oriented and knowledge based e-government project management platform. In *43rd Hawaii International Conference on System Sciences*. IEEE.
11. Yang, X., Wang, Y., Zhang, X., & Jin, J. (2008). *Research on the classification of e-Government information resources by categories-theme integrating*. IEEE. https://doi.org/10.1109/WiCom.2008.2155
12. WESTBIC: The Business and Innovation Centre for the West and North of Ireland (IE), Thessaloniki Technology Park (EL), ATLANTIS Consulting (EL) *ALBA (Greece), LINK MV (DE), Institute for Support and Competence Development Pro-Competence (DE), BILSE Institute (DE), The Multidisciplinary European Research Institute Graz (MERIG) (AT), Cyprus Institute of Technology (CY), Sofia University St. Kliment Ohridski (BG), National Council for Private SMEs in Romania (RO), TRAINing Material in ORganisational KNOWledge Management for European ORganisations & Enterprises, *A pilot Leonardo Da Vinci project* (2008).
13. Ping, Z. (2008). An integrated architecture for effective knowledge e-Government. In *International Symposium on Knowledge Acquisition and Modeling Workshop*. IEEE.
14. RDF Working Group—W3C, Resource Description Framework—RDF. http://www.w3.org/standards/techs/rdf
15. W3C official site: www.w3.org
16. Cyganiak, R., Wood, D., & Lanthaler, M. RDF 1.1 concepts and abstract syntax. *W3C Recommendation*, 25 February 2014. https://www.w3.org/TR/RDF11-concepts/#bib-WEBARCH
17. Brickley, D., & Guha, R. V. RDF schema. *W3C Recommendation*, 25 February 2014. https://www.w3.org/TR/RDF-schema/
18. Uschold, M., & Gruninger, M. (1996). Ontologies: Principles, methods and applications. *Knowledge Engineering Review, 11*.
19. O'Connor, M. *The semantic web rule language*. Stanford Center for Biomedical Informatics Research, Stanford University. http://protege.stanford.edu/conference/2009/slides/SWRL2009ProtegeConference.pdf
20. Sirin, E., & Parsia, B. (2007). SPARQL-DL: SPARQL query for OWL DL. In *Proceedings of the 3rd International Workshop on OWL: Experiences and Directions*.
21. Derivo Semantic Systems, SPARQL-DL. http://www.w3.org/2001/sw/wiki/SPARQL-DL
22. Protégé wiki page, DL Query Tab. http://protegewiki.stanford.edu/wiki/DLQueryTab
23. Uschold, M., & King, M. (1995). Towards a methodology for building ontologies. In *IJCAI95 Workshop on Basic Ontological Issues in Knowledge Sharing*, Montreal.
24. Gruninger, M., & Fox, M. S. (1995). Methodology for the design and evaluation of ontologies. In *Workshop on Basic Ontological Issues in Knowledge Sharing*, Montreal.
25. Hellenic Republic, Statistical Law No. 3832/2010 (2010). http://www.et.gr
26. Pierrakou, C., & Theocharis, S. (2018). Coordination of national statistical system—Experience from the Hellenic Statistical System (ELSS). In *Q2018 European Conference in Quality of Official Statistics, Statistics Poland & Eurostat*.
27. Official site of ELSTAT. www.statistics.gr
28. Official site of Protégé. http://protege.stanford.edu/about.php#citing
29. Hellenic Republic, Law 3492/2006 (2006). http://www.et.gr
30. Hellenic Republic, Law 3016/2002 (2002). http://www.et.gr

31. Ministry of Finance, Greece, Conducting manual of internal audit (2015). http://www.publicrevenue.gr/kpi/
32. COSO, Improving organizational performance and governance: how the COSO frameworks can help (2014). https://www.coso.org/Documents/2014-2-10-COSO-Thought-Paper.pdf
33. Official site of COSO: https://www.coso.org
34. Official site of The institute of Internal Auditors: www.theiia.org
35. Protégé Project, Stanford University. http://protege.stanford.edu/
36. ISO/IEC 27007:2011 Information technology—Security techniques—Guidelines for information security management systems auditing. http://www.iso27001security.com/html/27007.html

Chapter 5
Towards Open Data and Open Governance—Representation of Knowledge and Triplification of Data in the Field of the Greek Open Government Data

Abstract The subject of this chapter is the most specific field of the semantic representation of open data and its use in the context of electronic government. In particular, the relationship between e-Government, Open Government and open data is presented, while the advantages and expectations of opening government data are presented and analysed. In this context, the Greek government data are examined as a case study, and in particular the perspectives and critical points, while methodology and guidelines are proposed for their successful publication as open public data. In the context of knowledge representation and modelling, and as a case study, a specific ontology of the open data of the Greek repository is proposed and evaluated. In the context of the use of open data, a methodology/algorithm is developed and presented for the production of RDF triples from open data of the Greek public open data repository of the "Diavgeia program" as well as a methodology for the publication of open data in corresponding public open repositories. Finally, the opening and the publication of the e-Government ontology developed in a previous chapter as open data in a corresponding data repository, is presented as a case study.

Keywords Open government · Open data · Triplification · e-Government ontology

5.1 From e-Government Towards Open Government

The transformation of governance from the classical model to the model of the e-state is directed towards the adoption of principles and policies of the so-called "open governance". We are talking about a new system of governance, Digital Governance, which is essentially an extension of the original design for the e-model. The involvement of an administration in Digital Governance lies in the implementation of a series of actions related to the establishment of institutions to improve the regulatory framework governing the operation of the public sector to promote transparency, accountability and participation and to restore citizens' trust in the state and at the same time the necessary infrastructure to support them [1].

Openness and accountability concern all kinds of powers, Public Administration, Justice, the operation of state services and institutions, while they directly affect

S. Theocharis and G. A. Tsihrintzis, *Semantic Knowledge Modelling via Open Linked Ontologies*, Artificial Intelligence-Enhanced Software and Systems Engineering 4,
https://doi.org/10.1007/978-3-031-20585-9_5

the operation of the political system. Giving new possibilities to the consolidation of the constitutional principles and rights of the rule of law. The application of the principles of open government aims at the overall upgrading of the standard of living and the achievement of sustainable development in accordance with its basic principles. Particularly:

- In the sense of transparency, it means that citizens know and understand the functions of the administration, strengthening the procedures for the fight against corruption.
- The ability of citizens to control the actions and efficiency of the administration is linked to the concept of accountability of the administration and increase the level of trust in the administration and the government.
- The concept of participation means the ability of citizens to formulate together with the administration the various policies and any procedures that concern them.
- Open Government, in addition to being a form and method of good governance in terms of the functioning of public and public administration, also fully concerns the functioning of the basic institutions of the Republic, namely the Parliament, the Government and the political parties.

In addition to the new policies, new methods and technologies are needed to manage the data and information that is circulated and shared in the context of the completion of open government. A key technological pillar of this effort is an innovative system, the so-called "open data" along with the standards and technologies/applications that support their use. An equally remarkable effort is made in linking this data with the emergence of the so-called "Linked open data". Their importance, their uses as well as the methods used to manage them as well as their prospects, are the subject of this work. The development of new methods of retrieving information from the processing of huge databases that can be created from open data can enhance the effectiveness of this project. There is talk of big data and their methods of analysis to formulate standards and draw useful conclusions from them.

The reform from the e-Governance model towards the "Open-Digital Governance" is associated with the rapid increase in the volume of data published and traded through the Internet. All these data vary as to their usability and accessibility. In recent years, the governments of technologically advanced countries like the U.S., U.K. and other EU countries emphasise on the opening of government data to the public and at the same time on providing adequate tools for the consumption of their users [2]. Related to this is the initiative at the international level, called "Open Government Partnership-OGP"—and the relevant published guild—of a great number of states members from Europe and USA about the opening and sharing of government acts to the public. The OGP was established in 2011 as an international voluntary effort to foster more transparent, effective and accountable governments. Member governments embrace a high-level declaration of principles on transparency, participation, accountability, and innovation and develop their own individualised action plan, focused on local priorities for open government, developed through a local consultative process. Action plans should outline specific, measurable, actionable, relevant and time-bound ('SMART') commitments for advancing transparency, accountability and participation [3].

Under EU directives the Greek Government has already established freedom of publishing on the Internet a large amount of governmental data through an application called "DIAVGEIA" [4] (meaning clarity). This policy, through the improvement of transparency and image of public administration, can ensure the improvement of business and the economy, and eventually strengthen democratic institutions.

The data, as well as any product produced, are useful when they can be consumed. The concept of consuming data means the ability of the users to identify, process and generate information and knowledge. Thus, it becomes a challenge for the scientific community to find solutions for proper organising of data published on the web but also for their efficient and optimal use afterwards. Identification and processing of data, and production of information and knowledge from them, the recycling of data by other users are some of the issues that relate directly to the rights reserved by the data editor. Closed data is not accessible to everyone, therefore it can be used on a small scale and produce knowledge and benefits for a few. The so-called open operating data functions differently. These data have a particular licence of use and distribution, and if respected by the users, can generate knowledge for a large part of the population.

The production, publishing, connection and further processing of open data is a huge topic, with application in both public sector and citizens/businesses. Associated with the new trend of the Semantic Web it is expected to give further boost to e-Government. In this chapter we examine the prospects of opening up the public sector data to the benefit of the citizens as well as of public sector bodies.

The connection and consumption of open data have concerned the scientific community with numerous publications and the development of applications on open source and free software. On the Internet we found a large number of training documents regarding procedures for opening and linking data and a large number of publications. Examples include sites that focus on open data and related tools that support them.

As part of the European effort to open up public data and reuse them by machines and humans, the European Union presented on the 18/12/2012 the Open Data Hub for the European Union [5] decided by the European Commission on 12/12/2011 (2011/833/EE) in accordance with recent EU directives.

The OKFN, promotes and actively contributes to the development of the Strategic disclosure of open data, to creating a culture of openness by the publication of relevant handbooks [6] but also in building applications useful to people like CKAN and Open Spending Book [7].

The Greek Open Data Hub follows the EU Directive on Open Data and specifications met by the Open Data Portal. In its page, data are published and obtainable by anyone who wants to develop it, reuse it, connect it and create innovative services. In addition, data providers, application developers and the general public can use new features made possible by semantic technologies [8].

The website datahub.gr regards a project in constant evolution, which has been developed and maintained by the "Athena" Research Centre. There is currently a small number of open government data published on this site, but it is expected to become a digital knowledge hub where would it will be possible to detect, share, freely reuse and visualise open data provided by the public sector, businesses and citizens. This site runs on open source software and only applies open standards. The basis of the data list is CKAN, which has been developed and maintained by the Open Knowledge Foundation and by a community of volunteers, along with a number of extensions that provide advanced visualisation services. The current interface of the site and its functionality incorporate DGU, the expansion of data management developed for the needs of data.gov.uk [9].

5.2 Benefits—Perspectives from the Opening of Government Data

Adopting policies in the context of Open Government can bring many benefits to governments and their citizens [10]. In particular, the opening of public data can lead to the development of innovative applications and thus create new business activities. Public data can be the raw material for the production of information with added value as knowledge for all citizens as it promotes free sharing and reuse.

From the processing of data provided "openly" by the administration to the public, it is possible to produce important information and consequently knowledge in various areas of interest such as economics, administration, urban planning, etc. And this is because public open data may relate to various cases of public activity, such as financial data, geospatial administration, citizen safety issues, statistics and demographics, etc.

The core principles of Open Government are to ensure transparency and openness in the operation of state institutions. At the same time, they reflect the level of democracy that a country has, forming the framework for the effective participation of the citizens in the exercise and control of the work of the government. Open Government is expected to be a tool and guarantee for the full implementation of popular sovereignty and the democratic principle of the state, while guaranteeing the respect and implementation of human rights. In this way, the public's trust in the state is expected to be restored and its participation in the so-called "commons" will be increased.

We believe that the greatest benefit from the adoption of these principles is related to the fight against corruption, which in recent years has been of particular concern to national administrations in most countries [11]. Opening the data of the government and the bodies of the public administration to the public, acts as a deterrent to those involved who expect their private interest over the public, while at the same time providing the possibility of controlling the actions of the administration [12]. Especially in the case of public procurement, Open Government policies can be effective.

When the terms and procedures of the tenders, the pricing policies, the selection criteria of the suppliers or contractors of the projects as well as the procedures of monitoring the projects are made known to the general public, the opportunity is given for the effective control of the administration.

Opening up government data and then connecting them in order to reuse them, promises benefits for all parties involved in the future. The main benefits are:

- *Improvement of the state's function.* Opening up government data requires adjustment of structures, processes and functions across the public sector. If properly organised and achieved within a short period of time it will result in the improvement of the bodies themselves.
- *Transparency—Accountability.* The most important result is the achievement of transparency in the actions of the government. The publication of the acts of public administration, has the effect of improving both the image to the public and also the most consistent functioning of public institutions, through accountability imposed by the disclosure of government data.
- *Strengthen democracy.* Consequence of the above is to strengthen the role of citizens in public affairs, because of the opportunity to know and to directly control the actions of the government.
- *Social offer.* The free access of citizens in governmental primary information ("DIAVGEIA" program) or open government data will have a direct impact on society. On the other hand, the process of open government data by intermediate users and the development of relevant applications, can provide the simple Internet user a multitude of services that improve their daily lives and make his information easier and more reliable.
- *Improving economy—reducing cost.* The open access to government data and freedom of processing and reusing it under the Creative Commons licence, has immediate obvious financial results for citizens and businesses. The saving of financial resources for the acquisition of information is moreover a bet and at the same time an advantage for attracting knowledge-intensive enterprises. In any case, this is also a lever to increase entrepreneurship and help the exit of the economic crisis of recent years.

In particular, in the area of transparency of public administration operations, the expected benefits include:

- The wide disclosure of the terms and procedures for conducting public tenders, in a transparent and stable context, gives the impression from the beginning of an organised and objective process.
- The announcement of the results of the tenders and their complete documentation restores the image of the administration in terms of compliance with legality and accountability.
- The publication of the monitoring of the relevant contracts gives the opportunity to control the actions of the administration during the implementation of public works or the procurement of materials and services, preventing the most common

phenomenon of changing the terms of the announcement of tenders during the implementation of contracts.
- Finally, the organisation and publication to the general public of registers of suppliers, contractors and service providers and related pricing policies can act as a deterrent to the creation of closed circuits taking part in public tenders.

Corresponding benefits can be achieved in a large number of public administration activities that affect a large number of managed and involved employees, such as:

- Procedures for the selection, evaluation and development of the staff of public bodies,
- Staff mobility procedures between the various bodies,
- Procedures for selecting contractors and suppliers for public works, as the terms and procedures for the procurement and execution of public works are known in advance,
- Procedures for drawing up and monitoring budgets and the economic activities of public bodies in general,
- Business licensing procedures,
- Procedures for approving urban plans, etc.

5.3 The Case of Greek Open Government Data

The most distinctive effort of the Greek government to open up government data was the implementation of the program called “DIAVGEIA” (which means transparency) in 2010 [4]. Here we will present its main features and make a few remarks on the usability and potential of this program.

The “DIAVGEIA” program in accordance with the relevant provisions (Law 3861/2010 of the Greek state) aims to achieve maximum publicity of government policy and administrative activities, ensure transparency and consolidation of responsibility and accountability on the part of institutions exercising public power [4]. Free access to open government data can provide a meaningful citizen participation in social, economic and political life in the information society. The publication of government documents is an important step, which is to be completed by providing an open service that allows further processing and use of data, and search for specific transactions through a series of parameters as defined in the relevant site.

Through this program:

- The obligation to display the decisions of public administration bodies on the Internet was introduced for the first time in Greece.
- Citizens may have access from one point to all laws and decisions issued by all bodies of the close and broader public sector.
- Any decision regarding economic issues cannot be executed if it’s not posted on the relevant website. With the completion of the post each decision acquires a unique identification number (Number of Web Posting—AΔA), which certifies it.

- The modes of displaying and searching decisions per public bodies are provided through the central website of the National Printing Office operating at et.diavgeia.gov.gr.

In particular, according to the existing legal framework, among the data that can be posted on the Internet and that can be open data under specific licences (CC) are the following:

- Laws, legislative acts, Presidential decrees,
- Regulatory acts with the exception of the regulatory acts concerning the organisation, structure, composition, order, supply and equipment of the Armed Forces of the Country, as well as any other act, the disclosure of which causes damage to the national defence and security of the country,
- Circulars,
- Financial data for state expenditures and revenues, such as Budgets, accounts, balance sheets of state bodies,
- Contracts,
- Acts of appointment, acceptance of resignation, transfer, termination of employment of regular employees and employees of special categories, appointment of single-member bodies and establishment of collective management bodies of public bodies,
- Acts of appointment of committees, project groups,
- Tender announcements and results,
- Spatial data related to public areas,
- Spatial data related to local economic and business activities and residential development,
- The administrative division of the country,
- Data related to the safety of citizens and the fight against crime,
- The distribution of the population,
- Social data related to job, education and care opportunities.

The API of Open Data, as a standard feature, offers the possibility of deriving decisions and administrative acts which are posted under the "DIAVGEIA" by using open patterns. All data derived from the system are available under licence Creative Commons—Attribution. This means that anyone can use, reproduce and modify these data without any limitation other than the obligation to refer at the source and the explicit differentiation of secondary material that might be generated from the original material [4].

From the user's side, we note the following:

- The website of this program, offers through the interface, as an essential service, the search for documents based on filters concerning the ΑΔΑ (Number of Web Posting), the thematic area, category and other characteristics.
- The search services are provided only in Greek language, since the relevant decisions are only published in Greek.
- The number of published decisions are really big, so difficult to navigate if the user does not know the exact characteristics of the document he looks for.

- The large number of organisations which are obliged daily to publish their decisions, creates dysfunctions when accessing the website of the program mainly during working hours.
- There are no figures for the qualitative characteristics of the data published.
- No applications have been developed for the effective exploitation of the program data to the benefit of the public administration.
- The initiative for the consumption of open government data, seems to have been left to individuals or research institutes.

Apart from the "Diavgeia" program, the Greek Ministry of Interior and a large number of other bodies such as municipalities and regions use a specific repository, such as an instance of CKAN, in order to publish a large number of government data—datasets as open data [13]. Data.gov.gr is the central directory of public data that provides access to databases of Greek government agencies. The purpose of data.gov.gr is to increase access to high value, machine readable datasets by providing integrated services of cataloguing, indexing, storage, search and availability of public sector data and information, as well as online services to citizens and third party information systems.

Also it is worth mentioning that the Cadastral Agency and Mapping Association (OKCHE) [14] gives free access to all high-resolution orthophoto maps aerial photographs from 2010 taking on the area of the region. The range datasets, which was completed in December 2012, covering part of Central Greece and Salamis Island, as shown in the diagrams available orthophoto maps distribution.

The product consists of 1152 signs of the distribution 1:2500 of the reference system EGSA '87. In the first stage 558 signs were created, which were available since April 2012 in the second stage 394 signs were produced, which were available since June 2012 and in the third and last stage 200 signs which have been available since February 2013.

This particular project represents an important tool for all engineers, and contributes to the work of the Municipalities which will be able to significantly reduce the cost of their technical projects and studies. With digital orthophoto maps a new, modern and reliable background is created which every engineer or citizen has for free and works with on his computer, for any purpose, such as demarcations, control and tracking precision or arbitrary structures. The data are available under the terms of the Open Public Licence Geospatial Information (Version 1.0) which is available via the web [15]. Digital files of orthoimages are JPEG2000 with embedded geo-referencing; their name is based on distribution of plates 1:2500 EGSA '87. The orthoimages are available via FTP from the address published on the relevant website.

5.4 Critical Issues for Opening Government Data

5.4.1 Basic Rules

In the context of Open Government, the free movement and use of data and information is promoted. However, the sharing of useful open public data and their processing by stakeholders (scientific community and businesses) must be accompanied by respective rights of use for citizens and businesses so that they can deliver the expected development benefits.

The data that can be included in the so-called open data do not include, as is natural, a number of confidential or other classified documents circulating between public bodies. At the same time, the rights of citizens regarding the protection of their personal data must not be overlooked. The sharing of open data and information generated must be in line with the provisions of the General Data Protection Regulation [16] and any relevant national directives. For this reason, special security flaws are required to support the anonymization of data and the adoption of appropriate information systems security policies.

As already mentioned, the indiscriminate opening of data leads to a chaotic situation, and the storm of the citizen from information, which may eventually obscure the real picture and lead to the opposite of the desired result. It is therefore necessary to adopt appropriate policies and issue relevant directives so that the disclosure of data by the administrations of the organisations is limited to the necessary and critical size.

In recent years, a more systematic effort has been made for the most homogeneous and compact methodology of data opening with the aim of encouraging the opening of government data and their further use and processing. This methodology uses tools and methodologies from the Semantic Web. This is the basic idea that our work focuses on and especially the case of the "connection" of the data that is opened, the so-called "linked open data—LOD".

The successful implementation of open governance policies requires the provision of the required human and material resources as well as the training of those involved, both the staff supporting them and the executives who make the relevant decisions. Their support does not require a new type of infrastructure but perhaps an extension of the existing ones, as the needs for storing and distributing electronic content increase. As for the required management tools/applications, it is possible to either use the traditional electronic content management tools or to give incentives for the development of new innovative methods and tools within WEB3.org SemanticWeb that we will develop in the following chapters of this work. As already mentioned, the size of public data is already huge while it is constantly increasing with the result that traditional processing methods become inadequate. At the same time, new processing needs are emerging to draw useful conclusions and make relevant decisions. This issue is related to the emerging field of big data analytics, which is not part of this work.

Data that can be published and thus opened, have the following characteristics:

- Are not sensitive personal data of citizens or employees of the public sector
- Data whose publication is prescribed by specific national legislation, such as administrative acts published in the issue of the Government Gazette
- Individual administrative acts do not fall in the case of sensitive personal data
- General rules of law or circulars and instructions
- Statistical, economic or geographic features.

Not all the data held by the public administration are in a format that can be opened. Thus, data to be classified as open, must have the following basic characteristics [6]:

- Regarding availability and accessibility. The data must be easily accessible. They should be individual and independent and easily available through the Internet. The form of documents published should be such that anyone can access and edit them without the need of supplying a particular commercial software. For example, the publication of a document should not be done in image format. Such features should meet all the data available to the public administration no matter if they are going to be published widely or only to the internal network.
- Regarding the reuse and redistribution. The data should be available to the public under specific terms and licences. There should also be a provision for the combination of data with datasets from other bodies. In the case of government data, there are many cases where primary data from two or more bodies are related and produce important information. For example, if the data about supplies of a local hospital regarding the hospitalisation of patients (primary data of the Ministry of Health) are combined with the data regarding population and tax features of the area (primary data from the National Statistical Service of the Ministry of Finance) we will come to conclusions for possible waste of financial resources. To achieve this goal interoperability is essential at both levels: the one concerning communication applications that exchange data, and the other concerning the semantics of the data. The issue of common meaning and representation of data is already being treated by the development of the Semantic Web.
- Regarding global participation. The reuse and redistribution of the data should not only concern specific groups of users or operators. Data should be addressed to everyone, without any exceptions regarding status (public or private) or other quality characteristics (for example: data only for educational purposes).

Although the first steps in opening up government data are encouraging, there must be careful planning for the future. There should be a specific guide to rules and steps to maximise the benefit from the opening of data both to citizens and businesses and the public sector itself. The benefit should be evaluated based on measurable outcomes and quality characteristics.

Three main general rules are indicated when opening data, which in the case of public institutions, are adjusted as follows [6].

- Simplicity of data. Public bodies hold a vast amount of data, which differ in the various qualitative characteristics and their allocation according to the body. In the first steps of opening the data, these should be as simple as possible and few in quantity in order to quickly perceive the failures and try alternative actions. Given that public bodies move slowly in adopting innovations, thus adjustment problems can be overcome to innovations. In the following steps the operators will be enabled to select data with specific quality characteristics and increase the amount of data to be opened. In any case the possibility to combine data between different entities should be provided.

- Early and often involvement Public entities wishing to open their data, at an early stage should seek the involvement of users either to use or to support the reuse of data. The involvement of the interested parties ensures the successful delivery of the most relevant services.
  An important issue associated with the end users of the data is what they really want to have as information. In many cases end users want information in the form of summary, maps, tables, images etc. This information is based on the process of large volumes of primary data. This processing can be achieved by intermediate users, who will seek the primary data from the various bodies, combine them and undertake the delivery of information through relevant applications. So it is important to involve intermediate users in the process of opening up data.
- Arrangement of issues that discourage the opening of data. An important issue in the adoption of innovations is related to their acceptance by all the bodies involved. Any fears or concerns about the project can delay or cancel the entire effort. For this reason, we need to recognize and deal with these problems as soon as possible. In this direction, we need to develop an evaluation system of feedback from the first application of the relevant innovations.

5.4.2 Basic Steps for Opening Government Data

There are four main stages to convert the data to open.

Stage 1. Select dataset

This is the first step in opening the data. The three rules above should apply here as far as possible. It is likely to be repeated as a step in the future, depending on the feedback of the system and any problems which will emerge in the next stages.

Stage 2. Application of licence

Usually government data that can be published is not subject to copyright. However, all the data that is about to be opened, should be investigated for possible copyright. Also, the data deriving from public repositories should be available by licence similar to the Creative Commons licences. For example, anyone can use, reproduce and modify these data without any limitation other than the obligation to refer to the source and the explicit differentiation of secondary material that might be generated from the original material. Such a licence has the data published in the Greek administration program "DIAVGEIA".

Stage 3. Data availability

The dataset must be provided into useful forms with alternative ways in order to enable data processing for users with different characteristics.

Stage 4. Spotting

The dataset should be easy to detect. A good tactic would be a public website known as one single national depository, where the data will be classified according to specific criteria. Such an effort has been launched with the "DIAVGEIA" program of the Greek government. Highly efficient would be the cooperation and availability of data from intermediate users or communities that appeal to specialised teams, through which even most end users are linked.

In any case, the procedure for opening data requires special planning and compliance with specific rules. Particularly in the case of government data, special attention should be given in the following cases, which can lead to the discrediting and ultimate failure of the project.

- Bad metadata: incorrect use of metadata, will result in errors in references or mistakes when searching data or files.
- Errors in the licensing: such mistakes lead to failure treatment and further use of data from end users.
- Interoperability: the failure to achieve interoperability between the data opened the various public bodies, practically devaluing the project itself.
- Open data when combined with data from other operators may lead to the disclosure of personal data or other confidential information. It is possible that data which does not seem to affect the privacy of citizens, if properly combined with other data will lead to the disclosure and identification of certain individuals.

Special attention should be paid to the following points:

- The data intended to be opened by a public body should be completely relating to the responsibilities and the public it serves. The publication of more data, irrelevant to the organism, may cause conflicts with any parallel publications.
- Combination of services. In any case, the opening of the data should be combined with specific services to specific audiences.
- Intermediate users with added value for data. There are cases where the data is not directly usable by the final user. In this case, it is particularly important to find and cooperate with intermediate users who will be processed and enhance the value of government data.
- Constant, continuous and consistent disclosure of data. As an innovative policy, opening up government data should be backed up with consistency and continuity of services by the public sector. And this should be global and based on certain common guidelines for all public bodies.
- Restructuring the public sector. The exploitation of solutions offered by new technologies in computing, requires rapid responses and adaptations to the operation

and structure of the public sector. This, however, is associated with the necessary restructuring of public sector organisations, which in Greek reality generally progresses slowly.

- Interoperability. One of the most important issues when exchanging data and information between the two systems is the ability of every system to perceive the function of the other. On one hand, systems should be compatible with operating systems and data to be provided by common or compatible standards. So why is functional interoperability? On the other hand, it is particularly important that the exchange of data and information be shared with the notion of semantics, since it would avoid any misinterpretation or conflicts. When bodies use the same data with the same meaning, the search data will be successful and efficient.
- Data of high value for the individual. The task of opening up government data will be successful when the data is useful and efficient for the end user. Of course a definite end user does not always have to be a citizen or business. There is data that may not seem useful or of significant value to the citizens but are valuable within the public administration.
- Anonymising data. The opening of the data can lead to disclosure of personal data ending in violation of the privacy of citizens. A good tactic to avoid such phenomena is the generalisation of data. For example, instead of an organisation publishing a home address it could publish your zip code, instead of publishing the full date of an event (e.g. birth, death) publish the month or the year.

5.4.3 A Proposal for Linked Open Government Data

Nowadays it is essential to rationalise the distribution of staff in the public sector. There are services that have a large number of employees with a minimum workload and in contrast, bodies serving the public, with a small number of employees who have to deal with a heavy workload. With existing procedures, all bodies must register in a central body the vacant positions they want to cover with staff and transfer and in a later stage the central body that collects data on a national level, will proceed in filling these gaps, through a competition. The whole process is not expected to be completed soon.

In this case, we could use linked government data. The rationale behind the idea is the following. All public sector bodies keep records of their job positions and their staff. The relevant departments may publish by using open data the vacant positions, or the availability of their employees in the relevant access point. To connect the open data, they must be converted according to the RDF standards to acquire common semantics. Then by the use of an agent, the data will be recovered, processed and return relevant information to parties involved. The overall idea is the automated process of open data after their publication and until the final information of the services involved (Fig. 5.1).

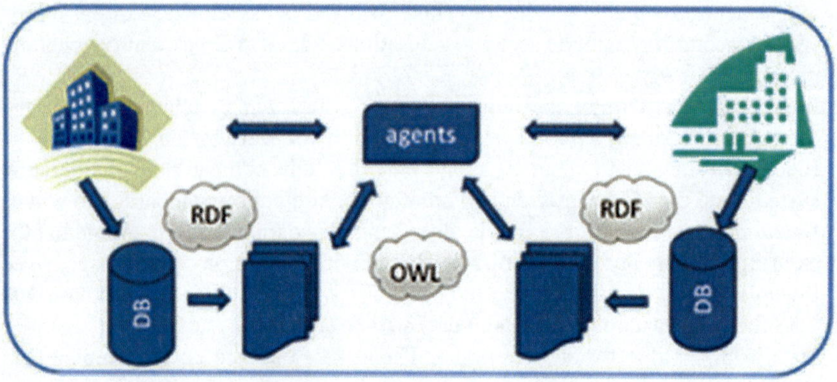

Fig. 5.1 Architecture for linked open government data

5.5 The Open Data Ontology—Our Proposal for the Case of the Greek Open Data Repository

5.5.1 Introduction and Motivation

The task of the transition from the classical model of governance into the e-Government model seems to have been accepted by the majority of citizens in every case it has been applied. However, what is required from citizens and businesses is a more responsible governance, consequent to the needs of modern society. The civil societies require on one hand the accountability of those involved in the governance of the countries and on the other hand the opportunity to gain knowledge from the data managed by public sector bodies and concern them. That way, we talk about open public data and the so-called open government. At a global level has been attempted through specific initiatives, such as the initiative Open Government Partnership—OGP [17] to support governments in transition. In Greece this requirement was met initially (from 2010) by publishing part of administrative and other public documents on the Internet ("Diavgeia Project" [4]) and then by the country's participation in the initiative OGP [17].

The opening of more and more datasets by institutions of the government, has made it difficult to search information and therefore to produce knowledge from them. It is therefore a clear need to develop systems to acquire knowledge from public data and especially those that can constitute open data. In addition another critical issue is the linking of open data in a way that supports the special characteristics and diversity of data published by various agencies worldwide.

The process of acquiring knowledge in a field of interest, requires initial registration, mapping and representation of the concepts and relationships between them. A key tool in this direction are the technologies supporting the Semantic Web and in particular ontologies and RDF representation of information. Thus, in this part we are dealing with the possibilities of acquiring knowledge through representation and management of concepts associated with public documents that can constitute open data, using the RDF representation of OWL ontologies and implementation of questions using the appropriate tools.

The open data and the tools for connecting and consuming it through the technologies of Semantic Web, seem to concern part of the scientific community. In particular by our bibliographical search we have distinguished tasks dealing with the representation and linking of ontologies describing open data as for example in [18] which presents an approach to the production of equivalent relations through research in the environment stating the equivalent classes in various entities. In addition in [19] we refer to the ontologies behind the Linked open data and present a system about finding schema-level links between LOD datasets. A view on the ontology of public administration is given in [20] which will form, according to the authors, the cognitive basis for developing Semantic Web applications in the field of e-Government. Just as important as the previous ones, [21] highlights the potential of Question Answering Systems within its approach. Finally, [22] presents a framework for constructing semantic model ontologies in OWL Web Service Standard for e-Government applications.

The experience of recent years by opening public data in Greece, mainly through "Diavgeia" [4] has demonstrated critical points in the direction of use, connection and availability of open data to the public. The main points we recorded are the following:

- Frequently organisations refer to the same concepts or entities but using different dataset.
- All providers do not use the same terminology for the thematic or types of decisions.
- As far as the structure of organisations is concerned, similar but not at all the same terminologies are used for departments or directories that share the same object.
- There have been several cases where the position of the subject of a decision is a lengthy document or the existing text is not appropriate.
- There is confusion about the correct completion of the thematic field and the type of decision.
- The search for a published decision on the existing search system is based on a key word or expression and takes place either by organisation or among all organisations but the results are still unrelated.

All the above create problems of efficiency and overall reliability of the system to search for information thus resulting in delays and the depreciation of the system, and have consisted of the motivation to work.

5.5.2 *The Ontology Implementation in Protége*

In this Book, we used the Protégé 4.3 for creating OWL ontology that describes the environment and the functioning of the Greek government. An attempt was made to include all aspects of the state and its functions as the so-called tripolar model which considers that the overall system consists of three basic pillars: State—Citizens—Business. The concepts institutions involved in this model are in the form of classes and subclasses of the ontology that we propose, while the features and their properties are represented by the object and data types properties. The Protégé enables us to link the individuals together with the help of object properties that have already been created for this purpose. The data types properties are used to represent the properties of the classes and their instances. Special information used as metadata are declared in the form of annotation properties in each class or instance separately. Each entity of the system structure is represented as a snapshot in the form of an individual associated with a particular class of the ontology. The structure of the ontology with the basics-classes is shown in Fig. 5.2.

In this part we did not fully present the properties in the whole of the classes of the ontology but just on the particular subsystem of the public documents. Here we have made an attempt to present a part of the ontology concerning the characteristics of administrative acts handled through "Diavgeia program". The corresponding part of Protégé showing their related properties are in Fig. 5.3.

In order to assess the correctness and consequence of the ontology and to control the behaviour of the model in terms of putting questions we have designated instances (individuals) and properties associated with a fact, that of elections for the local authorities in Greece. In particular we considered a scenario based on actual operating conditions of public administration based on the author's personal experience from years of working in the Ministry of the Interior of Greece. Initially the Ministry of Interior (class: MinistryOfInterior), publishes the decision of the Minister of Interior (individual: DecisionForLocalGovernmentElections) for conducting the forthcoming elections in which applicants are advised to submit nomination as candidate city councillors in specific combinations to the competent Court (class: FirstDegreeCourt). For our example we assume that the citizen (individual: Citizen1) submitted (object Property: isCreatorOf) an application (individual: ApplicationForApprovalInTheLocalGovernmentElections2014/object Property: hasRecipient) Court of First Degree Court of Athens (individual: FirstDegreeCourtOfAthens). The Court approves (object Property: issue, individual: ApprovalForApplicationOfCitizen1) the nomination of the person concerned and which announces (object Property: hasRecipient) later to the Ministry of Interior. With the election class (class: Elections) the Ministry of Economics is also involved (individual: MinistryOfEconomics) by approving (object Property: issue) the budget for the municipal elections (individual: BudgetForLocalGovernmentElections2014).

The concepts mentioned in this scenario and the properties of the link are represented in the ontology either as a class or object/data types property respectively. This scenario is visualised with the help of OntoGraf Tab of Protégé 4.3 and shown

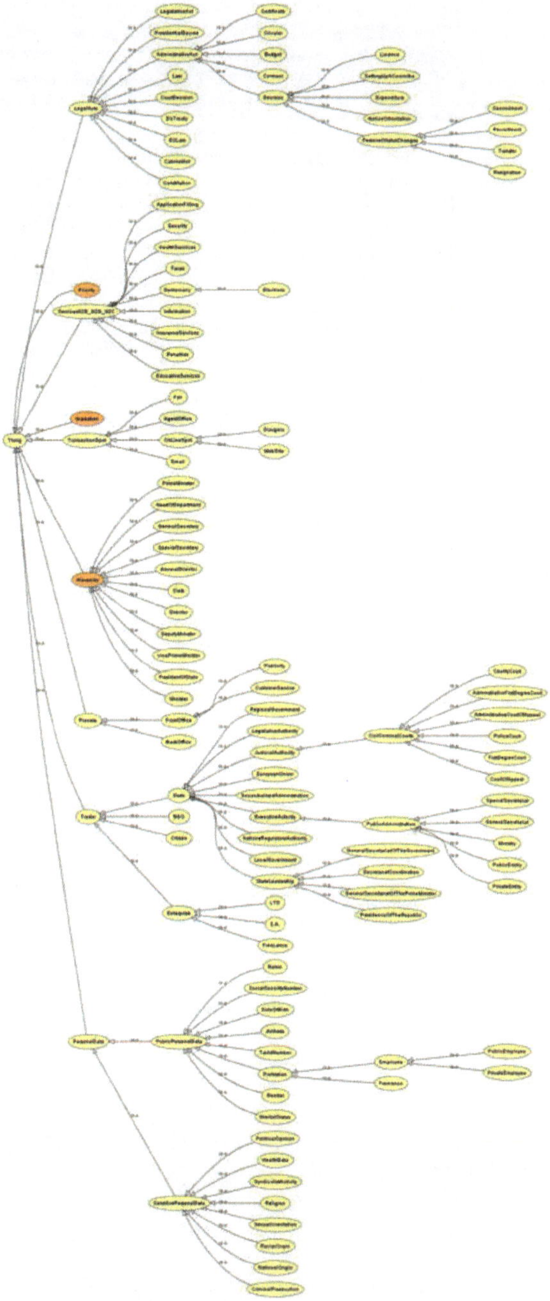

Fig. 5.2 The ontology hierarchy via the OWLViz view of Protégé

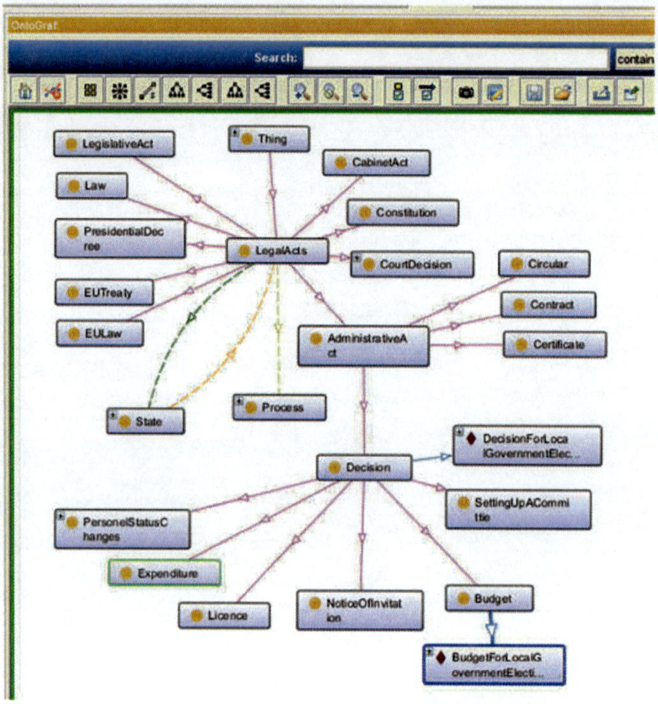

Fig. 5.3 Part of the ontology hierarchy via the OntoGraf view of Protege

in the Fig. 5.4. Relationships between individuals are represented by dashed lines and relationships between class and subclass or between class and individuals are represented by continuous lines of different colours depending on the relationship.

5.5.3 Ontology Evaluation

The Protégé 4.3 features two options for the evaluation of ontologies created by the user. As shown below there are two reasoners: the FaCT++ and HermiT. We activated both sequentially and tested the correctness and consequence of the ontology as far as covering the classes and taxonomy is concerned as well as any logical errors. These errors can easily be determined from the conclusions of the reasoners, concerning individuals and the relationships between them. As shown in Fig. 5.5, no problems occur.

As to the evaluation from experts, the ontology was given to judgement by senior executives of the Ministry of Interior, from whom we requested to make an assessment as to the completeness of the concepts included in the basic visualised shape. We

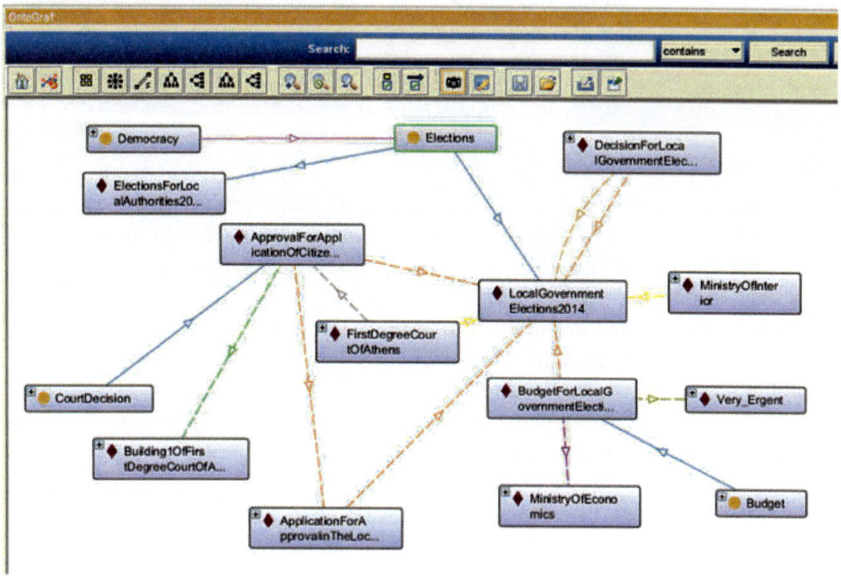

Fig. 5.4 The election class and subclasses in OntoGraf in Protégé

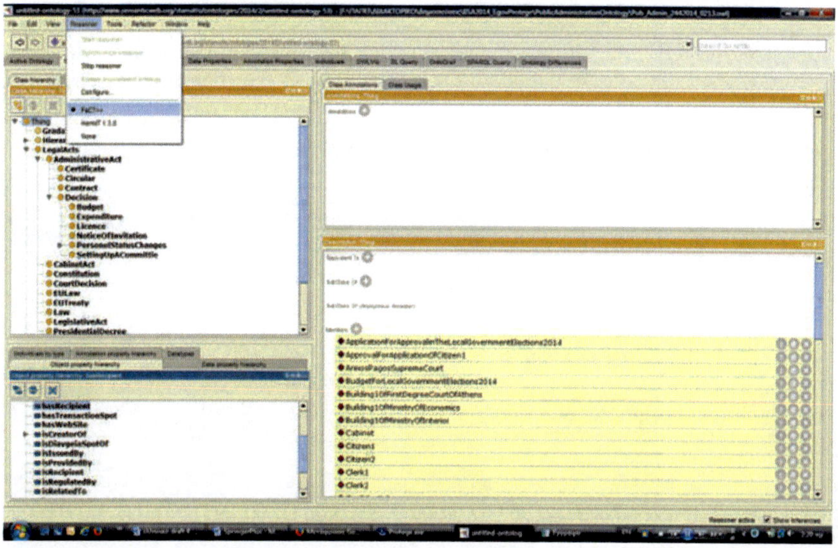

Fig. 5.5 Evaluation of the ontology within the reasoner

chose the method of interview as a more direct one and in order to quickly receive any comments. These interviews did not reveal any issues about correctness or omissions.

From the implementation of the ontology in Protégé 4.3 we can observe the following:

- The ontology represents the knowledge from the study of different datasets held at various public bodies. For example, in the scenario that we developed three entities are involved with the same object (municipal elections) maintaining different datasets. Protégé enables us to show all the three bodies at the same shape and also show the point at which they are involved simultaneously.
- It is common practice to use the "owl:sameAs" property for stating that another data source also provides information about a specific non-information resource. An "owl:sameAs" link indicates that two different URIs references actually refer to the same thing. Therefore, "owl:sameAs" is used to map between different URI aliases. For example the Fig. 5.6 from the Protégé 4.3 where we have stated that "individual:ElectionsForLocalAuthorities2014" is the same (owl:sameAs) with the "Individual: LocalGovernmentElections2014". As seen from the figure below, the properties of the first individual are the same as the ones of the second, in line with the conclusions given by the reasoner.
- The use of object/data types properties for connecting individuals in fact implements at a higher level the RDF triples hiding from the user the corresponding code RDF.
- The OntoGraf Tab of Protégé 4.3 gives a very useful feature to display the properties of various entities depicted. Once the trace of the mouse passes over the

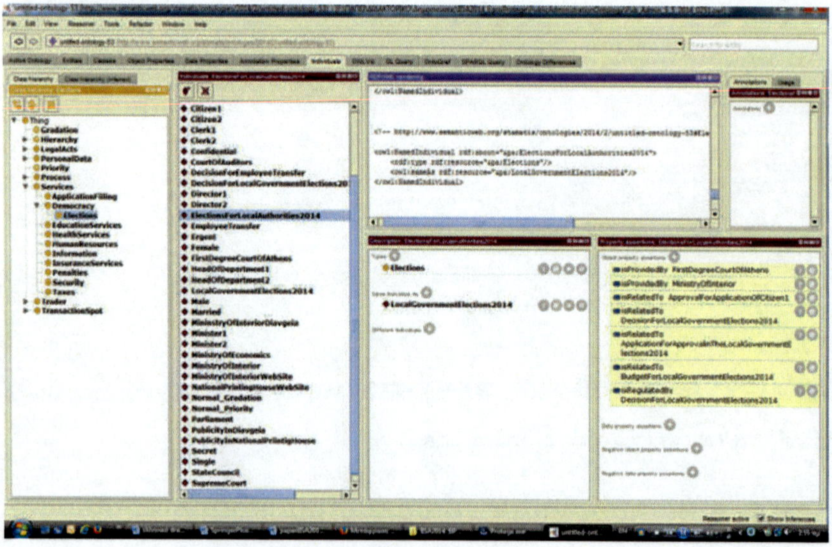

Fig. 5.6 The properties of individuals within the reasoner

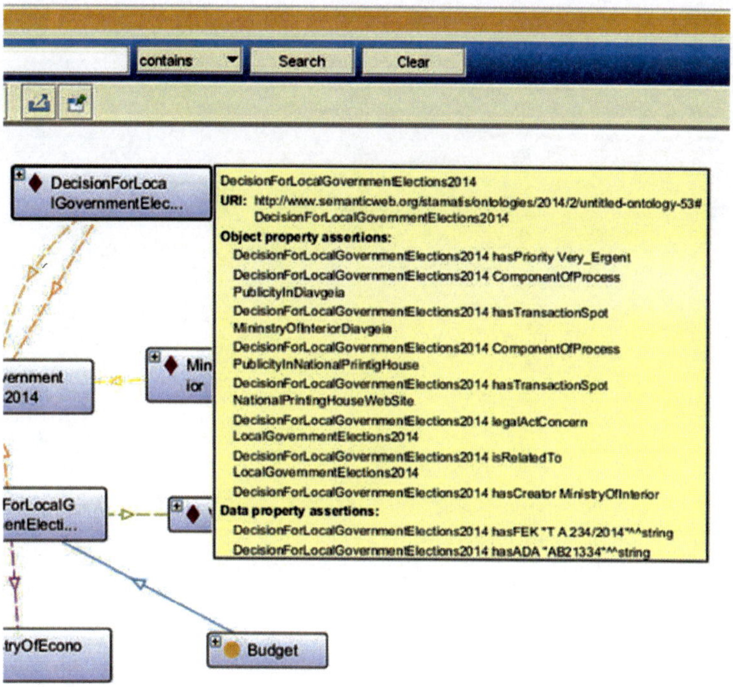

Fig. 5.7 The properties of the "individual:DecisionForLocalGovernmentElections2014"

corresponding object shape, a special context with all information related to a specific entity is shown (Fig. 5.7).

- In order to make it as easy as possible for client applications to process the posted data, you should reuse terms from well-known vocabularies wherever possible. You should only define new terms yourself if you cannot find required terms in existing vocabularies.
- When you cannot find good existing vocabularies that cover all the classes and properties you need, then you have to define your own terms. Defining new terms is not hard. RDF classes and properties are resources themselves, identified by URIs, and published on the Web, so everything we said about publishing linked Data applies to them as well.
- There is no possibility of handling ontologies in a language other than English. This creates a particular problem if we wish to use the ontology in real conditions, for example, to be used as a knowledge base for the development of knowledge management tools in the Greek language.
- The protégé 4.3 offers two excellent tools for information retrieval through queries. One tool concerns the OWL version of the ontology through the DL Query Tab and the other the SPARQL Tab for questions in the form of triplets of RDF. We

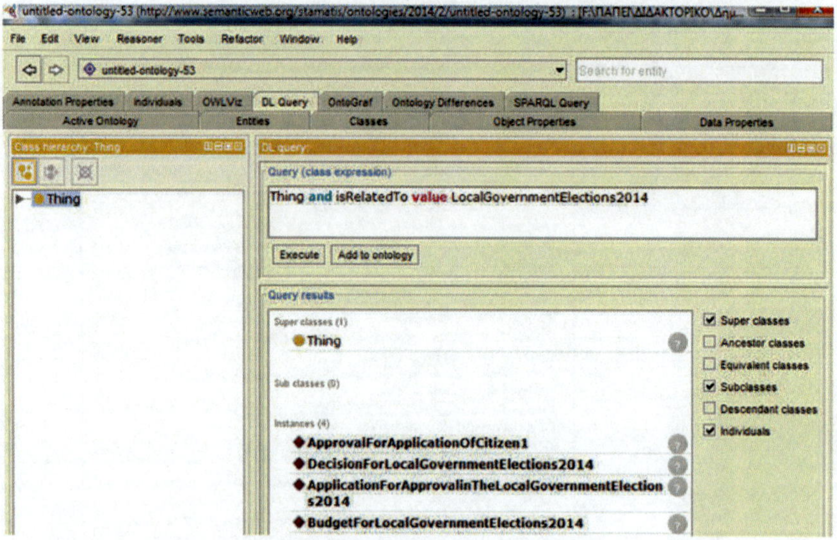

Fig. 5.8 The results of the query in the DL Query Tab

consider the first of the tools friendlier to the casual user and the second one most qualified. In this work we applied both tools in order to evaluate the consequence and correctness of the ontology and check any logical errors. Here we give screenshots relating to information on entities (class or individual) in connection with municipal elections (individual: LocalGovernmentElections2014). The questions were made in both tools and the results are shown in the Figs. 5.8 and 5.9.

- Similar to the search of all public documents published on the web-based program "Diavgeia" is the following question we made to SPARQL Tab Protégé of by using the property object "Property: hasADA" (Fig. 5.10). The query returns not only the name of the entity but also the corresponding number of "ADA".

5.6 Open Data Triplification—The Case of the Greek Open Data from the "Diavgeia Program"

5.6.1 *Introduction*

In recent years, the field of knowledge management has been enriched with smart technologies and methodologies of the Semantic Web. The global scientific community has adopted standards for the organisation and query semantic information such as RDF(S) and SPARQL while developing semantic technologies to organise, publish, exchange and recover data. The RDF stores have become sufficiently robust to support data volume reaching billions of records (RDF triples), while providing

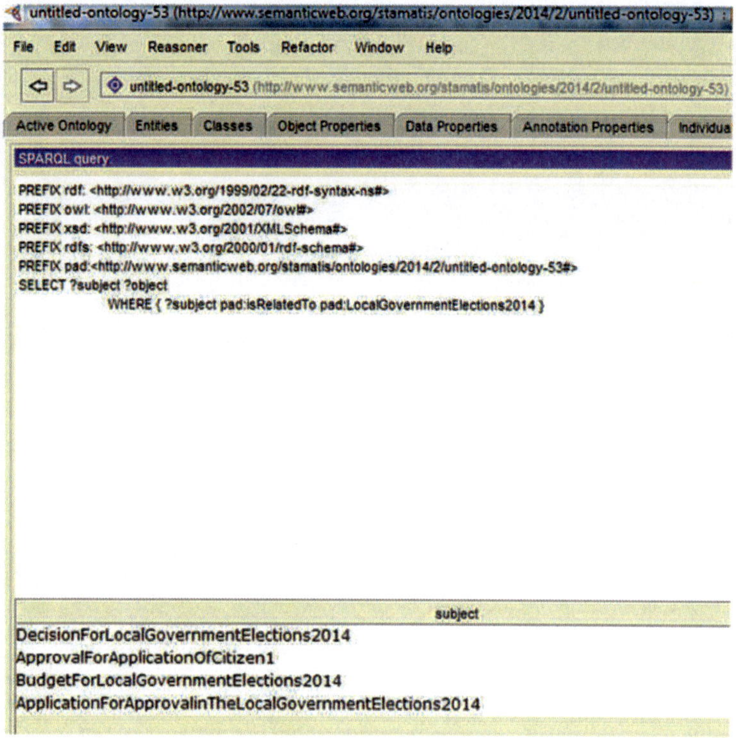

Fig. 5.9 The results of the query in SPARQL Tab

data management capabilities and query functions similar to those of traditional relational database systems (RDBMS).

We believe that the field of e-Government can be one of the main consumers of smart services based on Semantic Web. This is because the modern trend of open government is supported by the publication of "open data" and its further linking to Semantic Web technologies. By using Semantic Web technologies, expected improvements related to both parts of the overall system, i.e. the so-called "front-office" and "back-office".

The objective in this part of our work is to present our methodology for the production of structured information in the form of RDF triplets from open public data to achieve the subsequent connection to other similar datasets. To achieve this goal, it is also desirable to use already developed vocabularies and ontologies relevant to public administration and open government.

In particular, we developed and presented an algorithm of RDF triples production from tabular data (either in form or .CSV .XLS). In the baseline scenario we consider the case of recovery of open public data from the Greek government site www.diavgeia.gov.gr and then producing RDF triples by manual statement RDF schema through specific supportive application. At the site of the "Diavgeia program" are

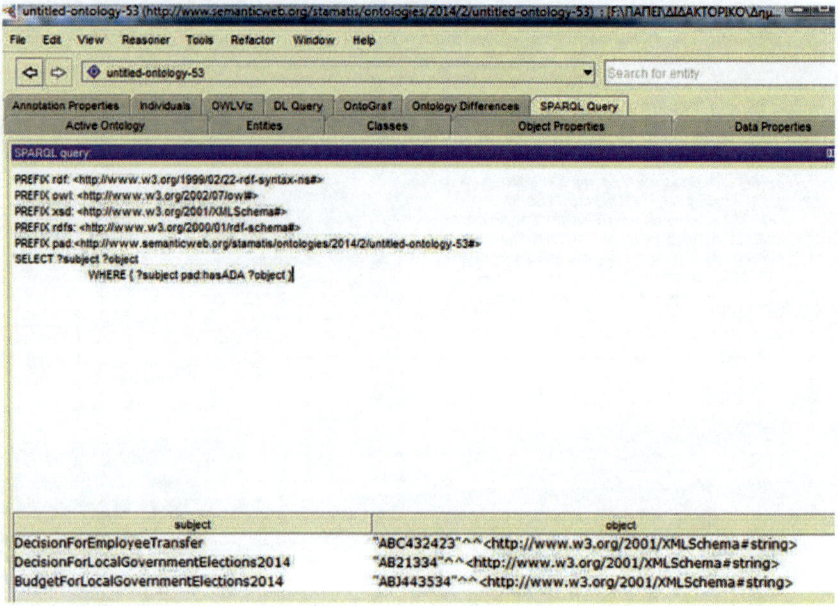

Fig. 5.10 The results of the query about the posted public documents in SPARQL Tab

posted almost all administrative acts of public administration in the last five years in Greece, in three formats: XML, JSON and XLS.

The completion of open public data for use in the open government requires that the science community faces two key challenges:

A. The definition of appropriate standards and common vocabularies that describe public information across the range of public sector activities in accordance with the protocols RDF and SPARQL.
B. The development of technologies for efficient storage and querying on semantic public data.

Regarding the first challenge, several attempts have been made by the Semantic Web community in order to consolidate the relevant standards. Greece has already made the first steps toward adopting standards for open public data involving the Greek government's OGP initiative [3] and the adoption of the principles of open government with a series of pieces of legislation, such as the "Diavgeia program" (according the Law 3861/2010 [23]).

In addition to the general standards of open public data we mention the adoption of specialised RDF standards like geospatial Basic GeoVocabulary of W3C [24] that allows the representation of points in WGS84, the GEORSS which provided support for more geospatial objects (lines, rectangles, polygons) and GeoOWL that was developed to provide a more flexible model for geospatial concepts.

Of particular interest for Greek data, is the first attempt to free distribution of geospatial data of the general public administration to all citizens of the country through www.geodata.gov.gr site [25]. This site is constantly updated with data from more and more entities of the Public Administration, and will be enriched with functionality, in order to control public administration and activate citizens in environmental protection. Today it hosts data from different ministries, the local Government and NGOs on various areas of activity of public administration such as urban plans, defining foreshore and coasts etc.

5.6.2 Relevant Tools

5.6.2.1 Open Refine (Google Refine)

The Open Refine [26] is an open source tool for cleaning, processing and converting data, their enrichment by using Internet services and their connection to databases such as Freebase. Google Refine The works on data lines with information organised in cells under columns. Data can be entered in a Google Refine Project from formats TSV, CSV, text separated by spaces or special characters, Excel, XML, JSON, RDF (RDF XML or N3) or Google Spreadsheets. After the user enters data, the tool enables the user to filter the data by column using facets, so that it can draw conclusions such as be able to see the distribution of values for a column that contains numeric values or the number of cells whose value matches a regex for columns containing strings.

The part of the processing, gives many different conversions on the project data using functions written in a special language Google Refine Expression Language (GREL) or Jython or Clojure languages. Reconciliation data can also be done with a database, such as Freebase or through APIs from Web Services. For example, the Google Geocoding API can be used for a column containing addresses to produce two other columns that contain the latitude and longitude of them.

Google Refine can extract the processed data in the formats TSV, CSV, Excel and HTML tables, while the processing steps can be extracted in JSON format and respectively be inserted into another Google Refine Project so as to resume in new data.

5.6.2.2 RDF Extension for Google Refine

Of particular interest are extensions that have been developed for their use via Google Refine, such as the RDF extension and the extension CKAN. In particular, using the RDF Refine extension for Google Refine export opportunities are added for the processed data in the RDF graph in format RDF/XML or TURTLE and the possibility of mixing them with a SPARQL endpoint. With the export functionality, you can determine the intended structure of the RDF data by drawing a template graph. The exporter iterates through the project rows, evaluates expressions in the template graph and produces an equivalent RDF subgraph per row. The final result is the merge of all the subgraphs. Terms from existing vocabularies such as ontologies, can be imported. The exporter attempts to retrieve and index the terms so that they can be suggested when adding classes and properties. Additionally, by using a dump from prefix.cc, URIs of popular prefixes are automatically suggested. Reconciliation is identifying multiple representations of the same real-world object. In the Semantic Web field, it is usually known as instance matching and refers to identifying equivalent resources in two RDF datasets.

Google Refine with RDF Refine extension was used for data conversion from the official site of Greek open public data Diavgeia program in RDF format.

CKAN Storage Extension for Google Refine allows data of Google Refine projects to be uploaded to CKAN Storage and connected to a package on a running CKAN instance (for example ckan.net). In its features, are included (a) Upload any combination of CSV table, RDF data and Google Refine operation history. (RDF data is only available if you have the RDF Extension) and (b) Link the uploaded files to an existing package or register a new one.

5.6.2.3 Protégé 4.3

In this work, we used the Protégé 4.3 to update our initial ontology for e-Government. Besides the creation of the knowledge base, this tool offers us the possibility of using concepts and properties declared as vocabulary for creating RDF triples to OpenRefine [26] (latest edition of GoogleRefine), as described below.

5.6.3 Triplification—The Case of Open Data by the "Diavgeia Program"

The most systematic and efficient processing of heterogeneous data may be achieved when such data may be transformed in the same format, without losing information. Furthermore it is desired to yield semantics in the various data and then the jointing between each. In this way interoperability of system data is increased. An interesting and very easy to use data format that serves the above is the RDF format triples. Converting data that are not in the form of RDF triples or else triplification is a particular process in a heterogeneous data integration system, and so far the bibliographic search showed the need for further specialisation. Of particular interest is the automated production of RDF files from various formats such the forms of .XLS, .CSV and .JSON which are some of the forms of open data in public administration.

The proposed conversion process is based on tabular data such as XLS or CSV. The output relates to the form of RDF/XML or Turtle, in order to reduce the file size produced in parallel with its complexity. Our interest concerns the additional attribution semantics to tabular data by generating RDF triples. That transformation

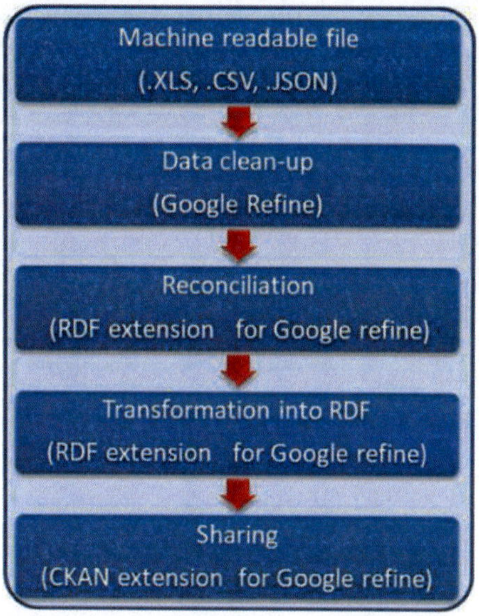

Fig. 5.11 The basics steps for tabular data integration

is necessary in order to lead to automated processes linking the original data with other similar data. The basic steps of the methodology are shown in the following Fig. 5.11.

In order to realise the transformation of data into RDF we need an RDF format. The proposed scheme is shown in Fig. 5.12. More specifically, we define the classes OpenPublicDataList and AdministrativeActs as well as properties arising from the header of each column. In the general case it is possible that a change of the relevant header must be made so that the respective properties correspond to the semantics we want to describe. The class OpenPublicDataList represents the data set and the class AdministrativeActs represents each line of the file. Each line is connected through the properties of the cells and each cell is associated with the value of the property through owl: hasValue. The data set and each row, column and cell are represented by a URI. So it makes sense to connect the cells representing the same entity of the natural world. For example, this is particularly important in sets of statistical data where the columns can be geographic parameters such as domain names, cities or countries, or as in our case, the key elements of administrative acts published freely online. In this case, connecting the cells means in fact connecting the URIs of the respective documents/administrative acts making it possible to combine different data sets (Fig. 5.12).

In order to convert tabular data into RDF we need to traverse them. With corresponding methodology we also make the conversion of data from relational databases into RDF triples. So a classical iterative structure is used to access each cell of the table (dataset). The formation of the URIs and triples is based on the Table 5.1.

According to the foregoing derives the conversion algorithm shown in Fig. 5.13. Initially (*step 1*) we introduce the necessary pairs prefixes-namespaces in order to be

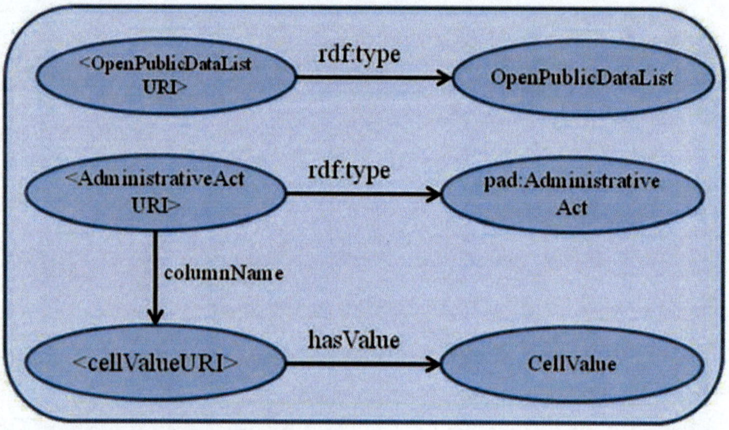

Fig. 5.12 Ontology/schema for modelling tabular data

Table 5.1 Prefixes and URI establishment

baseURI	http://127.0.0.1:3333/
URI-Prefix	baseURI+"tab/"+filename+"/schemaDefinition#"+"tab-"+filename
datasetURI	baseURI+datasetName
rowURI (subject)	datasetURI+"/rowId/"+rowId
predicateURI	vocabularyNameSpaceURI+columnName
cellValueURI	rowURI+columnName+"/"+cellValue

able to read the resulting RDF file. Then (*step 2*) created the properties of ontology from the column names. At this point it should be noted that the names of the columns may vary slightly compared to the source, in case of containing a character incompatible with the syntax where we use RDF/XML or Turtle. The column names are required to be at the forefront of the data source and should be appropriately stated in the introduction to the application conversion. *Steps 3 and 4* concern the formation of triples, according to the ontology-shaped dataset. Here we report that these steps (Fig. 5.13) also represents in an abstract way the more specific RDF conversion algorithm given as a recommendation of the W3C [27].

In the environment of Open Refine the procedure has been done manually via the relevant API that provides the tool as follows:

(a) As an initial step in the procedure we introduced from the website the "Diavgeia project" file concerning the key elements of administrative acts posted in format .XLS. This program provides the same information as the forms of XML and JSON.
(b) In Google Refine processing environment we were initially able to clean-up the cells of the file and with the assistance of facets to identify cells with similar contents. This is a very important step as we observed differentiation

```
•Declaration of prefixes-namespaces
•Rename the columns headers into the preferable properties /predicate names
•Define the properties/predicates according the headers
•Dataset define through RDF statement
•For line in File_Input
        •Establish Subject defining the row
        •For column in File_Input {
                • Establish the Predicate defining the column
                • Establish the Object defining the cell
                • Declare RDF statement defining the cell value
                }
        •Write the RDF representation of this row in the File_Output
```

Fig. 5.13 XLS to RDF conversion algorithm

concerning formatting (Uppercase–Lowercase) and use of synonymous words. This situation creates problems in linking the data with the next steps, so we felt appropriate to update the cells.

(c) Then, using the RDF extension, we made the declaration of the prefixes in a specially designed tab managed by the user (Fig. 5.14). Here we note the statement by the prefix "pad" concerning the introduction of the vocabulary of the ontology of e-Government that we have developed in our previous work [28]. The prefix refers to our local computer's internal address. At the same time we additionally declare other popular vocabularies like DublicCore (dc/dcterms) and DataCube (qb).

(d) The following steps are made in the environment of the processing of Google Refine as shown in Fig. 5.15. As described above in the general case, we have defined each line (row index URI) of the original table as pad Class: AdministrativeAct (the corresponding ontology mentioned above). Each line is essentially the individual respective class. The definition of triples involves determining the subject, the predicate and object. So in the environment of the processing of RDF extension (Fig. 5.15) the respective statements were made. In this tool we are given the ability to knock triples, meaning that some object is the subject of another RDF statement.

(e) As a final step in this processing, we export, using the corresponding option of the tool, the data format RDF/XML. Part of the relevant output is shown in Fig. 5.16.

Using the Protégé 4.3 we can have a visualised view of the RDF file generated by the above procedure. As shown in Fig. 5.17 class "`AdministrativeAct`" is set with members of the lines of the source file. Each member has the properties shown in the section Annotations View of this tool.

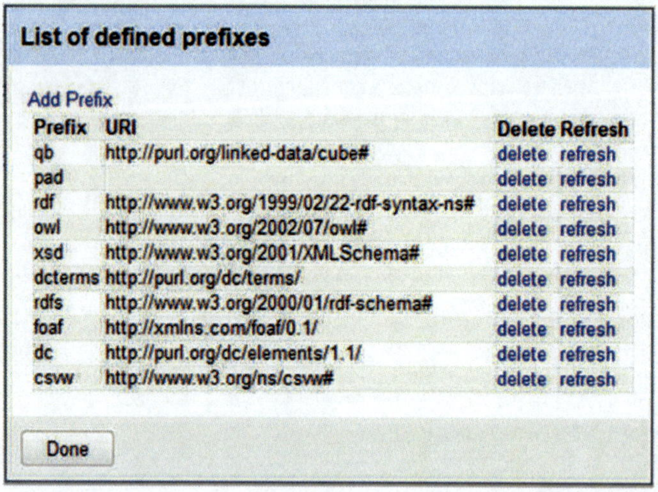

Fig. 5.14 Defining the necessary prefixes in Google Refine

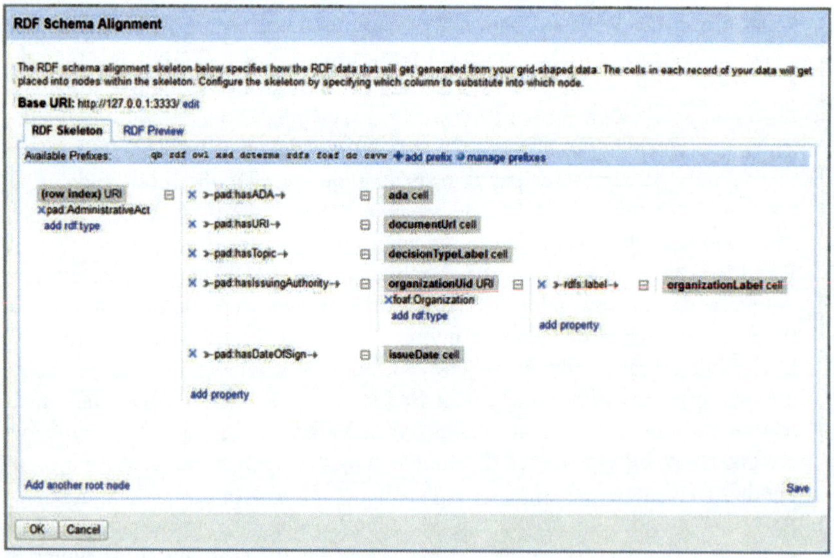

Fig. 5.15 The RDF triples definition in RDF extension

5.6.4 Conclusions

This work has presented issues concerning the application of Semantic Web technologies in e-Government and open data. This area is characterised by the huge number

```
<rdf:RDF>
<rdf:Description rdf:about="http://127.0.0.1:3333/0">
    <rdf:type
rdf:resource="http://www.semanticweb.org/stamatis/ontologies/
       2014/2/untitled-ontology-53#AdministrativeAct"/>
    <hasADA>Ψ634465ΦΘΕ-ΝΧ4</hasADA>
    <hasURI>https://diavgeia.gov.gr/doc/Ψ634465ΦΘΕ-ΝΧ4</hasURI>
    <hasTopic xml:lang="el">ΕΓΚΡΙΣΗ ΔΑΠΑΝΗΣ</hasTopic>
</rdf:Description>
<rdf:Description rdf:about="http://127.0.0.1:3333/100010874">
   <rdf:type
rdf:resource="http://xmlns.com/foaf/0.1/Organization"/>
   <rdfs:label xml:lang="el">ΥΠΟΥΡΓΕΙΟ ΕΣΩΤΕΡΙΚΩΝ ΚΑΙ ΔΙΟΙΚΗΤΙΚΗΣ
      ΑΝΑΣΥΓΚΡΟΤΗΣΗΣ</rdfs:label>
</rdf:Description>
<rdf:Description rdf:about="http://127.0.0.1:3333/0">
       <hasIssuingAuthority
rdf:resource="http://127.0.0.1:3333/100010874"/>
       <hasDateOfSign
rdf:datatype="http://www.w3.org/2001/XMLSchema#
        date">21/05/2015 03:00:00</hasDateOfSign>
</rdf:Description>
....
</rdf:RDF>
```

Fig. 5.16 The RDF statements in the output RDF file

of raw data that constitute the so-called open public data in various formats, with key features the non-compliance with a common vocabulary for the concepts and procedures used throughout the public sector, and the insufficient set of semantic tools for information retrieval. The technologies which we have introduced are intended to support the semantic approach in seeking information and subsequent interconnection in order to constitute linked open data. They are the infrastructure for the development of new intelligent applications that will use both citizens, and the public administration itself for the management of open public data. From the study which preceded us we were finally able to note the following conclusions.

- The basic tool we have used serves the production of RDF triples from the popular types of files, such as .XLS files and JSON, in a semi-automatic way.
- The statements of predicates and the subject and object as resources are being made very easily from the functional environment of RDF extension.
- The RDF statements can perform the semantic shape of a knowledge base that supports open data so these will be able to connect in the next step.

In our interests there is the fully automated .XLS and JSON files triplification process and the support of the connection of generated files from this process in order to constitute linked open data.

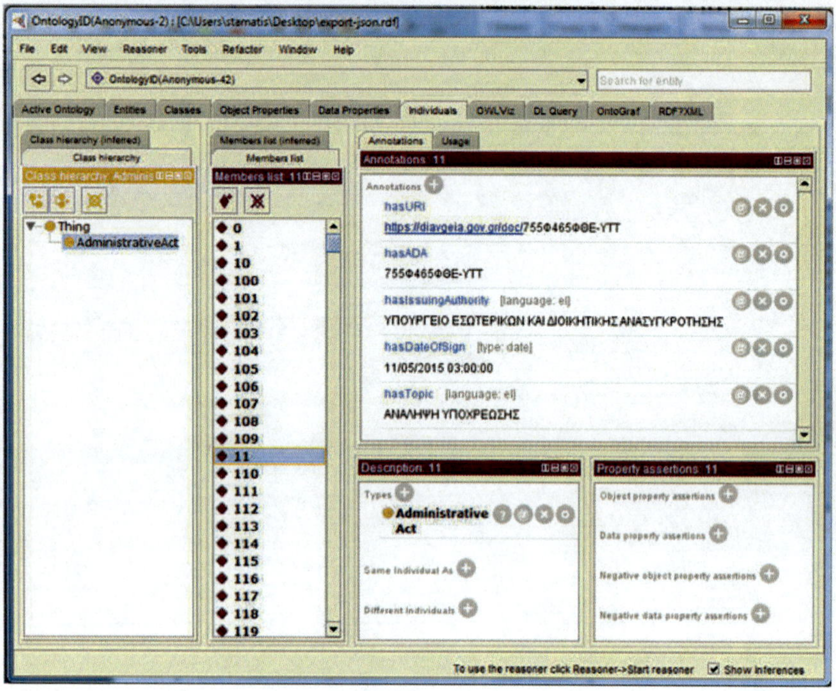

Fig. 5.17 Visualisation of the RDF output in Protégé 4.3

5.7 Open Data Repositories—The Case of the e-Government Ontology Publishing as an Open Ontology in CKAN

As mentioned above, the main use of ontologies developed in different areas of interest is on concept the alignment of the concepts-entities used in the various electronic files in order to achieve semantic interoperability between users and on the other hand the sharing of information with in order to reuse, adapt them to each user in order to produce knowledge. If this sharing is done in terms of open data the expected benefits will be multiplying in each of these respects. For this reason we have chosen to develop the ontologies presented in the previous chapters using Protégé as .owl files that follow the RDF/XML syntax and according to the general framework guidelines for publishing the open data described above. In this way our ontologies can be stored and published as open data in special repositories and retrieved with methods and tools that support RDF, JSON or more generally XML compliant documents.

In this Book we present as an example of the publication of an open ontology in the field of e-Government the "e-Government ontology" that we developed and presented in the previous paragraphs. This ontology as an open data-open ontology can be a special and innovative knowledge in the field of e-Government. This is made possible by the development of a suitable open data repository accessible to those interested through the operation of a relevant server and a corresponding database. Below we analyse the process of developing the special repository and the subsequent publication of this ontology.

After the relevant search we came up with the development of a CKAN [29] instance repository for the reasons mentioned below.

- CKAN is a tool for making open data websites. (Think of a content management system like WordPress—but for data, instead of pages and blog posts.) It helps you manage and publish collections of data.
- It is used by national and local governments, research institutions, and other organisations who collect a lot of data e.g. the central point of open data data.gov.gr.
- Once your data is published, users can use its faceted search features to browse and find the data they need, and preview it using maps, graphs and tables—whether they are developers, journalists, researchers, NGOs, citizens, or even your own staff.
- CKAN is open source software, with an active community of contributors who develop and maintain its core technology. CKAN is modified and extended by an even larger community of developers who contribute to a growing library of CKAN extensions.
- From the technology view CKAN is built with Python on the backend and Javascript on the frontend, and uses The Pylons web framework and SQLAlchemy as its ORM. Its database engine is PostgreSQL and its search is powered by SOLR. It has a modular architecture that allows extensions to be developed to provide additional features such as harvesting or data upload.
- CKAN uses its internal model to store metadata about the different records, and presents it on a web interface that allows users to browse and search this metadata. It also offers a powerful API that allows third-party applications and services to be built around it.
- CKAN has particular features, it is a fully-featured, mature, open source data portal and data management solution. CKAN provides a streamlined way to make your data discoverable and presentable. Each dataset is given its own page for the listing of data resources and a rich collection of metadata, making it a valuable and easily searchable data catalogue. The core functionality of CKAN provides a wealth of features and has over 200 community extensions which can fill almost any feature gap. We mention the following features as the most useful and basic as well.
  - API: The CKAN's Action API is a powerful, RPC-style API that exposes all of CKAN's core features to API clients.
  - Datastore: The CKAN DataStore extension provides an ad hoc database for storage of structured data from CKAN resources.
  - FileStore, that allows users to upload data and image files.

- Publish and manage data: An intuitive web interface allows publishers and curators to easily register, update and refine datasets.
- Visualization CKAN's data previewing tool has a host of powerful features for previewing data stored in the DataStore.

5.7.1 Methodology—Steps for the Development of the CKAN Repository

The development and use of a CKAN instance requires in addition to the installation of the CKAN package on the local computer, also the operation of a server that will host the relevant implementation and the necessary database. The easiest solution for installing the complete system is by using a suitable package system (operating system package installation) versus the manual process using the source system (installation from source). Whatever method one follows to install CKAN, special guidance is provided through detailed instructions provided by the official CKAN site [30].

We are not going to give here a detailed description of the technical steps, which are also described in detail on the official site of CKAN, but a critical reference to the basic steps and important points during the development of the repository and especially to those points that we consider not covered by the above instructions and we had to adapt more to our case.

Step 1. Selection of installation process depending on the available operating system

The installation of CKAN depends on the operating system that will host the relevant application. Since we had a local computer with Windows 10 at our disposal, the preferred installation method based on the relevant instructions is the installation from source (installation from source). However, because we considered this method to be more complex, we chose to install using a package, on a virtual machine, using the virtual box [31] in which we installed Ubuntu 16 as the operating system. During the development period of this application, CKAN version 2.8 was available, which was fully functional with version 16.04 of Ubuntu. However, in August 2020, CKAN 2.9 was released and is available, which is also supported by newer versions of Ubuntu.

Step 2. Install Ubuntu 16.04 on a virtual machine

In order not to bind our local computer with the parallel operating system, we chose to install Ubuntu 16.04 on a virtual machine that we developed through the Oracle VM VirtualBox. Note that due to the increased resource requirements of the overall system, it is necessary to install an appropriate version of VirtualBox that supports ×64 systems as well as to enable the virtualization function of the local computer, if possible. The virtual machine we installed on our local computer was supported with Ubuntu 16.04 [32] as this version proved to work better with the CKAN 2.8 system prerequisites mentioned in step 4.

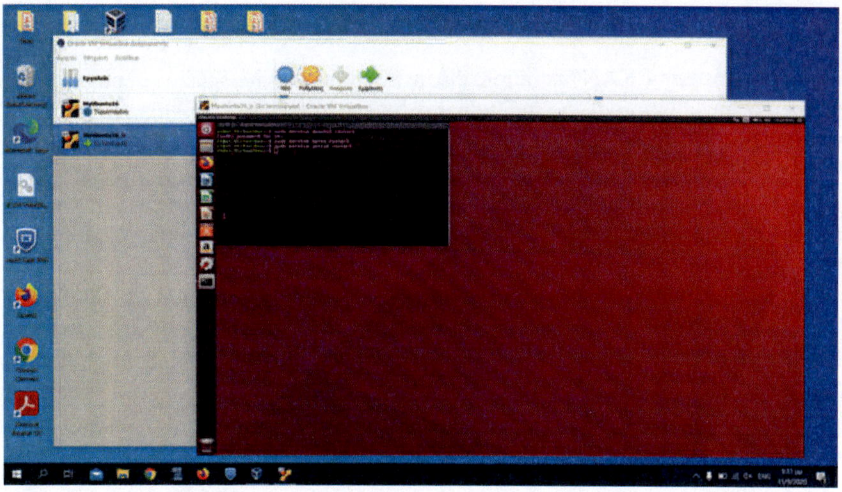

Fig. 5.18 Starting apache2 and nginx in Ubuntu 16 into VirtualBox machine

Step 3. Installation of the CKAN package 2.8

Based on the instructions mentioned above, the download and installation of CKAN was completed immediately and without system failures. Note that before downloading and installing the basic CKAN package, you need to install and start apache2 and nginx. In some cases, such as Ubuntu 18, the download, installation, and startup of apache2 and nginx must be done in distinguishable step-by-step instructions to avoid any dysfunction not observed with Ubuntu16 during downloading through one command of these applications. Also, in our case, it was necessary to pause apache in order to complete the nginx installation correctly. Then the restart of the two servers was completed without any problems (Fig. 5.18).

Step 4. Installation of prerequisite software

As mentioned above, the operation of CKAN requires an additional database and search platform. For this reason, in addition to the Step 3 installation, PostgreSQL and Solr-Jetty installation were required. Based on the relevant instructions, the installation of the PostgreSQL database and the Sorl-Jetty platform was completed without malfunctions or failures. We note that it is important to correct/complete the CKAN configuration files which depend mainly on whether the servers and the database operate on the same or different machines and on the absolute addresses that respond to the relevant calls. In any case, the installation of JAVA and the relevant configuration in the configuration files of the installed applications of step 3 are required. We note that the incorrect configuration of the environment variables creates system malfunctions-failures.

Step 5. Update the configuration and initialization of the database

In order to start the CKAN operation, the instructions require specific configurations per step of installation of the individual software in the basic configuration file of the CKAN package as well as the initialization of the database and the definition of the database administrator and separately the CKAN administrator. These steps are important for the functionality of the system as the system seems to start without the appointment of an administrator, however in practice it seems that skipping this step leads to malfunctions. Finally, we consider it necessary for the easier use of the system to install the provided extensions such as DataStore extension and FileStrore. These extensions are linked to the ability to upload files that we manage locally so that they can then be published and made available to system users online. In particular, the CKAN DataStore extension provides an ad hoc database for storage of structured data from CKAN resources. Data can be pulled out of resource files and stored in the DataStore. When a resource is added to the DataStore, you get:

- Automatic data previews on the resource's page, using the Data Explorer extension
- The DataStore API: search, filter and update the data, without having to download and upload the entire data file.

The DataStore is integrated into the CKAN API and authorization system while is generally used alongside the DataPusher, which will automatically upload data to the DataStore from suitable files, whether uploaded to CKAN's FileStore or externally linked.

The final format of the CKAN configuration file that includes all of the above configurations is shown in Appendix A.1.

Step 6. Access to the application

Upon completion of this step, using the operating system browser, a visit to http://localhost (or http://127.0.0.1) returns the CKAN homepage as shown in the Fig. 5.19.

This initial screen remains indicative of the proper operation of the system, as the operation of CKAN requires the connection to an account with administrator privileges. Failures to configure or create system user roles may result in incorrect CKAN operation. Please note that this home screen can be adjusted in terms of logos and titles as well as enriched with additional hardware by the CKAN administrator. In our case the home screen after connecting to CKAN returns our system status as shown in Fig. 5.20.

5.7.2 *Opening Data Through CKAN: The Case of Publishing-Opening e-Government Ontology*

In order to publish our data as open data, through CKAN it is first necessary to create a user with rights to upload the data according to the relevant instructions. The

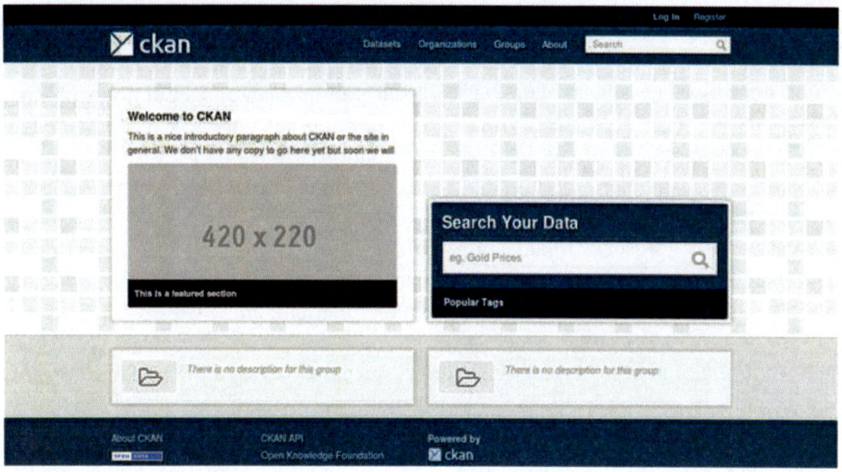

Fig. 5.19 "Welcome" CKAN screen

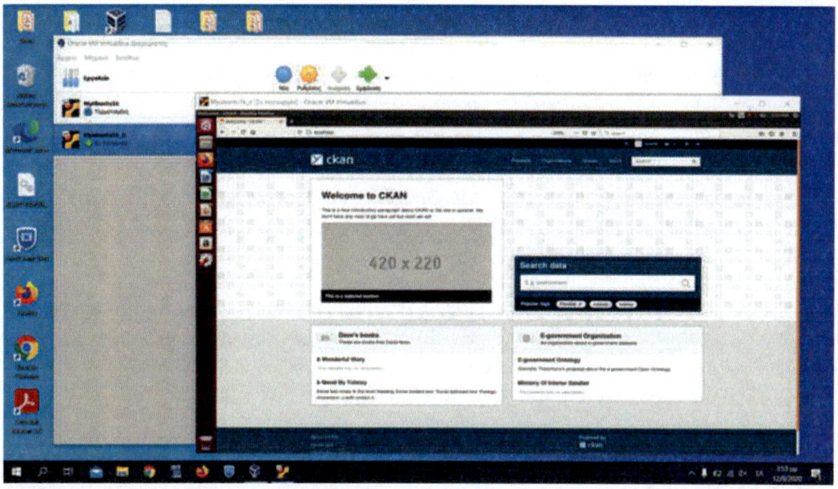

Fig. 5.20 Login to CKAN as a system administrator

connection is then made to either the role of this user or the role of administrator. For the needs of our work, we did not create ordinary users while we limited ourselves to the definition of database administrator and CKAN administrator through which we performed the following.

Step 1. Creating a new organisation

A necessary step for the publication of data is the creation of an "organisation" in which the data are published as datasets based on a specific procedure and logic.

Organisations are the primary way to control who can see, create and update datasets in CKAN. Each dataset can belong to a single organisation, and each organisation controls access to its datasets. Datasets can be marked as public or private. Public datasets are visible to everyone. Private datasets can only be seen by logged-in users who are members of the dataset's organisation. Private datasets are not shown in dataset searches unless the logged in user (or the user identified via an API key) has permission to access them.

When a user joins an organisation, an organisation admin gives them one of three roles: member, editor or admin. A member can view the organisation's private datasets while an editor can do everything a member can plus (a) Add new datasets to the organisation (b) Edit or delete any of the organisation's datasets and (c) Make datasets public or private. An organisation admin can do everything as editor plus (a) Add users to the organisation, and choose whether to make the new user a member, editor or admin (b) Change the role of any user in the organisation, including other admin users (c) Remove members, editors or other admins from the organisation (d) Edit the organisation itself (for example: change the organisation's title, description or image) (e) Delete the organisation.

In order to create our organisation we followed the steps below:

1. From the "Organisations" link at the top of the starting page we select the "Add Organisation" button below the search box.
2. CKAN displays the "Create an Organisation" page.
3. We enter the name for the organisation, and a description for the organisation's home page. Description and the image URL are optional in this step.
4. CKAN creates the organisation via the "Create Organisation" button and after this displays its home page. Initially, of course, the organisation has no datasets.

We note that when we created the organisation, CKAN automatically made us its "Admin". From the organisation's page we can see an "Admin" button above the search box. From this selection CKAN displays the organisation admin page. This page has two tabs: (a) *Info*—Here you can edit the information supplied when the organisation was created (title, description and image) (b) *Members*—Here you can add, remove and change access roles for different users in the organisation. Note: you will need to know their username on CKAN.

In this case, we generated as administrator "e-Government Organisation" adding the relative description as shown in the figure below (Fig. 5.21).

Step 2. Dataset creation and management

As we mentioned before, for CKAN purposes, data is published in units called "datasets". A dataset is a parcel of data—for example, it could be the general censuses or statistics for a region, the spending figures for a government department, or temperature readings from various weather stations. When users search for data, the search results they see will be individual datasets. We mention that on early CKAN versions, datasets were called "packages" and this name has stuck in some places, specially internally and on API calls. Package has exactly the same meaning as "dataset".

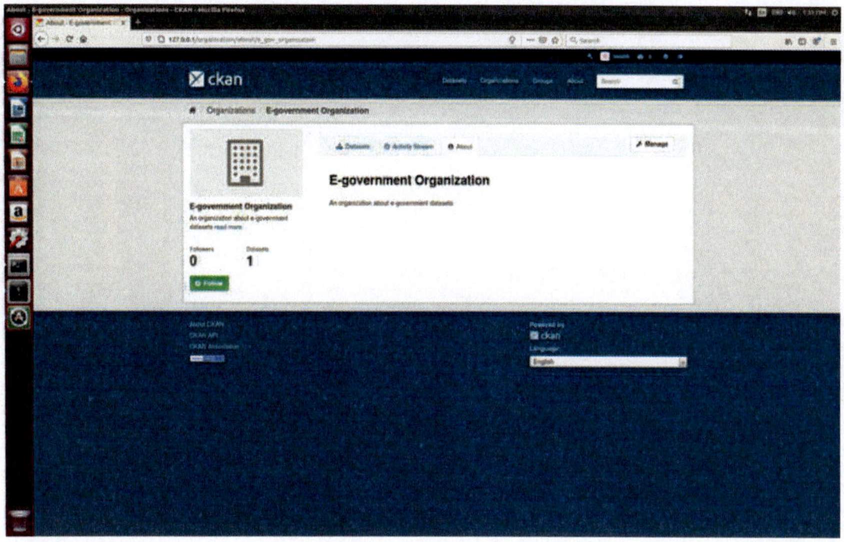

Fig. 5.21 The e-Government Organisation published in CKAN

A dataset contains two basic things:

- Information or "metadata" about the data. For example, the title and publisher, date, what formats it is available in, what licence it is released under, etc.
- A number of "resources", which hold the data itself. CKAN does not mind what format the data is in. A resource can be a CSV or Excel spreadsheet, XML file, PDF document, image file, linked data in RDF format, OWL ontology, etc. CKAN can store the resource internally, or store it simply as a link, the resource itself being elsewhere on the web. A dataset can contain any number of resources. For example, different resources might contain the data for different years, or they might contain the same data in different formats.

In our case we chose to create two different datasets into the organisation we created in the previous step and to which they will refer in the following. The first is intended for publishing metadata of the archive of decisions of the Ministry of Interior. For this reason we created the dataset entitled "Ministry of Interior Dataset" with the relevant description shown in the figures below (Figs. 5.22 and 5.23).

The second concerns the publication of the e-Government ontology as shown in Fig. 5.24.

These datasets were created through the administrator account we created and the relevant menu "datasets" offered by CKAN. Alternatively we could create these datasets, through the menu options provided by the editing organisation that we have already created through the management menu provided by the relevant selection on the organisation page. We note that at each stage, either during creation or after, it is possible to edit the dataset in all fields such as title, description, licence that

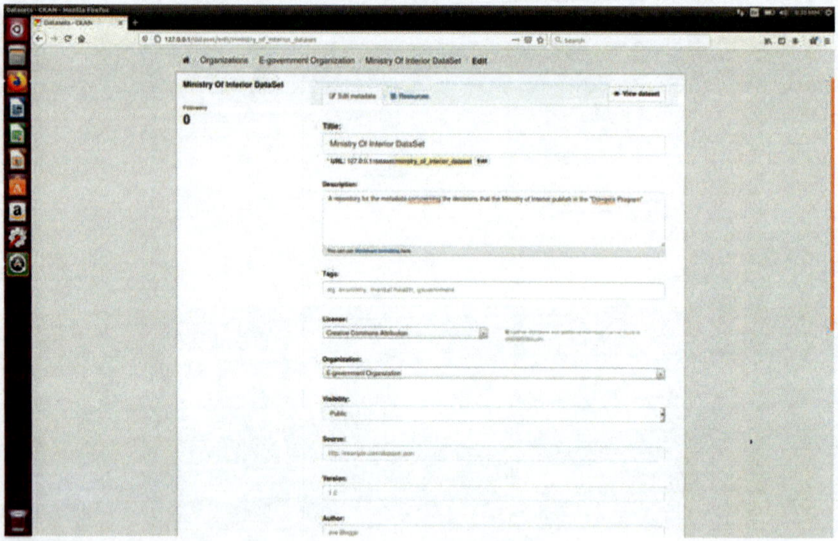

Fig. 5.22 Ministry of Interior DataSet creation in CKAN

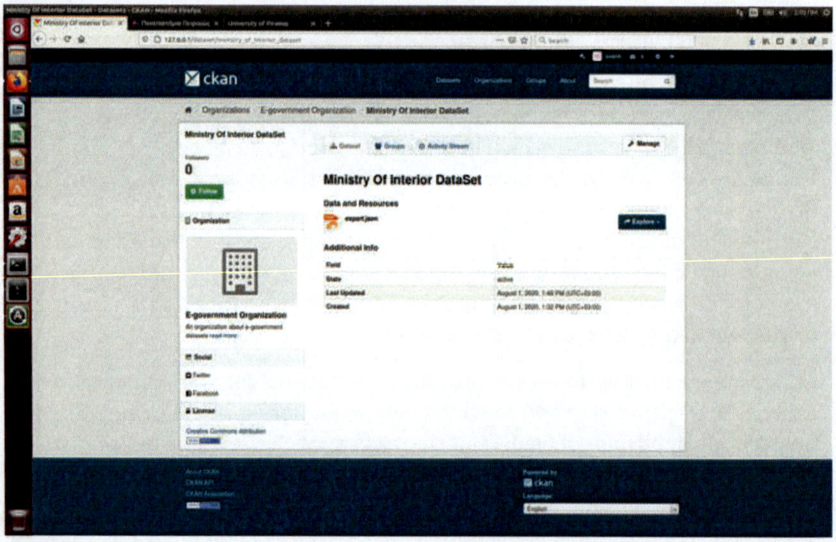

Fig. 5.23 The "Ministry of Interior DataSet" dataset page in CKAN

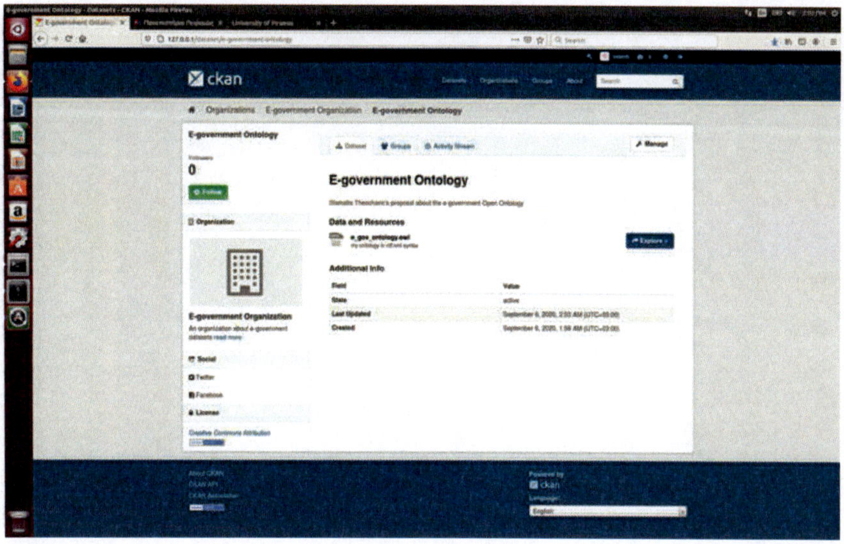

Fig. 5.24 The "e-Government ontology" dataset page in CKAN

accompanies it. In order to complete the relevant process we must select the "Add Data" link at the bottom of the relevant page. By default, the only required field on this procedure is the dataset title. However, it is good practice to include, at the minimum, a short description and, if possible, the licence information. You should ensure that you choose the correct organisation for the dataset, since at present, this cannot be changed later. You can edit or add to the other fields later. The process of creating the dataset is completed in the next step by uploading and publishing the data in the form of resources, as shown in the following figure (Fig. 5.25) which concerns the creation of the dataset for the ontology of e-Government.

Step 3. Publication of data

CKAN treats data (in its physical-electronic form) as resources (either as files or as links to data) that refer to specific datasets. To achieve this, in addition to the relevant electronic file, appropriate metadata must be declared such as the distinctive title, relevant description, creation and revision dates as well as the URL based on which they will be searched. Also, in the creation phase it is stated whether the data will be private (visible only to the local user) or public (visible as open data). The above is done as follows.

A. In case the corresponding dataset has not been created yet, we are in the second phase of creation by uploading the data in the form of resources. This is done as follows:

    1. At the end of the previous step and in order to complete the creation of the dataset, the addition of data is pending. At the bottom of the screen, select

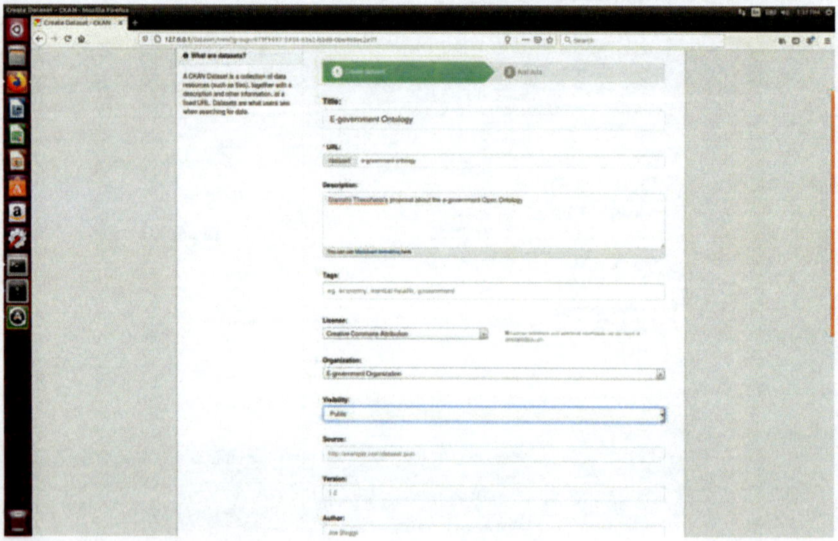

Fig. 5.25 Uploading data in the e-Government ontology dataset in CKAN

the "Add Data" button to go to the relevant page. On this page we can add one or more resources as files from our local computer or as links to the corresponding data.

2. Adding other information on the page. CKAN does not require this information, but it is good practice to add it:

   - *Name*—a name for this resource, e.g. "Population density 2011, CSV". Different resources in the dataset should have different names.
   - *Description*—a short description of the resource.
   - *Format*—the file format of the resource, e.g. CSV (comma-separated values), XLS, JSON, PDF, etc.

3. If you have more resources (files or links) to add to the dataset, select the "Save and add another" button. When you have finished adding resources, select "Next: Additional Info".
4. CKAN displays the "Additional data" screen.

   - Visibility—*a Public dataset is public and can be seen by any user of the site. A Private dataset can only be seen by members of the organisation owning the dataset and will not show up in searches by other users.*
   - Author—*The name of the person or organisation responsible for producing the data.*
   - Author e-mail—*an e-mail address for the author, to which queries about the data should be sent.*
   - Maintainer/maintainer e-mail—*If necessary, details for a second person responsible for the data.*

- *Custom fields*—If you want the dataset to have another field, you can add the field name and value here. E.g. "Year of *publication*". Note that if there is an extra field that is needed for a large number of datasets, you should talk to your site administrator about changing the default schema and dataset forms.

  Note: Everything on this screen is optional, but you should ensure the "Visibility" is set correctly. It is also good practice to ensure an Author is named.

5. Select the 'Finish' button. CKAN creates the dataset and shows you the result.

Following this procedure we uploaded to CKAN the .owl file concerning the ontology of e-Government as shown in the following figure (Fig. 5.26) in the dataset "e-Government ontology" and the .JSON file concerning metadata from its page DIAVGEIA program in the dataset "Ministry of Interior DataSet" as shown in figure (Fig. 5.27).

B. In case we want to add data to an existing dataset then:

1. We select the dataset to which we want to upload our data.
2. Using the "manage" button we choose to manage the already open dataset.
3. Through the tab "Resources" we choose either to upload a specific file from our computer or to provide a specific link for retrieving data remotely.

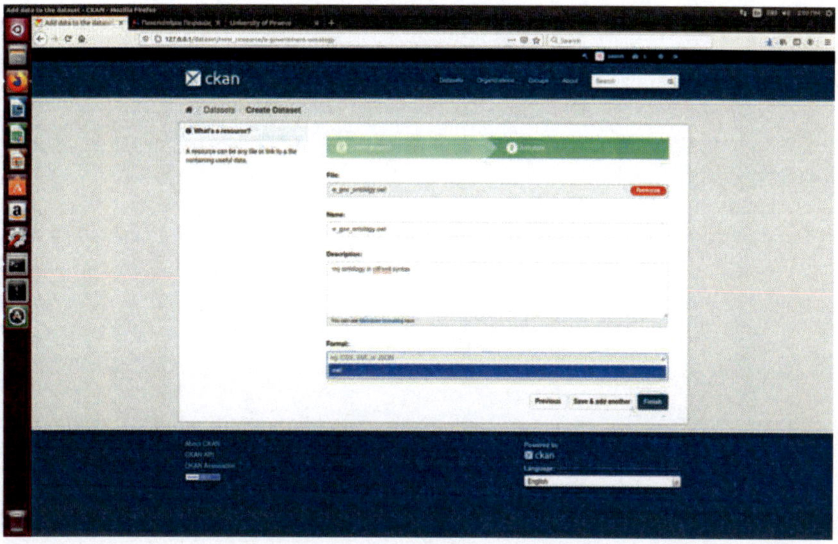

Fig. 5.26 Uploading the .owl file with the e-Government ontology in e-Government Organisation in CKAN

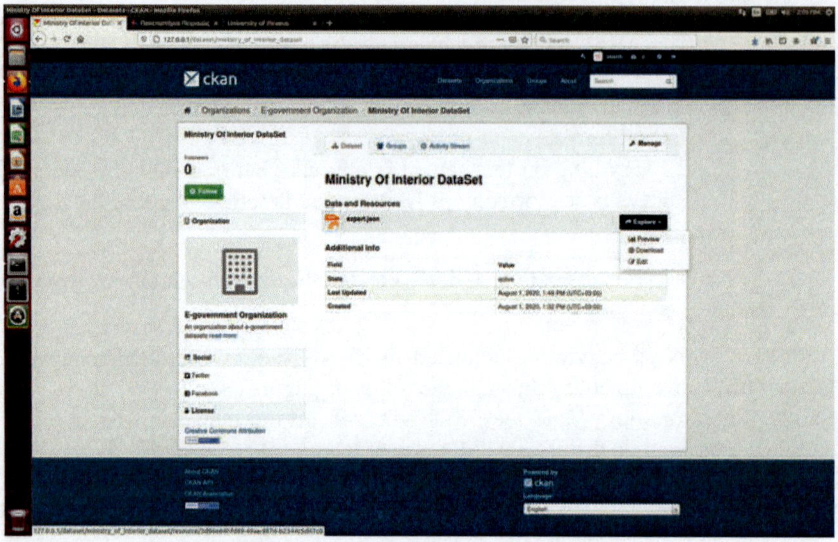

Fig. 5.27 Uploading the .JSON file in the e-Government Organisation in CKAN

4. Fill in the remaining metadata elements for the source we upload and select the relevant button to end the process according to the procedure of case A described above.

In each case it is possible to edit the resources (add, delete, edit) by selecting through the corresponding page. The page of each resource gives the general information concerning it, as well as the possibility to display a preview, the downloading of the resource etc.

Through the "preview" option we display the content of the resources that we have already uploaded to CKAN as shown in the following images (Figs. 5.28, 5.29 and 5.30). Finally, we note that if this data has been classified as public, then it is accessible in the same way by the ordinary user of CKAN as long as he selects the CKAN home page provided that the relevant ports (port forwarding) of the communication system are open via the Internet.

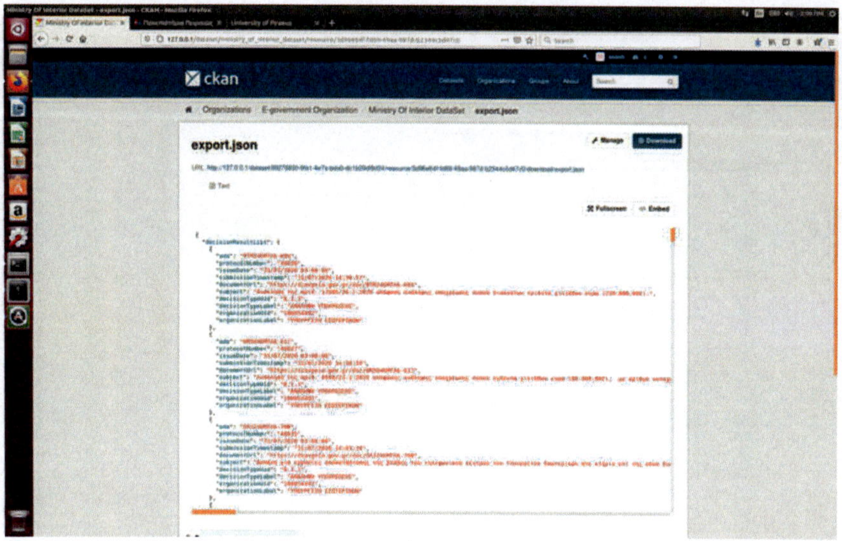

Fig. 5.28 Preview of the .JSON file we uploaded in the e-Government Organization via the CKAN page

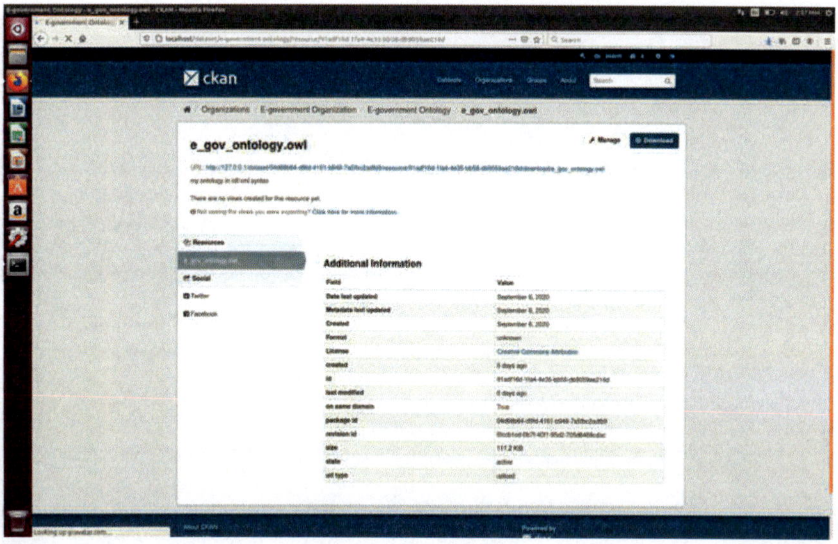

Fig. 5.29 Preview of the .owl file concerning the e-Government ontology we uploaded in CKAN

Fig. 5.30 The result of the access in the .owl file we uploaded as the e-Government ontology in CKAN

References

1. OECD. (2016). Digital government strategies for transforming public services in the welfare areas. http://www.oecd.org
2. The Greek Open Data Hub, official site: https://okfn.gr/the-greek-open-data-hub/
3. Open Government Partnership Initiative, official site: www.opengovpartnership.org/
4. Hellenic Republic, The official site of "DIAVGEIA program" of the Greek Public Administration: www.diavgeia.gov.gr
5. European Union Publication Office, The official portal for European data: https://data.europa.eu/en
6. Open Knowledge Greece, Open data handbook. http://opendatahandbook.org/guide/el/
7. OKFN/Greece Working Groups, Open Spending Book. https://github.com/openspending
8. OKFN/projects: https://okfn.gr/what-we-do/projects/
9. datahub.gr/official site: http://datahub.gr/data
10. OECD official site: www.oecd.org
11. G20/OECD Compendium of good practices on the use of open data for Anti-corruption. www.oecd.org
12. The World Economic Forum site: https://www.weforum.org
13. The official central directory of the Greek public data: www.data.gov.gr
14. Cadastral Agency and Mapping Association site: http://www.geodata.gov.gr
15. Cadastral Agency and Mapping Association, Open Geospatial Information Public Use License. http://www.inspire.okxe.gr/images/Files/Licences/Licence_OKXE_BY_v1.0.pdf
16. European Commission, Data protection in the EU. https://ec.europa.eu/info/law/law-topic/data-protection/data-protection-eu
17. The Open Government Partnership Initiative site: www.opengovpartnership.org
18. Parundekar, R., Knoblock, C. A., & Ambite, J. L. (2010). Linking and building ontologies of linked data. In *The Semantic Web—ISWC 2010, Lecture Notes in Computer Science* (Vol. 6496, pp. 598–614).

19. Jain, P., Hitzler, P., Sheth, A., Verma, K., & Yeh, P. (2010). *Ontology alignment for linked open data* (pp. 402–417). https://doi.org/10.1007/978-3-642-17746-0_26
20. Goudos, S. K., Peristeras, V., Loutas, N., & Tarabanis, K. (2007). A public administration domain ontology for semantic discovery of e-Government services. In *2nd International Conference on Digital Information Management, ICDIM 2007*. IEEE. https://doi.org/10.1109/ICDIM.2007.4444233
21. Najmi, E., Hashmi, K., Khazalah, F., & Malik, Z. (2013). Intelligent semantic question answering system. In *IEEE International Conference on Cybernetics (CYBCO)*. IEEE.
22. Vincent, J., Dombeu, F., Huisman, M., & Szpak, Z. (2011). A framework for semantic model ontologies generation for e-Government applications. In *The Fifth International Conference on Digital Society, ICDS*.
23. The official site of National Printing House of Greece: http://www.et.gr
24. W3C Incubator Group, W3C Geospatial Vocabulary, *W3C Incubator Group Report* (2007). https://www.w3.org/2005/Incubator/geo/XGR-geo-20071023/
25. geodata.gr official site: http://www.geodata.gov.gr
26. Openrefine official site: https://openrefine.org
27. W3C, W3C recommendations. http://w3c.github.io/csvw/csv2RDF/
28. Theocharis, S., & Tsihrintzis, G. (2014). Ontology development to support the Open Public Data—The Greek case. In *Proceedings of IISA 2014*. IEEE. https://doi.org/10.1109/IISA.2014.6878820
29. CKAN project official site: https://ckan.org
30. Open Knowledge International and contributors, Installing CKAN from package. https://docs.ckan.org/en/2.8/maintaining/installing/install-from-package.html
31. VirtualBox official site: https://www.virtualbox.org/
32. Ubuntu official site: https://ubuntu.com

Chapter 6
Production and Publication of Linked Open Data: The Case of Open Ontologies

Abstract The subject of this chapter is linked open data, issues related to their production and publication as well as the more special case of open ontologies. In particular, key themes, concepts and methodologies are examined as a theoretical infrastructure of the field of linked open data. It also examines the basic steps for publishing linked open data and the key technologies and tools used to produce and publish it. As a basic process in this context, triplification is examined and as a special case a methodology and an algorithm is proposed for the production of RDF triples from the data that can be retrieved from the relational database maintained as part of the operation of the Greek National Registry. The case of RDF serialisation from JSON data is also examined and as a special case study the production of RDF data from the open data retrieved by the "Diavgeia program". A methodology and an algorithm for publishing linked Data is then proposed and a case study on publishing an open ontology in the form of linked open data is presented.

Keywords Linked open data · Open ontologies · Data transformation · Triplification

6.1 Linked Open Data

6.1.1 Semantic Web, Linked Data and Linked Open Data—The Fundamental Rules

Semantic Web and linked data, as concepts, are closely connected. Semantic Web, as the Web of Data, started as an initiative of the World Wide Web Consortium—W3C [1] expanding the existing Web. Specific tools and technologies, such as RDF, OWL, SPARQL etc., that have been developed, support the automated discovery, access, processing and sharing of data by humans and computers as well. However, in order for the Web of Data to become a reality, it is necessary to use a standard method for accessing and managing the huge amount of data on the existing Web. In addition, unlike a clean data collection, relationships among data must also be

S. Theocharis and G. A. Tsihrintzis, *Semantic Knowledge Modelling via Open Linked Ontologies*, Artificial Intelligence-Enhanced Software and Systems Engineering 4,
https://doi.org/10.1007/978-3-031-20585-9_6

accessible in order to achieve the automated discovery of other data. The collection of these interrelated datasets on the Web is the fundamental idea behind the so-called linked data.

The term linked data is not about a new specification but sets out the basic principles governing the publishing, connecting and sharing structured open data via the Web. It is essentially a method of creating links between open data so that both humans and machines can automatically discover other data or information based on the semantics assigned to them by their creators. This connection concerns their relationship in the real world or their seemingly "irrelevant" information pieces. Many times the same concept is used in different verbal ways (vocabulary) in different sources, resulting in inefficient information search. The success of the project is based on wide acceptance and use of technologies used (HTTP, URIs) to achieve effective retrieval of information between websites in a way processable by computers. The fundamental principles introduced by Tim Berners-Lee [2] and are summarised in the following points:

1. Use of URIs as identifiers for objects of the natural world. As objects we consider either abstract concepts derived from our experience or intellect or even the individuals-persons themselves.
2. Use of HTTP URIs as identifiers in order to easily search by humans and to obtain descriptions of the identified concepts or real world resources as well.
3. Provision of useful information based on standards such as RDF and SPARQL etc. when someone looks up a URI.
4. Linkage between URIs in order to enable transition and hence finding new information.

As data does not always have to be open, Tim Berners Lee in 2010 [2] in order to encourage data holders, and in particular public administrations, to publish linked data as open, adds another rule to the already published principles for creating linked data. This rule applies to open source content and is intended to describe linked open data. In addition, it developed a five-point evaluation system to achieve the ultimate goal, namely linked open data—LOD, which is presented below:

1. The data must be available on the Web, in any form, with relatively open permission, in order to be considered as Open Data (*1 star*).
2. Data must be published in a structured, machine-understandable format. For example, tabulated data should be published in editable formats by applications instead of publishing scanned data e.g. in .pdf format (*2 stars*).
3. The data must be published in a structured, and machine-understandable format, which will be non-proprietary. For example tabular data should be published in open .csv format instead of .xls corresponding to a commercial application (*3 stars*).
4. In addition to the above, the data uses the W3C instructions (RDF, SPARQL) to identify objects so that users can refer to them (*4 stars*).
5. In addition to the above, the data must be linked to other data in accordance with the rules of linked data (*5 stars*).

6.1.2 *The Way That Linked Data Work*

The linked open data is a method of publishing structured data whose main objective is to connect seemingly "irrelevant" information to each other by their real relationship or their semantics. The success of the project is based on the wide acceptance and use of technologies used (HTTP URIs) so as to achieve effective retrieval of information between websites in a manner processable by humans and computers as well. The Semantic Web and consequently the linked data were not developed to change the way people manage and understand data out of the web, but to automate these processes through machines with the ultimate aim of assisting the user to connect information available and thus facilitate their processing.

The dominant idea of linked Data is the application of the existing architecture of the Internet in sharing structured data worldwide as an extended version of the Web. For this reason the current form of the Internet based on the concept of URIs (Uniform Resource Identifiers) as a global mechanism of unique identification of objects and to the HTTP protocol (Hypertext Transfer Protocol) as a global mechanism of to access online documents and HTML language as the dominant format for the submission of data. Furthermore, there is the idea of linking web documents—files that can be stored in different locations (servers). The links between documents convert the scattered content to a global information space. The URIs are not used for the identification only of digital content, but also as identifiers of people, places, or even abstract concepts. Thus, it is possible for a person, a public organisation or even an institution to be identified on the Web. For example, an organisation, e.g. the Ministry of Interior, can be identified through the URL of its official website (e.g. http://www.ypes.gr) or a person with the URL of his personal public page. Abstract concepts can also have their own URI such as the concept "person" may be determined by the URI http://xmlns.com/foaf/0.1/person. Various institutions or properties attributed to specific concepts can be defined through specific URIs in specially formulated vocabularies or ontologies found in publicly accessible URLs as well.

The HTTP protocol is the widespread mechanism of access to online content. In the current form of the Internet HTTP URIs are used to uniquely identify all forms of available information. Therefore, linked data encourages the use of HTTP URIs to identify the entities of the physical world. In fact, the HTTP protocol makes it possible to search for these entities.

The third principle of linked data requires the representation of structured information in a common format. This format is usually the Resource Description Framework—RDF that is a model based on the logic graph as we mention in the previous paragraph.

The connection of the digital data constitutes the fourth principle of linked data. Thus, the general direction is not only the connection between online documents and between different types of objects. In contrast to existing unstructured and without semantic content Internet connections, the links between the entities identified by URIs become important. This means that links can describe relationships of the natural world. For example, it is possible to connect a man and an organisation in order to describe the working relationship between the man and the organisation. Links to linked data are called RDF links in order to be separated from the links between the online documents. Therefore, in line with the global information space of the current Internet, we are led to the global data space.

As we mentioned before, by incorporating HTTP into the resource naming schemes of Web accessible resources, the same linking mechanism used for retrieving a Web Page (and other information resources) also extends to the process of retrieving a structured description of a resource in the Semantic Web. This process is quite similar to accessing a SQL recordset via an ODBC or JDBC data source name (DSN) (Fig. 6.1). Nevertheless there are fundamental differences between linked data and SQL CLI based data access such as [3]:

- Linked data URIs name/identify individual records while SQL data access DSNs are scoped to record containers such as Database, Table, View, or Stored Procedure names.
- Linked data URIs facilitate untethered data access that transcends database vendor, database engine, operating system, programming language, application server, application development framework, and other infrastructural confines, unlike ODBC, JDBC, and other data access mechanisms.
- The illustration below highlights linked data's parallels with ODBC in relation to DSN based data access.

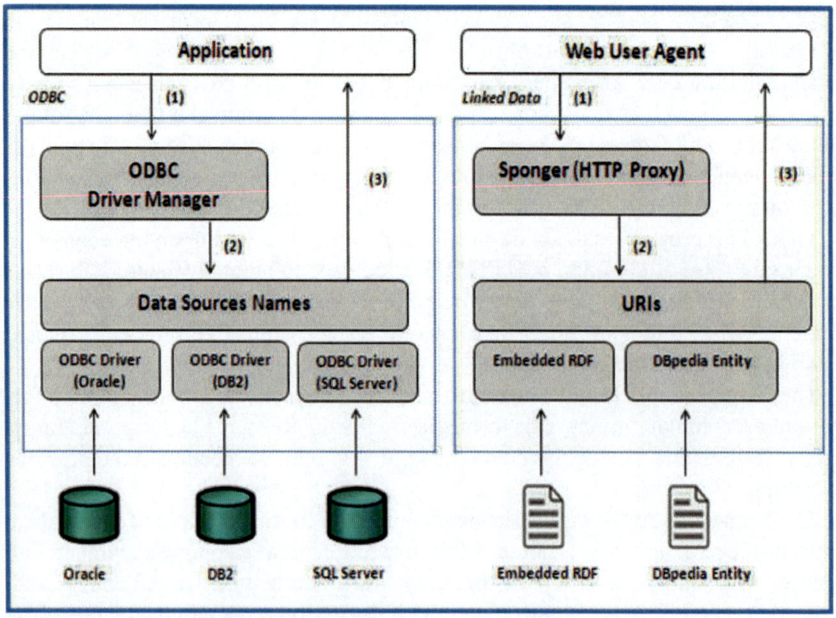

Fig. 6.1 Linked data and ODBC similarities [3]

6.1.2.1 Information and Non-information Resources

The World Wide Web is an information space in which the objects in focus are known as resources. Resources are identified by unique Uniform Resource Identifiers (URIs) based on the HTTP schema as discussed above. However, a URI alone is not enough to make a resource accessible on the Web. For this reason, a means of access is necessary. HTTP determines how such access can be made and identifies the various participants in the process.

The URI provides the visitor of the resource with the necessary information to enable the visit and retrieval of its data. In some cases the connection will be direct and additional information will be available directly through the resource access associated with that particular URI. In other cases, the link may be more indirect. We are talking about two main categories of resources: Information Resources and Non-Information Resources.

The relationship between a resource and its Web presence that supports access to it is important. For example, if an administrator of a page makes a statement about a resource, its Web presence should work in a way that supports that claim. This is usually achieved by including a link to the respective website which is usually accompanied by relevant text as a resource statement. However this process concerns humans as these statements on the existing web are perceived only by humans and not by machines.

6.1.2.2 Information Resources: Resources Whose Substance Is Information

Many resources available on the Internet are in fact documents that provide natural representations of complete information. For example, a document on the Web can provide a description of a public body, person, or object such as a planet or a car. URIs allow documents to be ambiguously identified as resources on the Web, while their presence on the Web allows them to access and retrieve the information they embed. Such retrieval is immediate in the event that the resource consists of a set of information that can be transmitted over the Internet via appropriate messages. These resources are referred to as Information Resources.

In order for the information of an Information Resource to be transmitted through the Web, it is necessary to represent it through defined formats. HTTP, for example, specifies that information should be displayed as an octet stream. In addition, the requesting person or system for resource-related information may also place restrictions on representations that are considered acceptable, such as natural language. Whether the requested representations are actually available for a given resource depends on the nature of the materials provided to support its online presence. However, we note that the architecture of the existing Internet clearly separates the concept of a resource from that of the representations that can be provided by its presence on the Web.

By the term "representation", we mean a form of information related to a resource that can be transmitted over the Internet. For example, for a video resource, two or more representations may be provided depending on the resolution supported by the video. Another example concerns the Greek, English or French version of an information note concerning the provision of a service by a public body. The process by which an appropriate representation is selected, based on the constraints on the application, is known as content negotiation.

6.1.2.3 Dynamically Produced Representations

Static documents—whether physical documents or multimedia files—are not the only source of information that corresponds to information resources. There are cases where information related to certain resources is provided by computer systems. Typical examples are systems that interact with users when submitting data in electronic forms to public or banking organisations or systems that periodically produce up-to-date information such as the weather or the stock market. Such systems are able to retrieve data from sources that do not have their own Internet presence. They can also perform calculations to assimilate the information that will eventually be returned to the applicant in an appropriate representation. These dynamic systems produce representations that, although not pure documents, can be considered dynamic by creating documents in response to the requests they receive. Just as with real-time document resources, representations returned by such systems may need to meet the constraints set out in the original request.

6.1.2.4 Websites as Information Resources

The most common type of information resource on the Web is the "well-known" website. As we have mentioned in one sense the Web is a huge set of linked websites. The pages themselves usually contain references and links to other pages—resources. They are often rendered in ways that allow users to interact using links. Enabling a link may cause the URI resource to which it refers to be rendered available to the user. However, the idea of enabling links to navigate between and within pages is so natural to web users that most of them do not distinguish between a resource, the representation that crosses the web, and the final version they perceive.

Pages often contain other types of references to resources such as page-based performance information or additional media. However users usually do not interact with these reports, which indicate resources that are not the websites themselves.

Although the term Website is associated for most users with a rather static entity that provides information, nowadays websites have dynamic features allowing access to a huge variety of functions, from buying a book to controlling a machine remotely. Increasingly, organisations are making applications available to their users as sets of web pages with which they can interact.

6.1.2.5 Non-information Resources: Resources Whose Substance Is Not Information

The vast majority of resources on the Web as mentioned above are information resources. Representations of these resources are available through appropriate interaction with their Web presence. However, there is increasing interest in using URIs to identify resources whose substance is not information, i.e. non-information resources specifically in the light of Semantic Web and linked data. The operation of a non-information resource must be differentiated from the operation of the information resources in order not to create misinterpretations or malfunctions when accessing them. For example, it is important that non-information resources do not respond with representations if they are present on the Internet. The question of the purpose of the Internet presence for non-information resources as well as the circumstances under which they should have one is discussed below.

In an attempt to describe a concept such as e.g. a Ministry or something more tangible like a country or a planet, in semantic terms using RDF statements through a relevant file, specific problems arise. For example, dedicating a relevant URI as a unique identifier for this concept raises the issue of the presence of a non-information resource on the Web. That is, access to this resource through this URI may not have the same results as in the case of an information resource. Consequently, we need to address the issue of using URIs for information resources that describe the possible properties that characterize the original concept and any other concept associated with it. This issue is related to the so-called dereferencing that we analyse below. In other words, the issue is the "discovery" of descriptive documents, which are information resources present on the Web, and which themselves refer to an RDF document identified by a specific URI. The solution is given by the analysis shown in the following figure (Fig. 6.2): a non-information resource has a unique URI identifier, is linked to a specific RDF document located at a certain URL while its information representations are given through an understandable HTML document by the user.

6.1.3 Linking Information Resources to Other Resources Through Redirection

From the above, we conclude that it is extremely useful to be able to correlate Information Resources with other types of resources. However, it should be noted that representations of such Information Resources are not representations of the resources to which they are linked. In fact, relevant Information Resources convey information about the concepts to which they relate through Non-Information Resources.

Information Resources related to an Information-free resource have their own URIs as they are separate resources and provide representations. They may have other uses besides providing additional information about the non-information resource.

Fig. 6.2 Non-information resource analysis

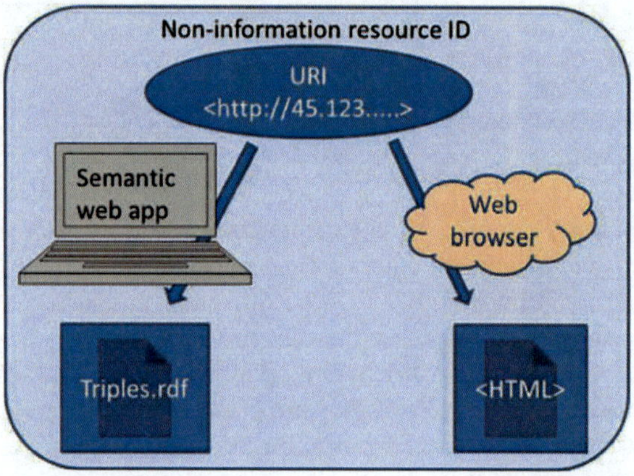

When such resources are linked to a resource without information, we need a means of clearly identifying the correlation. This is achieved through HTTP redirection which provides a means of showing this correlation with the use of an appropriate Internet presence.

6.1.3.1 Internet Presence for Non-information Resources

The presence on the Internet, for an Information Resource is responsible for the return of appropriate representations on access, as we have seen. We have also seen that the Internet presence for a resource without information should not return representations upon access. In order not to create misunderstandings when accessing a non-information resource, the appropriate URI should be returned, from which the appropriate information about the original resource can be retrieved, on the contrary to accessing an information resource. This presupposes the exchange of appropriate HTTP messages and the presence of appropriate information resources in the indicated URI.

6.1.3.2 Use of HTTP Redirection to Represent Correlations

The representation that is returned when there is Internet access for a resource consists of two parts. One part is the data, the other part the metadata. For example, when accessing an HTML web page, the data is the part that contains the markup and related instructions from which the rendered version is created. Metadata contains the information in the HTTP headers that are part of the response. These metadata include a specific HTTP response code. A specific value of this response code forms the mechanism by which HTTP itself can indicate correlations.

The HTTP 303 response code, named "See Other", indicates that there is no representation of the URI being accessed, but relevant information may be available.

The important thing in this response is the return of a URI that can be accessed when searching for relevant information. This type of request redirection to a different URI is exactly the behaviour we need to indicate that there may be additional information associated with a non-information resource. If the Internet presence for a resource without information is set to have this behaviour, the information can be associated with the resource without misleading individuals or systems that have access to it.

In this process, however, there is no guarantee that the URI returned to a 303 "See Other" response will actually provide an access representation. Access to the returned URI is a separate function from accessing the original resource without information. It could be that the returned URI leads to the need for further redirection actions before a representation is retrieved. Indeed, the returned URI may not be accessible. On a practical level, both behaviours are non-optimal, as they increase network circulation without necessarily allowing a representation to be retrieved [4].

The following is a good practice: Authorities, as URI's owners, may create HTTP URIs for non-information resources as well as for information resources:

- If a URI identifies an information resource, the URI owner should create a Web presence for it. The Web presence should provide representations of the resource when accessed.
- If a URI identifies a non-information resource, the URI owner may create a Web presence for it. The Web presence should respond by returning information about an associated information resource, related to the resource. The URI of this associated information resource should be indicated using the redirection mechanism based on the HTTP 'See Other' response code, 303.
- A URI owner providing an information resource associated with a non-information resource should avoid the need for additional redirection operations after the original 'See Other' response. In particular, the URI returned in the 'See Other' response should be able to provide representations of the associated information resource.

6.1.3.3 Content Negotiation

The usual HTML browsers, with their basic settings and without the use of any extensions, display the RDF representations in the form of RDF code and in some cases "download" them as RDF files without displaying them. This is neither useful nor understandable to the regular user. Therefore, presenting a correct HTML representation in conjunction with the RDF representation of a resource helps a person understand the resource that corresponds to a URI. There is talk of so-called content negotiation. This means the process of selecting the best representation for a given response when multiple representations are available. The process starts when an HTTP client sends an HTTP header that states the representation of the type of representation it is waiting for in response. If the headers indicate that the client prefers HTML, then the server can generate an HTML representation. If the client prefers RDF, then the server can generate RDF.

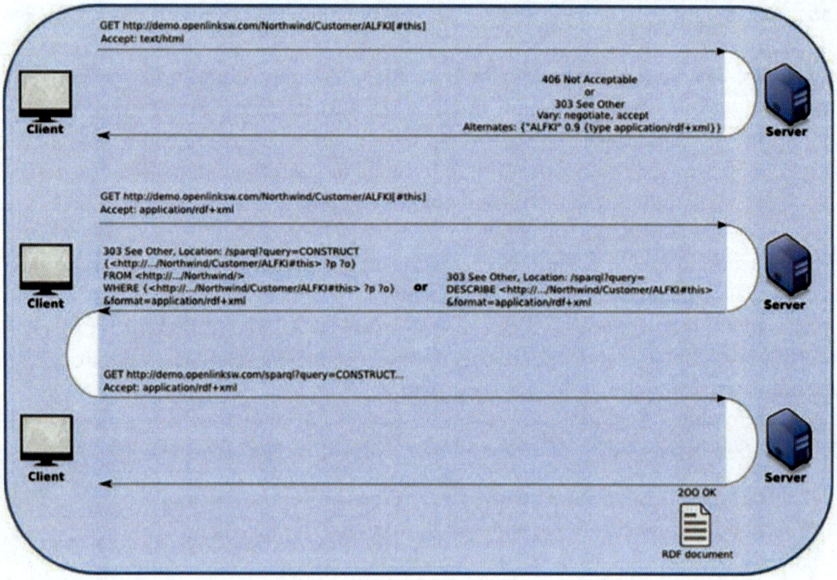

Fig. 6.3 Linked data access mechanics [6]

Content trading for non-information resources is implemented in the following way. When the client visits a URI that identifies a non-information resource, the server sends a 303 redirect to an information resource suitable for the client. Therefore, a data source often serves three URIs associated with non-information resources. The following figure (Fig. 6.3) shows how visiting an HTTP URI that identifies a non-information resource uses content trading [5]:

1. The client performs (sends) an HTTP GET request on a URI that identifies a non-information resource. If the client is a browser supporting linked data and would prefer an RDF/XML representation of the resource, it sends the "`accept:application/RDF+xml`" header along with the request. On the other hand HTML browsers would send the "`Accept:text/html`" header.
2. The server recognizes the URI that identifies a Non-Information Resource. As the server cannot return a representation of this resource, it responds using the HTTP response code 303-See Other and sends the client the URI of an information resource that describes the Non-Information Resource. In the case of RDF sends: "RDF …. content location".
3. The client now asks the server to receive (GET request) a representation of the information resource by requesting application/RDF + xml again.
4. The server sends the RDF/XML document to the client which contains a description of the vocabulary URI of the original resource.

6.1.4 Basic Steps for Publishing Linked Open Data

In order to publish our data as linked open data it is necessary to follow specific rules which are coded in the following steps [7–9].

Step 1. Understand the rules/principles of linked data and linked open data

It is important as a primary step to fully understand from the beginning of the process the basic principles/rules as well as the evaluation of the 5 stars mentioned above. Among these principles we note the particular importance of using appropriate HTTP URIs for all resources to be used and not only for documents as well as the use of RDF data representation. This will mainly serve to save resources, effort and time for the effective completion of the publication of the data.

Step 2. Understand and Analyse your data

The first of the practical steps of the process concerns the understanding and analysis of the data. It is crucial to take a deeper look at your data models, your metadata and the data itself. Get an overview and prepare a selection of data and information that is useful for publication. For this reason it is necessary to define the key entities-entities between the data, any properties they have and how to connect to other entities. For example, concepts such as: people, places, multimedia, documents, abstract concepts, etc. must be defined. It is also important to specify the data that is stored or is to be stored in databases. Such data are usually lists of products, services, places, people, etc.

Step 3. Setup your infrastructure

As mentioned above, the whole process involves information and non-information resources that must be accessible through the Web. Resources such as HTML documents, multimedia files (audio, video, video, etc.), RDF files, ontologies, vocabularies, etc. must be "uploaded" and served by respective servers. It is crucial to decide early on the technological infrastructure (networks, capabilities, capacity, operating systems, applications, etc.) that will support the project both in the development of data and in their maintenance.

Step 4. Clean your data

Data and information that comes from many distributed data sources and in several different formats (e.g. databases, XML, CSV, Geodata, etc.) require additional effort to ensure easy and efficient modelling. This includes ridding your data and information of any additional information that will not be included in your published data sets.

Step 5. Model your data

Since the focus is on the data and the format it must have in order to be able to link to other data, the process of modulating data located either in relational databases or in other types of resources is particularly important. Data modelling is a process that can be time consuming for specialised data sets and at the same time require the assistance of expert analysts. For this reason, linked data has been criticised for the effort and resources required to model. However, this investment is offset by the benefit to the community from the openness and ease of discovering information related to the concepts of interest each time.

The relevant procedure includes the following steps:

5.1. Make or download copies of raw data from their source.
5.2. Understanding data, recording objects/concepts and their properties as well as how they relate to each other.
5.3. Creating a design that includes the objects-concepts of interest and the lines that connect them, indicating the correlations and the properties that characterise them. In this step it is important to search for any published relevant vocabularies or ontologies and reuse the terms that fit.
5.4. Naming of objects-concepts using the appropriate URIs, taking into account the factor of evolution or change of data over time. It is important for the use of URIs to use the domain which is directly controlled by the production of the linked data.
5.5. Performing tests on the assumptions of the schema with the assistance of experts in the field to which the data belong. There may be a need to revise the original shape so it is a good practice to use appropriate design tools.

Step 6. Choose appropriate vocabularies

The general principle of vocabulary and ontologies must be applied in this process: "Reuse instead of reinvention". In other words, before developing new concepts through a new ontology or vocabulary, a search should be made for dedicated repositories for existing similar concepts or ontologies that need to be reused. However we must note the value of the validity of the ontology or vocabulary being reused. Usually such resources are published but not updated or not sufficiently documented. Through the use of URIs and the use of valid vocabularies, curators and data publishers can publish information more efficiently and reduce the cost of data integration.

Some vocabulary guidelines for consideration in your project:

- To name things, use RDFs: label, foaf: name, skos: prefLabel
- To describe people, use FOAF, vCard
- To describe projects, use DOAP
- To describe Web pages and other publications, use dc: creatoranddc: description
- To describe an RDF schema/vocabulary/ontology, use a VoID description
- To describe addresses, use vCard
- To model simple data, use RDF, RDFS, custom vocabularies
- To model existing taxonomies, use SKOS.

Step 7. Convert your data to RDF

After completing the schema for the data model, the next practical issue is to convert the source data into a representation of linked data or otherwise serialisation in RDF.

Considering the various RDF serials, such as Turtle, N3, RDF/XML, we choose the most appropriate one for each case, as there is no serialization better than another.

Since three conversion categories are mentioned: (a) Automatic conversion, sometimes called triplification, (b) Partial scripted conversion and (c) Modelling by human and subject matter experts, followed by scripted conversion, we choose the one that suits our needs in combination with the original data set. It should be noted, however, that although automatic conversion is a good practice that helps large data sets, it does not always produce high quality results. In any case, the views of experts in the field and the relevant data model that has been developed should be taken into account.

A best practice is to validate a representative sample set of your data format after converting it into one or more of the RDF serialisation formats. RDF validation helps to avoid unnecessary errors when the data is loaded into an RDF database.

Step 8. Interface tests and server access

Before the final publication and promotion of the data, the settings of the servers as well as the ontologies, vocabularies and related resources should be tested with appropriate tools in order to ensure the correct process for the content negotiation. We note that especially in the case of SparqL Endpoints the tests should concern access to datasets both locally, i.e. both inside the local machine that hosts the data and from the external environment or by using appropriate tools such as the Vapour Validator, or W3C RDF Validator either with remote machine access. These links ensure optimised data processing and integration for data (re-)use and allow for the creation of new knowledge from your data sets by putting them into a new context with other data.

Step 9. Specify licences

Publishing existing content as linked Data means still requires specification of an appropriate licence. Software licences for data is a complex issue and organisations dedicated to software licensing, such as Creative Commons, is a valuable resource for learning about the issues and choosing the appropriate licence (see Appendix A.3). Proper licensing ensures broad and effective re-use of data provided as linked open data.

Step 10. Publish and promote your LOD

In order for linked open data to be easily discovered and reused it is necessary to publish and promote it through appropriate channels. This means that the data in their electronic form must first be copied to a publicly accessible server. As a next and final step, ways to promote the dataset should be sought. Alongside other ways of promotion it is a great idea to add your LOD sets into the LOD cloud [10], a visual presentation of LOD sets by providing and updating the meta-information about your data sets on the data hub [11]. Remember to always provide human-readable descriptions of your data sets to make the datasets "self-describing" for easy and efficient re-use.

6.2 Linked Open Data Tools

In order to produce linked open data in the field of Public Administration for the needs of this work, we used specific tools which we briefly present below. Their common feature is that they are either free software or software provided as free under specific licences on a case by case basis. The way in which these tools were used as well as the results from their use are analysed in the following paragraphs.

6.2.1 Apache Jena and Apache Fuseki Server

Apache Jena is a free and open source Java framework for building Semantic Web and linked data applications [12]. The framework is composed of different APIs such as RDF API, TDB, Fuseki and OWL Ontology API, interacting together to process RDF data and OWL ontologies. Particularly:

- RDF API: Interact with the core API to create and read Resource Description Framework (RDF) graphs. Serialise your triples using popular formats such as RDF/XML or Turtle.
- TDB is a component of Jena for RDF storage and query. It supports the full range of Jena APIs. TDB can be used as a high performance RDF store on a single machine. A TDB store can be accessed and managed with the provided command line scripts and via the Jena API. When accessed using transactions a TDB dataset is protected against corruption, unexpected process terminations and system crashes.
- As we outlined above, there are various different ontology languages available for representing ontology information on the Semantic Web. They range from the most expressive, OWL Full, through to the weakest, RDFS. Through the Ontology API, Jena aims to provide a consistent programming interface for ontology application development, independent of which ontology language you are using in your programs.
- Apache Jena Fuseki is a SPARQL server. It can run as an operating system service, as a Java web application (WAR file), and as a standalone server in Windows systems or in Linux as well. Fuseki comes in two forms: a single system "webapp", combined with a UI for admin and query, and as "main", a server suitable to run as part of a larger deployment, including with Docker or running embedded. Both forms use the same core protocol engine and same configuration file format.
- Fuseki provides the SPARQL 1.1 protocols for query and update as well as the SPARQL Graph Store protocol. It is tightly integrated with TDB to provide a robust, transactional persistent storage layer, and incorporates Jena text query.

In this work we used the Fuseki server to develop a SparqL Endpoint to host and share our ontologies and our linked data with other applications. In our case we developed Fuseki as a standalone server which provides a SPARQL server that can use TDB for persistent storage and provides the SPARQL protocols for query, update and REST update over HTTP.

In order to use Fuseki it is necessary to install it on our local computer after downloading the corresponding file from the official repository. After decompressing the file and copying the generated folder to the desired location from the root location, we run either the batch file: "`fuseki-server.bat`" or the command in the formata of: "`fuseki-server:[--mem|--loc=DIR] [[--update]/NAME]`", where "`/NAME`" is the dataset publishing name at this server in URI space. As a first step for using fuseki we visit the address heard by the respective server: "`http://localhost:3030`". As a key feature of the system we mention the ability to handle RDF graphs with storage either in memory, which means the graph is lost in case the server is turned off or locally through the TDB database, which allows the graph to be retained when the server is closed. Another feature is the ability to query SparqL queries on the stored graph through which the graph can be revised, the RDF file checked in conjunction with the SparqL Endpoint function.

Detailed examples as well as comments from the use of fuseki are presented in the following paragraphs.

6.2.2 Eclipse RDF4J Framework

Eclipse RDF4J is a powerful Java framework for processing and handling RDF data [13]. This includes creating, parsing, scalable storage, reasoning and querying with RDF and linked data. It offers an easy-to-use API that can be connected to all leading RDF database solutions. It allows you to connect with SPARQL endpoints and create applications that leverage the power of linked data and Semantic Web.

RDF4J offers two out-of-the-box RDF databases (the in-memory store and the native store), and in addition many third party storage solutions are available. The framework offers a large scale of tools to developers to leverage the power of RDF and related standards.

RDF4J fully supports the SPARQL 1.1 query and update language for expressive querying and offers transparent access to remote RDF repositories using the exact same API as for local access. Finally, RDF4J supports all mainstream RDF file formats, including RDF/XML, Turtle, N-Triples, N-Quads, JSON-LD, TriG and TriX.

A. Core databases

RDF4J offers a set of database implementations out of the box.

- The RDF4J **Memory Store** is a transactional RDF database using main memory with optional persistent sync to disk. It is fast with excellent performance for small datasets. It scales with the amount of RAM available.

- The RDF4J **Native Store** is a transactional RDF database using direct disk IO for persistence. It is a more scalable solution than the memory store, with a smaller memory footprint, and also offers better consistency and durability. It is currently aimed at medium-sized datasets in the order of 100 million triples.
- The RDF4J **Elasticsearch Store** is an experimental RDF database that uses Elasticsearch for storage. This is useful if you are already using Elasticsearch for other things in your project and you want to add some small scale graph data. A good use case is if you need reference data or an ontology for your application. The built-in read cache makes it a good choice for data that updates infrequently, though for most use cases the Native Store will be considerably faster.
- On top of these core databases, RDF4J offers a number of functional extensions. These extensions add functionality such as improved full-text search, RDFS inference, rule-based reasoning and validation using SHACL/SPIN, and geospatial querying support.

B. Third party database solutions

The core RDF4J databases are mainly intended for small to medium-sized datasets. However, RDF4J-compatible databases are developed by several third parties, both open-source/free and commercial, and they often offer better scalability or other extended features. Because these triplestores are compatible with the RDF4J APIs, you will be able to switch your project to a different database with a minimal amount of code changes. In this work we used the Openlink Virtuoso RDF4J among several third party solutions. The Openlink Virtuoso RDF4J Provider is a fully operational Native Graph Model Storage Provider for the Eclipse RDF4J Framework, allowing users of Virtuoso to leverage the Eclipse RDF4J framework to modify, query, and reason with the Virtuoso quad store using the Java language.

C. RDF4J server and Workbench

RDF4J Server is a database management application: it provides HTTP access to RDF4J repositories, exposing them as SPARQL endpoints. RDF4J Server is meant to be accessed by other applications. Apart from some functionality to view the server's log messages, it doesn't provide any user oriented functionality. Instead, the user oriented functionality is part of RDF4J Workbench. The Workbench provides a web interface for querying, updating and exploring the repositories of an RDF4J Server.

In order to use both the RDF4J server and Workbench you will first need to download the RDF4J SDK. Both RDF4J Server and RDF4J Workbench can be found in the war directory of the SDK. The war-files in this directory need to be deployed in a Java Servlet Container such as the Apache Tomcat. After you have deployed the RDF4J Workbench webapp, you should be able to access it, by default, at path `<http://localhost:8080/RDF4j-workbench>`. You can point your browser at this location to verify that the deployment succeeded.

Examples and results from using the RDF4J workbench in order to hanld our ontologies in the domain of e-Government are presented in the following paragraphs.

6.2.3 Pubby

Pubby is a linked data Frontend for SPARQL Endpoints [14]. It can be used to add linked data interfaces to SPARQL endpoints or triples stores developed with several RDF servers. Much data in the format of RDF or OWL files that include semantic information lives inside triple stores and can be accessed only by sending SPARQL queries to a SPARQL endpoint. Thus the ordinary user must know the relevant query protocols, have access to special query engines and be able to interpret the triplets that are usually returned by the relevant queries. It is also hard to connect information in these stores with other external data sources. Pubby allows a wide variety of existing RDF browsers (e.g. Disco, Tabulator, OpenLink Browser), RDF crawlers (e.g. SWSE, Swoogle), and query agents (e.g. SemWeb Client Library, SWIC) to access the data.

There are two main reasons for setting up a Pubby site.

- Pubby is a linked data Server. It can make the dataset linked data compliant. This means, when you put a URI of a particular resource into a browser, you get back information about this resource, either in HTML format or in RDF format depending on the capabilities of the browser. This makes the relevant information part of the Web, rather than simply allowing access to it only through SPARQL.
- Pubby is an RDF Browser. This means it creates a nice and simple web site that allows browsing through your RDF dataset. It puts a simple user interface over your dataset. You can explore the dataset by following the links between resources.

How Pubby Works

Pubby is implemented as a Java web application. It makes it easy to turn a SPARQL endpoint into a linked data server. Pubby is designed to provide a linked data interface to the RDF data sources that live in the triple stores through the web (Fig. 6.4).

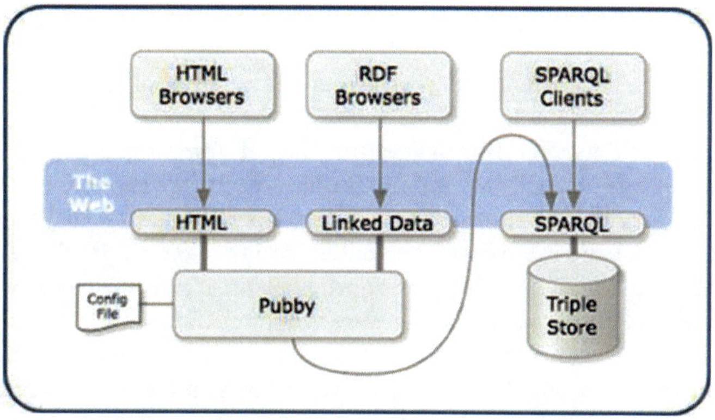

Fig. 6.4 How the Pubby works [14]

According to the RDF format, resources are identified by URIs as we mentioned before. The URIs used in most SPARQL dataset are not dereferencable, meaning they cannot be accessed in a Semantic Web browser, but return "`404 Not Found`" errors instead, or use non-dereferenceable URI schemes.

When setting up a Pubby server for a SPARQL endpoint, you must configure a mapping that translates those URIs to dereferenceable URIs handled by Pubby. If your server is running e.g. at `<http://myserver.org:8080/pubby/>`, then the particular resource URI above might be mapped to: `<http://myserver.org:8080/pubby/resource>`.

Pubby will handle requests to the mapped URIs by connecting to the SPARQL endpoint, asking it for information about the original URI, and passing back the results to the client. It also handles various details of the HTTP interaction, such as the 303 redirect required by Web Architecture, and content negotiation between HTML, RDF/XML and Turtle descriptions of the same resource.

Pubby installation is fairly easy. It first includes downloading the relevant file from the Pubby repository and then unzipping it and copying an image of it to the Apache Tomcat ROOT folder. The point that requires attention and relevant controls for proper operation is the configuration file, through which the dereferencing mentioned above takes place.

The configuration of the configuration file for the needs of our work is included in Appendix A.4. Examples, results as well as comments from the use of Pubby are presented in the following paragraphs.

6.2.4 Apache Tomcat

In order to host applications such as Fuseki, RDF4J and Pubby we used Apache Tomcat 8 as a friendly and stable HTTP web server. Apache Tomcat is a webcontainer which allows to run servlet and JavaServer Pages (JSP) based web applications. The Apache Tomcat software is an open source implementation of the Java Servlet, JavaServer Pages, Java Expression Language and Java WebSocket technologies [15]. It also provides by default a HTTP connector on port 8080, i.e. Tomcat, can also be used as an HTTP server. But the performance of Tomcat is not as good as the performance of a designated web server, like the Apache HTTP server.

The Java Servlet, JavaServer Pages, Java Expression Language and Java WebSocket specifications are developed under the Java Community Process.

The Apache Tomcat software is developed in an open and participatory environment and released under the Apache License version 2. The Apache Tomcat project is intended to be a collaboration of the best-of-breed developers from around the world. Apache Tomcat software powers numerous large-scale, mission-critical web applications across a diverse range of industries and organisations. Some of these users and their stories are listed on the PoweredBy wiki page.

Apache Tomcat can be installed in both Windows and Linux systems but with different procedures. Web applications provided via Tomcat can be managed through `<http://localhost:8080/manager/html>`.

The standard deployment format for web applications is a .war file. If you create a war application just put this application into the webapps folder. The next time tomcat starts it will unpack the war and make the application available. According to this method we created both fuseki and RDF4j server in order to establish our SparqL Endpoints that we analyse below.

Web applications may require external libraries. Typically, web applications contain their own libraries but if you want to make certain libraries available for all applications you can put them into the folder "lib" and a subfolder below "lib". These libraries are then available for all web applications.

Tomcat also contains a HTTP connector which can be used to serve static HTML pages. The standard directory which will be served is below the *Tomcat* `webapps/ROOT` installation directory.

6.2.5 D2R Server and D2RQ Mapping Language

In this work, we used and in this section present the D2R Server as a tool for publishing linked data views of relational databases. The D2R Server [16] is a tool for publishing the content of relational databases in the Semantic Web. The content of the database is mapped into RDF data with a declarative mapping that specifies how the resources are identified and how they produced the values of properties in the content database. Based on this mapping, the D2R Server allows in RDF and HTML browsers to browse the non-RDF database contents and also allows applications to submit queries to a database using the SPARQL query language of submission. The server receives requests from the web and transforms them "on the-fly" to corresponding SQL queries for submission to the database. Figure 6.5 illustrates schematically the operation architecture of D2R Server [16].

D2RQ Mapping Language

The D2R Server uses D2RQ mapping language to capture mappings between database shapes and RDFS or OWL ontologies shapes. A D2RQ mapping determines how resources are identified and how the values of properties in the content database are produced. An example of the structure of a D2RQ map file is shown in Fig. 6.6. The central object in D2RQ is ClassMap. A ClassMap represents a mapping from a set of entities described in the database to a class or group of such classes of resources. Each ClassMap has a set of PropertyBridges, which determine how the resource descriptions are created. The values of the properties can be created directly from the database values or by using "molds" or translation tables. The D2RQ supports conditional mappings at ClassMap and PropertyBridge, mapping n: m relations and handling highly normalised table structures where entity descriptions extend to several tables.

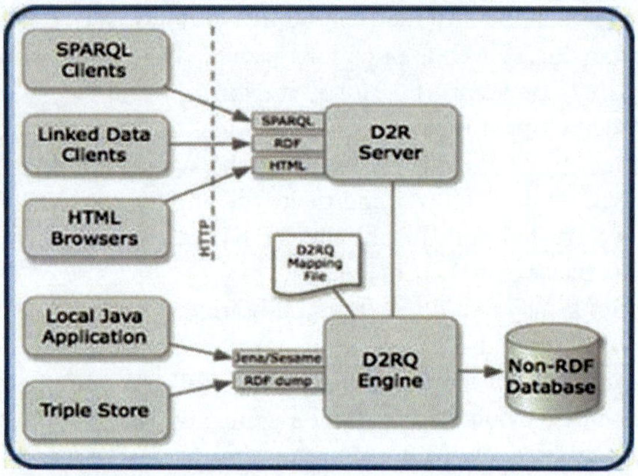

Fig. 6.5 Operation architecture of D2R Server [16]

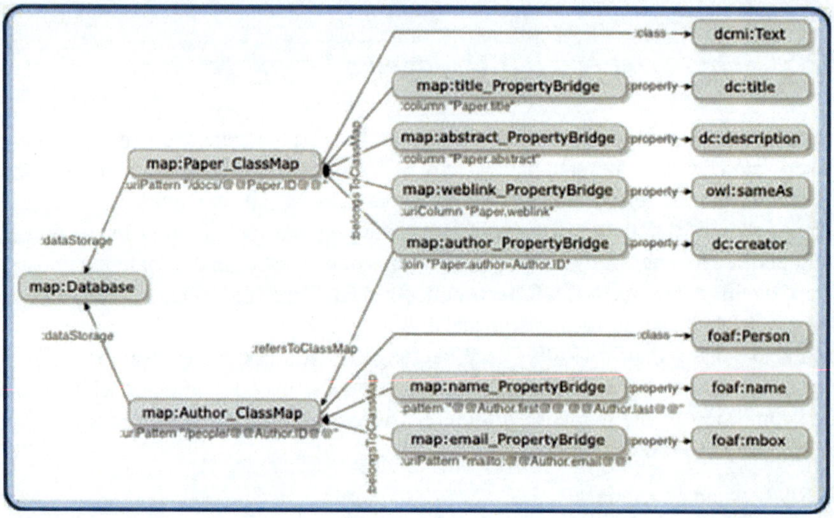

Fig. 6.6 The general structure of a D2RQ map [17]

The D2R Server includes a tool that automatically generates a D2RQ file mapping the structure of a database table. The tool generates a new RDF vocabulary for each database using the names of tables as class names and column names as property names. The mapping may be adjusted subsequently, replacing the automatically generated terms with terms of familiar RDF vocabularies [17].

6.2.6 OpenLink Virtuoso Server and Virtuoso Open Source Server

OpenLink Virtuoso is the first CROSS PLATFORM Universal Server to implement Web, File, and Database server functionality alongside Native XML Storage, and Universal Data Access Middleware, as a single server solution. It includes support for key Internet, Web, and Data Access standards such as: XML, XPATH, XSLT, SOAP, WSDL, UDDI, WebDAV, SMTP, SQL-92, ODBC, JDBC, and OLE-DB. Virtuoso currently supports the following Operating systems—Windows 95/98/NT/2000, Linux (Intel, Alpha, Mips, PPC), Solaris, AIX, HP-UX, Unixware, IRIX, Digital UNIX, DYNIX/PTX, FreeBSD, SCO, MacOS X [18].

Virtuoso is a revolutionary, next generation, high-performance virtual database engine for the Distributed Computing Age. It is a core universal data access technology set to accelerate our advances into the emerging Information Age. It provides transparent access to your existing data sources, which are typically databases from different database vendors.

Through a single connection, Virtuoso will simultaneously connect your ODBC, JDBC, UDBC, OLE-DB client applications and services to data within Oracle, Microsoft SQL Server, DB/2, Informix, Progress, CA-Ingres and other ODBC compliant database engines. All your databases are treated as a single logical unit.

The diagram below depicts how applications that are built in conformance with industry standards (such as ODBC, JDBC, UDBC, and OLE-DB) only need to make a single connection via Virtuoso's Virtual Database Engine and end up with concurrent and real-time access to data within different database types (Fig. 6.7).

Further still, Virtuoso exposes all of its functionality to Web Services. This means that your existing infrastructure can be used to support Web Services directly without any hint of replacement.

Virtuoso Open Source Server

At core, Virtuoso is a high-performance object-relational SQL database [20]. As a database, it provides transactions, a smart SQL compiler, powerful stored-procedure language with optional Java and .Net server-side hosting, hot backup, SQL-99 support and more. It has all major data-access interfaces, such as ODBC, JDBC, ADO .Net and OLE/DB.

Virtuoso has a built-in web server which can serve dynamic web pages written in Virtuoso's web language (VSP) as well as PHP, ASP .net and others. This same web server provides SOAP and REST access to Virtuoso stored procedures, supporting a broad set of WS protocols such as WS-Security, WS-Reliable Messaging and others. A BPEL4WS run time is also available as part of Virtuoso's SOA suite.

Virtuoso has a built-in WebDAV repository. This can host static and dynamic web content and optionally provides versioning. The WebDAV repository is tested to interoperate with WebDAV clients built into Windows XP, Mac OSX and others and makes Virtuoso a convenient and secure place for keeping one's files on the net.

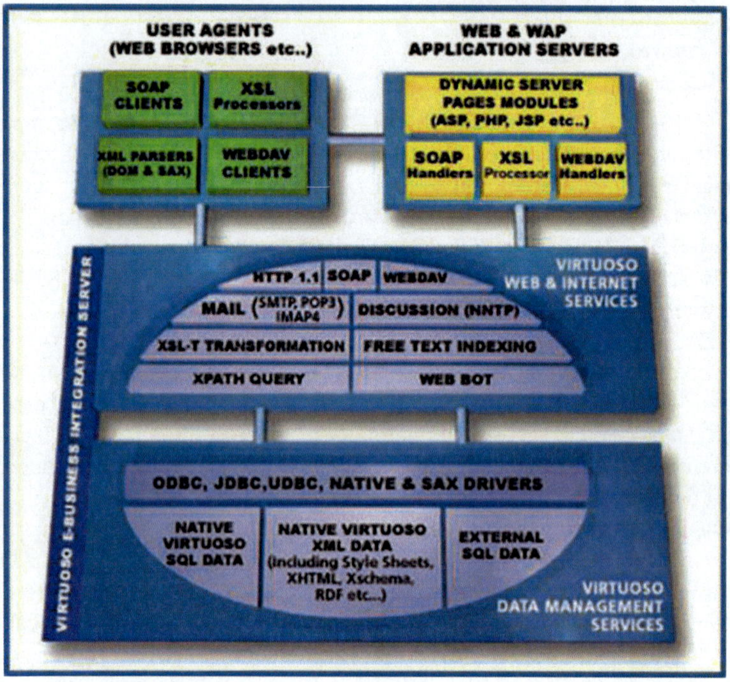

Fig. 6.7 OpenLink Virtuoso product architecture [19]

Further, Virtuoso provides automatic metadata extraction and full text searching for supported content types.

OpenLink Virtuoso supports SPARQL embedded into SQL for querying RDF data stored in Virtuoso's database. SPARQL benefits from low-level support in the engine itself, such as SPARQL-aware type-casting rules and a dedicated IRI data type. This is the newest and fastest developing area in Virtuoso.

We mention that the Virtuoso Conductor comes with a special open licence for use up to 30 days and the Virtuoso Opensource Server comes with an open licence without time limit. Examples and results from using Virtuoso and particularly the ODS Briefcase in order to handle our ontologies in the domain of e-Government are presented in the following paragraphs.

6.2.7 Comparison of the Basic Tools

Using the tool comparison application offered by DB-Engines, we present below comparative elements of the three basic tools we used in this work. DB-Engines is an initiative to collect and present information on database management systems (DBMS). In addition to established relational DBMS, systems and concepts of the growing NoSQL area are emphasised [21]. The DB-Engines Ranking is a list of DBMS ranked by their current popularity. The list is updated monthly. The most important properties of numerous systems are shown in the overview of database management systems. You can examine the properties for each system, and you can compare them side by side (Table 6.1).

6.3 Triplification—The Case of Production of RDF Triples from Data in Relational Databases in National Municipal Registry

6.3.1 Motivation

Nowadays, the majority of citizens and businesses use Internet applications in order to have access to primary electronic data or information published on various websites. Moreover, most of the public sector has digitised raw data held and has isolated Internet users (citizens, businesses, and public organisations) as open data, processed at relative governmental portals. In most cases, information retrieval is achieved with the help of mainly relational databases and related applications supporting access to these databases through the existing Internet.

At the same time, the interest of the scientific community focuses on further development of the Internet and the evolution of the Semantic Web. Initially, the Semantic Web is expected to contribute to smarter access and manage information handled on the Internet through new technologies and development of corresponding new applications. The overall vision is that of the transition from the current Internet of static pages into a dynamic network of providers of services that automatically discover the information sought, taking into account the semantics of the concepts referred to by the user. This will benefit in improving the quality of the answers given to specific questions-searches in each case. Thus, the research interest is focused on developing tools and methods based on the development and management of common vocabularies and new standards for representing information and knowledge produced.

Our interest focuses on the opportunities that can provide the technologies of the Semantic Web in the further development of e-Government and support of open government. The field of e-Government can be one of the main consumers of smart services based on the Semantic Web. This is because the modern trend of open government is supported both by the publication of "open data" to the public and also by the connection of digitised data and information in the back-office of public administration. By using Semantic Web technologies, it is expected that the search and data sharing between stakeholders will improve by the use of common terminology and semantics.

More specifically, in this part we examine the usefulness of processing the data stored in existing relational databases with the help of Semantic Web technologies.

Table 6.1 The comparison of basic tools from db-engines [21]

Name	Apache Jena—TDB	RDF4J	Virtuoso
Description	A RDF storage and query DBMS, shipped as an optional-use component of the Apache Jena framework	RDF4J is a Java framework for processing RDF data, supporting both memory-based and disk-based storage	Virtuoso is a multi-model hybrid-RDBMS that supports management of data represented as relational tables and/or property graphs
Primary database model	RDF store	RDF store	Graph DBMS Native XML DBMS Relational DBMS RDF store Search engine
Secondary database models			Document store
DB-engines ranking trend chart	Score 2.94 Rank #102 Overall #2 RDF stores	Score 0.57 Rank #220 Overall #9 RDF stores	Score 2.57 Rank #110 Overall #5 Graph DBMS #3 Native XML DBMS #54 Relational DBMS #3 RDF stores #10 Search engines
Website	jena.apache.org/documentation/tdb/index.html	RDF4j.org	virtuoso.openlinksw.com
Technical documentation	jena.apache.org/documentation/tdb/index.html	RDF4j.org/documentation	docs.openlinksw.com/virtuoso

(continued)

Table 6.1 (continued)

Name	Apache Jena—TDB	RDF4J	Virtuoso
Developer	Apache Software Foundation	Since 2016 officially forked into an Eclipse project, former developer was Aduna Software	OpenLink Software
Initial release	2000	2004	1998
Current release	3.15.0, May 2020		08.03.3316, March 2020
License	Open Source	Open Source	Open Source
Cloud-based only	No	No	No
DBaaS offerings (sponsored links)			
Implementation language	Java	Java	C
Server operating systems	All OS with a Java VM	Linux OS X Unix Windows	AIX FreeBSD HP-UX Linux OS X Solaris Windows
Data scheme	Yes	Yes	Yes
Typing	Yes	Yes	Yes
XML support			Yes
Secondary indexes	Yes	Yes	Yes
SQL	No	No	Yes

(continued)

Table 6.1 (continued)

Name	Apache Jena—TDB	RDF4J	Virtuoso
APIs and other access methods	Fuseki Jena RDF API RIO	Java API RIO Sail API SeRQL Sesame REST HTTP Protocol SPARQL	ADO.NET GeoSPARQ L HTTP API JDBC Jena RDF API ODBC OLE DB RDF4J API RESTful HTTP API Sesame REST HTTP Protocol SOAP webservices SPARQL 1.1 WebDAV XPath XQuery XSLT
Supported programming languages	Java	Java PHP Python	.Net C C# C++ Java JavaScript Perl PHP Python Ruby Visual Basic
Server-side scripts	Yes	Yes	Yes

(continued)

Table 6.1 (continued)

Name	Apache Jena—TDB	RDF4J	Virtuoso
Triggers	Yes	Yes	Yes
Partitioning methods	None	None	Yes
Replication methods	None	None	Chain, star, and bi-directional replication Multi-source replication Source-replica replication
MapReduce	No	No	Yes
Consistency concepts			Immediate Consistency
Foreign keys			Yes
Transaction concepts	ACID	ACID	ACID
Concurrency	Yes	Yes	Yes
Durability	Yes	Yes	Yes
In-memory capabilities			Yes
User concepts	Access control via Jena Security	No	Fine-grained Attribute-Based Access Control (ABAC) in addition to typical coarse-grained Role-Based Access Control (RBAC) according to SQL-standard. Pluggable authentication with supported standards (LDAP, Active Directory, Kerberos)

Our objective is to develop ontologies based on common vocabulary for describing each field of interest and then to feed them into the knowledge base of existing data. This is particularly important because until now a huge volume of digitised data stored in relational databases has been developed and their depreciation and their replacement by other technologies may prove not to be economic in terms of time and money.

Motivation for this work was provided by problems with the existing system when searching data from the database of the national population register used by municipalities of the country and the Ministry of Interior. Our interest is mainly focused on exploitation of the data held in existing relational databases to extract information based on dynamic queries. An example of such a query is the search for the nearest relatives of someone, excluding his/her immediate family. Existing database systems that support the management of public information are currently using relational databases. Searches in these databases are usually based on standardised and inflexible questions that have been designed beforehand and may not be personalised/individualised [22]. Their configuration and adaptation to new needs usually requires the assistance of managers/developers of the database. The process is time-consuming and uneconomical. Another serious problem is the compliance of local databases in municipalities with the central database in the Ministry of Interior. Although procedures of information synchronisation are provided by the central database of the Ministry, there is still a possibility of the existence of conflicting, incomplete or incorrect data in two or more public bodies.

The situation described above can be handled by technologies of the Semantic Web in which information is encoded in the RDF triples format and can be accessed by software-based questions in SparqL. This way, we can implement intelligent and dynamic questions that seek information based on specific properties and rules and not simple data control carried out in traditional relational databases. Our interest focuses on making full use of existing databases and data and not redesigning a system from scratch.

The number of Semantic Web tools is widening and there is already great interest in developing data conversion technologies from traditional forms into formats that can be managed by the technologies of the Semantic Web. These technologies are mainly based on programming languages like Java and Python. From the published work on the conversion of data from relational schemas in RDF triples, we indicatively present the following.

In [23], the authors analyse the various ontology construction approaches from relational databases and identify the advantages and disadvantages of these techniques, so that an enhanced and efficient approach can be proposed. The authors have performed detailed comparative analysis of various ontology construction techniques from relational databases (RDB) based on database schema analysis (metadata, cardinality restrictions and datatype information) and stored data (through data mining). The authors in [24] propose a particular process that integrates a design rationale approach with a triplification strategy. The process supports the reuse of standard RDF vocabularies recommended by W3C for publishing datasets and automatically collects the entire rationale behind the ontology design, using a specific vocabulary. In [25], the authors present advanced features of RDB2OWL mapping specification language that allows expressing RDB-to-RDF/OWL mappings in a concise and human comprehensible way. Relative with the previews is [26] which describes a practical approach about the transformation from relational database to RDF/OWL format. In [27], an Eclipse plug-in is introduced that supports the standard RDB to RDF Mapping Language (R2RML) to produce Direct Mappings in RDF. On the other hand, [28, 29] survey current techniques, tools and applications for mapping between Relational Databases (RDB) and the Resource Description Framework (RDF). In [30], a direct mapping is defined from relational data to RDF. This definition provides extension points for refinements within and outside of this document. This document has been reviewed by W3C Members, software developers and other W3C groups and interested parties and is endorsed by the Director as a W3C Recommendation. It is a stable document and may be used as reference material or cited in another document. W3C's role in making the Recommendation is to draw attention to the specification and to promote its widespread deployment. This enhances the functionality and interoperability of the Web.

Finally, in [31] links are provided to applications that are essential references of thematic data conversion to relational databases in RDF triples, among which the application D2RQ which we used in this study.

6.3.2 The Case of the National Municipal Registry

As already mentioned above our interest focuses on finding information on semantics using data from existing databases. This is because the retention of data in relational databases is the most widespread solution in most cases of public data regardless of their size and type. Their main disadvantage is that the stored data does not have the semantics of the original information so the questions submitted to databases are static and non-personalized to the needs of each user. In contrast, the transformation of this data into a format that supports the performance semantics will support the presentation of dynamic queries personalised to each user.

In this study, we examined the case of the National Population Registry kept by the Ministry of Interior and which is accessible by the country's municipalities. The system that supports the management of features of Registry has the architecture shown in Fig. 6.8.

According to this architecture a central database is maintained in the Ministry of Interior while databases are also maintained in the municipalities. It is the responsibility of the municipalities to synchronise local databases with the central database kept by the Ministry. More specifically, the Ministry of Interior hub also provides all the Web-Services to public bodies and citizens calling its services electronically via a web form that is installed in the portal of the Ministry of Interior. The server host has the ultimate oversight of the system and checks every kind of transaction of the citizens with the Population Register. The data of all the country's citizens are stored in a central database which communicates with the application of the Ministry

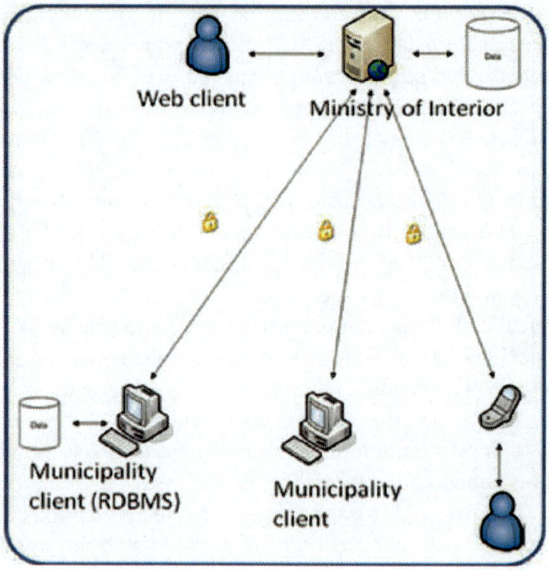

Fig. 6.8 The architecture of Municipality Registry RDBMS

of Interior server. The central database operates independently of the existence of other databases in the various municipalities-clients and does not require any kind of communication with these bases. It is the responsibility of the municipalities to update their information by periodically calling for updates and downloading service provided by the Ministry of Interior. In this scenario, all transactions start with the initiative of the customer who makes a request for a service from the Ministry of Interior and receives in response the results of the application.

Experience to date of the databases on specific areas of e-Government, such as the system introduced earlier, presents serious problems. Specifically, there are problems related to compliance with common vocabulary and common rules in the management of records in local bases. Thus, for example problems are commonly reported while seeking cases of double surnames or duplicate names of citizens. Also, using sometimes Latin and sometimes Greek characters or the abbreviations of people's names are commonly reported. For example instead of the full name "Constantine" someone appears as "Con/ne" while elsewhere the shorter version "Costas" is used. Such problems pose a risk of malfunction of the existing system.

Another challenge is the search for a citizen's family tree in order to find his/her closest relatives other than the direct relatives in his/her immediate family. The current system gives authorised users the ability to view only register status events that concern them and choose the register status documents which they can print either individually or en masse. According to these prints, the Municipal Registry employee can take the necessary actions in the Population Registers, such as deleting deceased from population registers, removing newly-weds from family register, doing neonatal entry in the family register, etc. [22]. Thus, in order to find the family relationships between citizens, it is necessary to present specially designed questions to existing databases to find possible links between members of different family registers to all the country's municipalities. The search is based on the entries and deletions from the family register which unfortunately does not record any family relationship except that of spouses and children. Thus, parallel tables must be maintained in order to follow changes in Population Registers. This, however, results in storing unnecessary information or duplicates. Conversely, as we shall see below, the process of finding immediate family members of any degree with the assistance of the ontology is treated with a single DL-query or a corresponding SparqL query on the knowledge base maintained by the ontology.

6.3.3 Triplification—Steps and Methodology

The most systematic and efficient processing of heterogeneous data may be achieved when such data is transformed in the same format, without losing information. Furthermore it is desired to assign semantics to the different data and to connect them. This increases the interoperability of the system data. An interesting and particularly easy to use data format that serves the above is the formatting to RDF triples. The conversion of data not in RDF ("triplification") is a particular process in a heterogeneous data integration system. Of particular interest is the automated production of RDF files of different formats such as .XLS, .CSV and .JSON, which are some of the forms of open data in public administration, but also from tabular data or data stored in relational databases.

The proposed conversion process is based on data stored in databases such as MySQL or other supporting SQL [32]. The output relates to the form of RDF/XML or Turtle, in order to reduce the file size produced in parallel with the complexity. Our interest concerns the extra yield of semantics to the data through the production of RDF triples. The basic tools of the methodology are shown in Fig. 6.9.

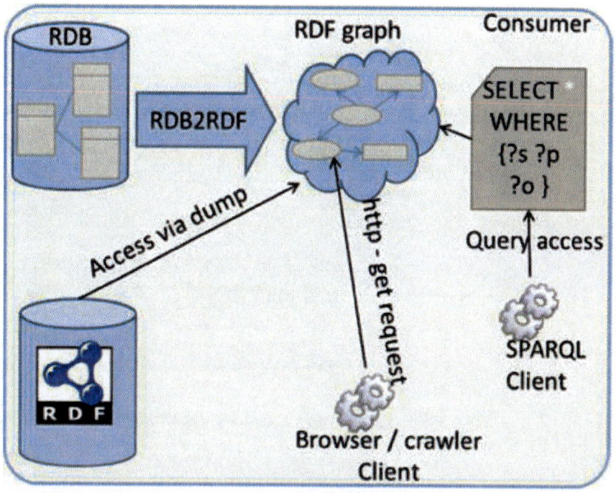

Fig. 6.9 The basics tools for RDB data triplification

While the basic steps of the process are:

1. Data analysis and Data Model
2. Relational DataBase Creation
3. RDF Graph Creation.

   a. Publishing tool selection
   b. Definition and assignment of URI's
   c. Concept schemes definition
   d. Defining vocabulary
   e. Data linking
   f. Access in data using a simple browser.

In the following paragraphs, we present the first steps in order to "build" a simulation of the database of the Ministry of Interior and the corresponding ontology and subsequently to transform relational data into semantically linked data. A diagram of the conversion process is shown in Fig. 6.10.

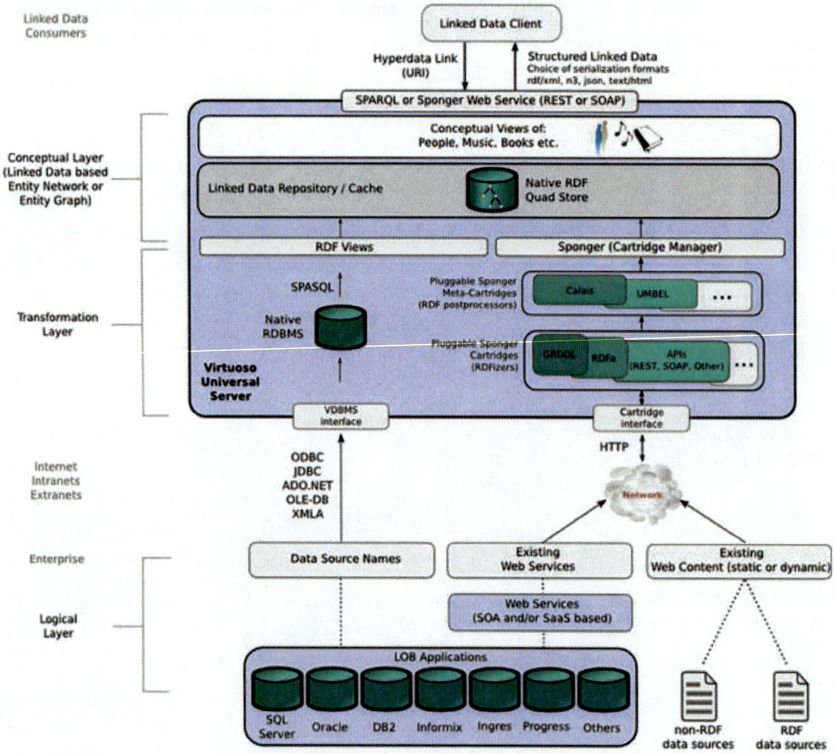

Fig. 6.10 Linked data generation via RDF middleware (RDFizers) [33]

6.3.3.1 Creating a Relational Database

In the environment of MySQL (in particular, the MySQL Workbench tool), we created a simulation of the database of the national population register with a model containing the key elements of the register, so that we could evaluate the process of conversion to RDF triples. The key elements were grouped into respective tables which present the following:

- Citizen, for the essential elements of each citizen
- Municipality, for the details of each municipality of the country
- Family_register, for the elements of family-owned register of each citizen
- Family_register_alterations, to monitor changes in family register, and
- Marriage, for storing information relating to marriages of citizens.

In the graphic environment of the tool we created the appropriate relations that appear in Fig. 6.11 in order to correlate the data of the tables above. Then, we fed the database with "synthetic" data and extracted the base data into a dump file.

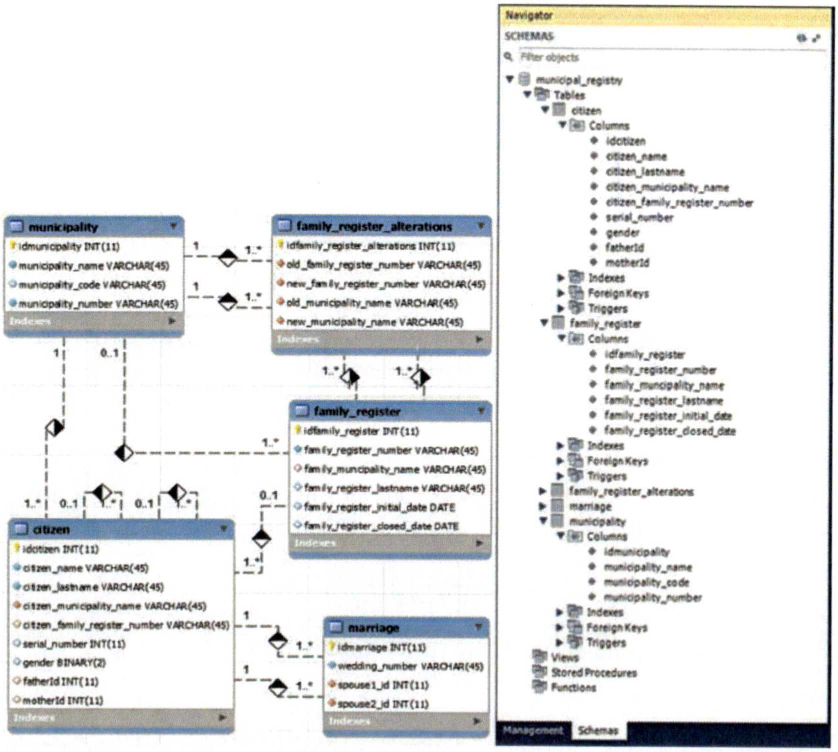

Fig. 6.11 The EER diagram of RDB municipal registry data

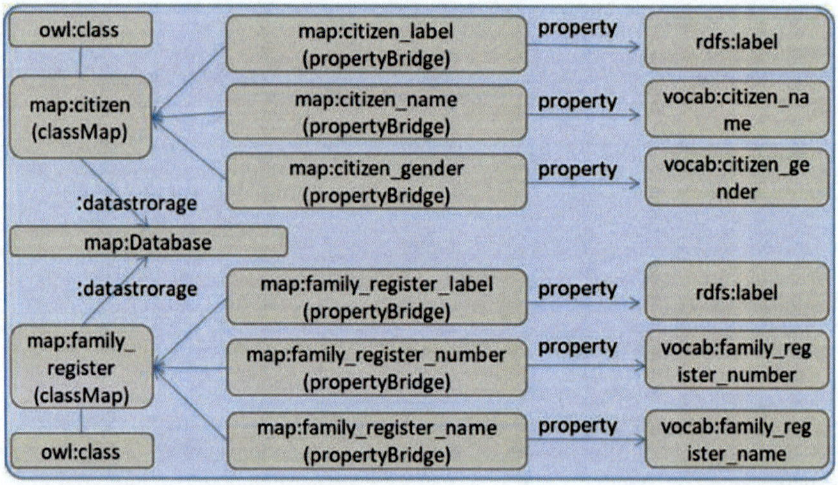

Fig. 6.12 The structure of the D2RQ map—the case of the National Municipality Registry

6.3.3.2 Triplification Using D2R Server

To recover the data from our relational database and convert them to RDF triples, we used the package D2RQ-0.8.1, which works directly with MySQL data and provides among other tools the generate-mapping and D2R-server. The generate-mapping tool creates a D2RQ mapping file by analysing the schema of an existing database. A part of the schema we used in this case is presented in the figure below (Fig. 6.12).

This mapping file, called the default mapping, maps each table into a new RDFS class that is based on the table name while it maps each column into a property based on the column name. This mapping file can be used as-is or can be customised. The command that we used to activate the tool was provided by the command line as: `<generate-mapping -o outputFileName.ttl -u root -p SqlServercode jdbc:mysql:///MySQLdatabaseName>`. Then, we activated the D2R-server with the command `<d2r-server -p 8080 outputFileName.ttl>` (Fig. 6.13) and with the help of a simple browser we went at <http://localhost:8080/snorql>. Results of the process are shown in Figs. 6.14 and 6.15.

6.3.3.3 Design—Implementation of the Ontology of the Population Register

To address the difficulties mentioned above by approaching relational databases in the Protégé 4.3 environment in the context of this work, we created an ontology in order to demonstrate the retrieval of information concerning family relationships between citizens through appropriate DL queries. Specifically, we created a basic

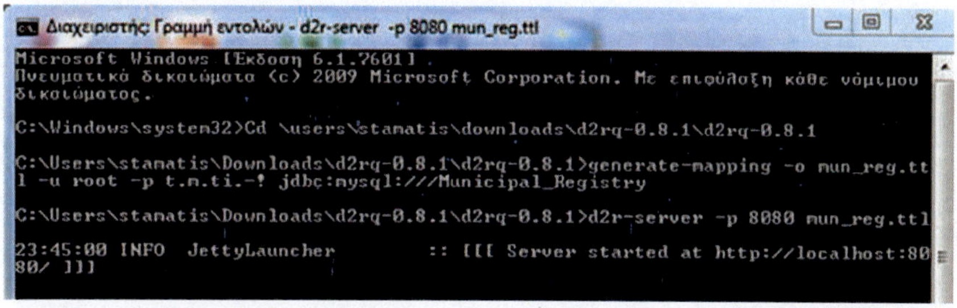

Fig. 6.13 Calling the d2r mapping tool

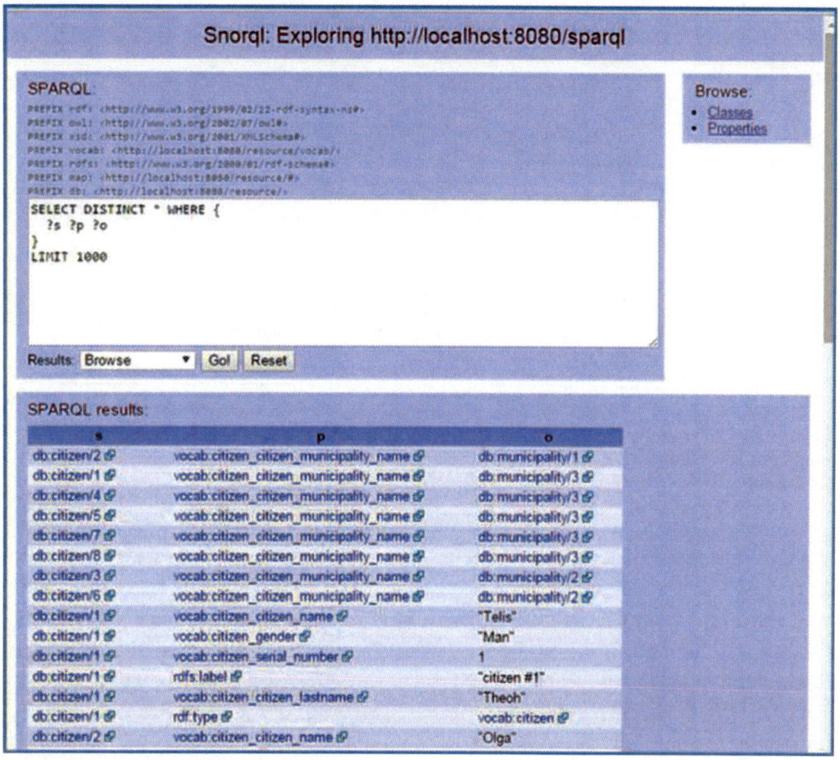

Fig. 6.14 The export of D2R Server in out browser within SparqL query

ontology (as a simulation of an integrated ontology to be presented in the following work) with three basic classes—similar to those of the relational database. We then provided our ontology with the necessary SWRL rules shown in Fig. 6.16 and the properties shown in Fig. 6.17. We see that a very simple rule of natural language that expresses "if X has Y as a parent and Y is a sister of Z, then Z is an aunt of X" has been translated in SWRL to declare when someone "Z is an aunt of X" with the

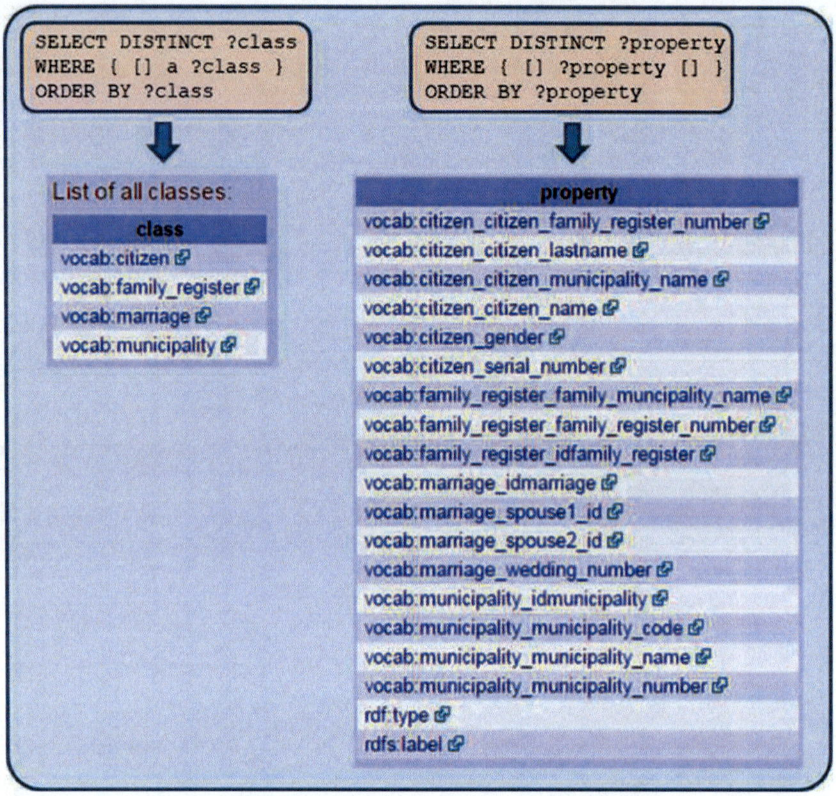

Fig. 6.15 List of all classes and their properties created via the D2R Server

statement "`hasParent (? x,? y), hasSister (? y,? z) - > hasAunt (? x,? z)`". Similar are the other statements of ontology. As shown at the tab of the properties of the ontology, the family relationships that interest us in this work but also the fact of the existence of family register for one person, have been implemented in the form of properties with suitable domains and sets of values.

In order to test the correctness and consistency of our ontology, we fed it manually with synthetic data and then submitted the ontology to be evaluated by the activation of reasoners available. We also submitted specific questions to the DLQuery Tab. As an example we refer to the query in natural language "who has MariaTheo as an aunt?". In the environment of the DL query, we implemented the statement "`Citizen and hasAunt value MariaTheo`" and its results are shown in Fig. 6.18. Clearly, the ontology is relevant.

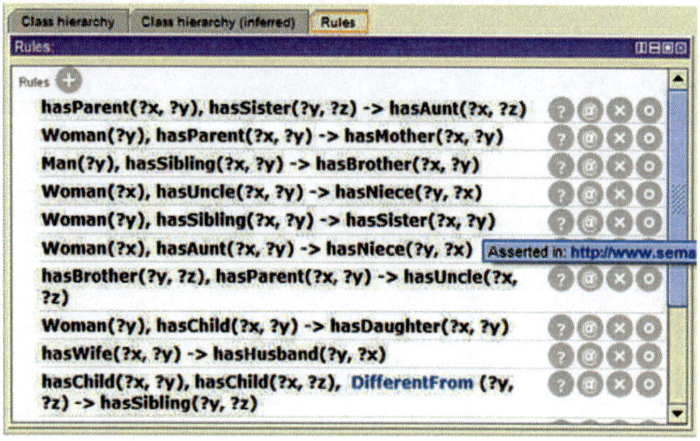

Fig. 6.16 SWRL rules for Municipality Registry ontology in Protégé 4.3

Fig. 6.17 Object properties of Municipality Registry Ontology in Protégé 4.3

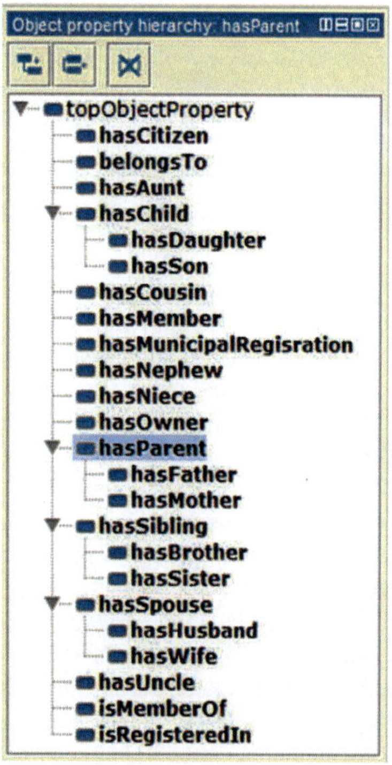

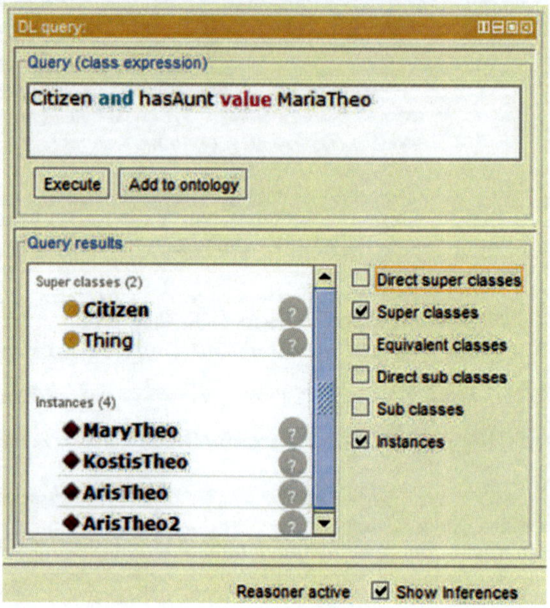

Fig. 6.18 Results from a DL query in the Minicipality Registry Ontology in Protégé 4.3

6.3.4 Conclusions and Future Work

The two main approaches to data management, namely relational databases and Semantic Web technologies, have been implemented simultaneously and shown to complimentarily operate in managing public data.

The two technologies manage data differently. Relational databases manage the data simply by checking their content while returning data instances (snapshots of their content) to the user, but do not recognize or support their semantics. Another difference relates to the knowledge that can be generated from the stored data. Knowledge acquisition from a relational database or a knowledge base is often associated via the submission of relevant questions through appropriate interfaces. The answer to a question on the Semantic Web, is not just data control (as in databases), but it also requires the implementation of complex considerations that run a huge number of classes and instances. The classes and instances that are stored in an ontology are linked via stated rules and constraints that describe specific events and roles during transfer from natural language. For this reason, it is particularly important to combine the facts and concepts that are declared in an ontology so conclusions about the correctness or otherwise of another event can be drawn or new unexpected events can be produced. This process is described as reasoning and is one of the main differences between databases and knowledge bases of the Semantic Web. Of particular interest is the sub-process of reasoning concerning the production of new events in the ontology, referred to as inferencing, which are added to existing events for future use.

But the fact that there is a large amount of data on the web stored in relational databases which may be obsolete, induces a high cost in redesigning a new system from scratch. It is therefore very important to produce semantic data using the already stored data in whatever format they are. To achieve the enrichment of the Semantic Web with real data, one should first map the data in relational schemas into the elements of each ontology.

Our future plans include the development of an application to automatically convert the output conditions of the RDF file D2R-server in conditions arising from mutually agreed vocabulary which will be described by a corresponding ontology and which will be associated with knowledge base which will be supplied automatically from existing relational databases. This work is currently underway and will be presented elsewhere in the near future.

6.4 RDF Serialisation from JSON Data—The Case of JSON Data in Diavgeia.gov.gr

6.4.1 Introduction

In recent years more and more governments of developed countries including the United States and Britain, participate in the establishment and promotion of open government through the initiative Open Government Partnership (OGP) [34]. This initiative displays a new system of governance, which is essential to strengthening governments' accountability and transparency in the management of public resources by public officials. One component of this system is associated with the publication of governments' and public institutions' acts. The publication is done by using special permits in the logic of free distribution and sharing. The basic tool of this component is open data. The opening of the data produced, maintained and shared by public sector bodies is achieved through the adoption of common standards of publication and modelling such as the JSON, CSV, etc. This operation is related to the new trend of the Internet called linked open data. The connection of open data, requires the use of specific tools for the Semantic Web and a specific sequence of steps, which have already been described by the mastermind of linked open data, Tim Berners Lee [35]. This connection allows the crossing of a graph of the data published on the web in an automated manner by means of suitable applications and browsers.

The next step after selecting the dataset and theoretical modelling in the form of triples "subject-property-object" according to the RDF model, is followed by the practical part of the procedure which involves the production of RDF triples. In this step, we implement the infrastructure for the effective connection of open data with appropriate programming procedures and tools of the web as the URI. The conversion depends mainly on the initial form of published or stored open data, for example in JSON or other tabular data e.g. CSV, XLS or relevant databases. Due to the enormous volume of data and the plurality of public sector bodies that publish data, it is perhaps the most important part of the login process and for this reason, it also constitutes the subject of this part of the Book.

The next steps of the connection process of open data include adding links to other data sets, in order not to present the RDF data as isolated graphs and succeed their availability to users through queries host-points.

The aim of this part is to highlight the advantages of formatting data in JSON. The JSON data have the advantage of being easily understood by the users/programmers and easily accessed by software applications that can be developed in various programming languages.

Two key issues are being examined in this work. One concerns the JSON data connection with the use of JSON-LD and the other one concerns the development of a methodology for the production of structured information in the form of RDF triples by open public data found in JSON format so as to achieve the subsequent connection to other similar datasets. To achieve this purpose we used the Google Refine and its extensions as well as the already-developed vocabularies and ontologies on public administration and open government. As a case study, we examined the recovery of open public data from the Greek repository [36] of public data by means of an application developed in the Python language, and then we examined the triples RDF production by the development of relevant algorithms.

More specifically, in the second part of the work, we report, in summary, to works relating to the subject of this work, while the third part is the presentation of data modelling on JSON and JSON-LD forms. In the fourth part we refer to data conversion using relevant algorithms from the form of JSON into RDF format while we also present a specific application in a case study of open data from the Greek repository of open public data "Diavgeia" [36]. In the fifth part we respectively present the RDF data generation tool from JSON-LD. The last part presents the conclusions from the comparison of the two cases and also references to future work are made.

The issue of the management of huge volumes of data found on the web in the form of open data, can be treated effectively with the Semantic Web and its technologies. The main goal of the Semantic Web is to enable the computer systems to identify, and share linked data as easily as possible. The machines are very efficient at finding data, but cannot perform the semantics that the user attributes. A large part of the scientific community works in this direction, making publications on related patterns and methodologies on crossing, data conversion, semantic enrichment and development of vocabularies and ontologies.

Our previous work also deals with similar topics. Specifically, in [37] we had presented the technologies used for the implementation of open data, like for example the RDF triples, the RDF formats, the languages OWL and SPARQL. We had also presented the ontology for public data and we had created a relevant knowledge base in Protégé. This was considered as a critical point, since the scope of public information is quite complicated as information is coming from various organisations using different datasets and there is no commonly accepted dictionary for meanings and associations between them. In [38] we had also presented the conversion of tabular data in RDF format, targeting the semantic annotation of data stored in open data files such as those found in XLS or CSV format. For this reason, we dealt with the first main part of the relevant procedure regarding the clean-up of the source file and the production of RDF triples thereafter. For its implementation we used the RDF extension of Google Refine and for the visualisation of the results we used Protégé.

The latest W3C Editor's draft in [39], defines the procedures, rules and algorithms to be applied when converting tabular data such as CSV and JSON into RDF. Tabular data may be complemented with metadata annotations that describe its structure, the meaning of its content and how it may form part of a collection of interrelated tabular data. This document specifies the effect of this metadata on the resulting RDF.

Author in [40] describes a syntax that can be used to write RDF graphs for presentation and editing purposes. This paper proposes an alternative means of serialising RDF triples using JSON. RDF/JSON is a textual syntax for RDF that allows RDF graphs to be completely written in a compact form. This means the format is quick and easy to read and write. This new serialisation format is developed in response to a demand from a wide range of users, who do not have an XML, Notation3 and Turtle background.

Finally, [41] presents a RESTful Web Service based on the HTTP protocol that translates between different serialisations. In addition to its core functionality, author's proposed solution provides features to accommodate frequent needs of Semantic Web developers, namely a straightforward user interface with copy-to-clipboard functionality, syntax highlighting, persistent URI links for easy sharing, cool URI patterns, and content negotiation using respective HTTP headers.

6.4.2 JSON Versus JSON-LD

6.4.2.1 The JSON Format

JSON (JavaScript Object Notation) is a lightweight data-interchange format. On one hand, it is easy for users/programmers to read and write and on the other it is easy for machines to parse and generate. It is based on a subset of the JavaScript Programming Language, Standard ECMA-262 3rd Edition—December 1999. Facebook, Twitter, Google are among them who use JSON to restore and manipulate their data [42]. JSON is a text format that is completely language independent but uses conventions that are familiar to programmers of the C-family of languages, including C, C++, C#, Java, JavaScript, Perl, Python, and many others. These properties make JSON an ideal data-interchange language. For example, Python provides particular JSON parsers and through them, the programmer is able to handle the JSON data as the Python dictionaries.

JSON is built on two structures:

- A collection of name/value pairs: In various languages, this is realised as an object, record, struct, dictionary, hash table, keyed list, or associative array.
- An ordered list of values: In most languages, this is realised as an array, vector, list, or sequence [43].

These are universal data structures. Virtually all modern programming languages support them in one form or another. It makes sense that a data format that is interchangeable with programming languages also be based on these structures.

In JSON, they take on these forms:

- An object is an unordered set of name/value pairs.
- An object begins with "{" (left brace) and ends with "}" (right brace).
- Each name is followed by ":" (colon) and the name/value pairs are separated by "," (comma).

As an example we present the result from the web page of the Greek public administration repository of open data in JSON format in Fig. 6.19. The particular result is provided by the relevant API of "Diavgeia Project" in www.diavgeia.gov.gr.

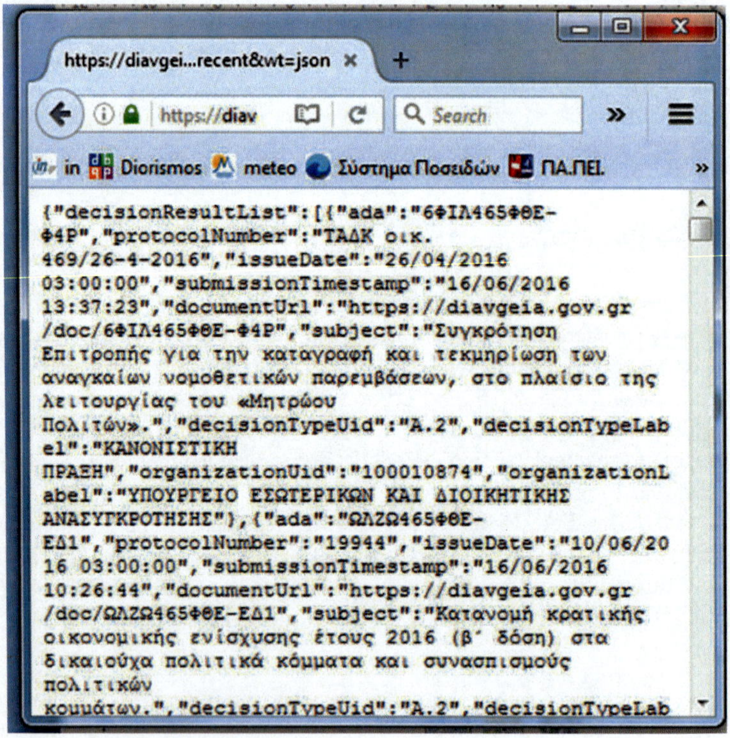

Fig. 6.19 The JSON open data output from "Diavgeia" project

Fig. 6.20 The proposed Python code for parsing JSON data from the URL of diavgeia.gov.gr and the relevant results

We can parse these data according the code that we propose in Python in the Fig. 6.20.

6.4.2.2 The JSON-LD Notation

The data notation with JSON-LD is a faster way to deliver annotating information in search engines more clearly and more simply. By its use, it is possible to connect simple JSON data to form linked data. It is a notating binding data format that allows easy integration of data in a web programming language tag. Unlike microdata, the JSON-LD runs in background and is not visible and therefore does not need configuration for the output in web pages. With this notation we exploit the fact that the search engines can detect the code of web pages easily and thus be able to understand the websites' data more quickly. The results are shown in Fig. 6.21.

Among the advantages of using JSON-LD we refer the following:

- Complexity—allows depiction of complex information in a piece of code, which is easy to understand by search engines.
- Based on variables—To enter new data in the future, simply add a new corresponding variable.
- It saves a lot of time—The optical notation is easy to fail and is difficult to deal with if there is no relevant experience with CSS.

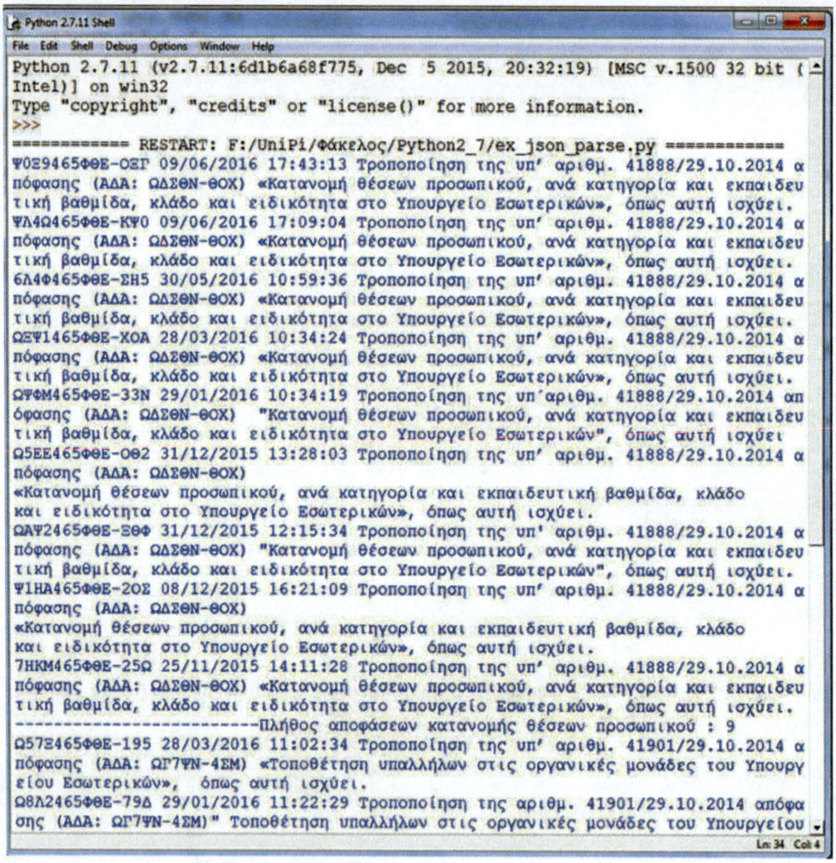

Fig. 6.21 Results from Python IDLE

- It is easy to include the comments and the reviews of your website—keeping the style for the current ratings, without having to make changes in the CSS file.

For example, previous JSON section from the output of diavgeia.gov.gr by the addition of new features such as "`@context`", "`@type`", "`@id`" etc., the JSON-LD could be made as follows in Fig. 6.22, obtaining thereby semantics shown in Fig. 6.23 lacking from the simple JSON format.

```
{
"@context": "http://schema.org",
"doc_id":{"ada":"ΩΠΟΙ465ΦΘΕ-Τ4Γ","protocolNumber":"2501/10/76-α",
"issueDate":"27/02/2015 02:00:00","submissionTimestamp":"27/02/2015
13:40:58", "@type":"@id"},
"documentUrl":"https://diavgeia.gov.gr/doc/ΩΠΟΙ465ΦΘΕ-Τ4Γ",
"subject":"Προσωρινή ρύθμιση κυκλοφορίας στη Ν.Ε.Ο. Αντιρρίου -
Ιωαννίνων","decisionTypeUid":"A.2","decisionTypeLabel":"ΚΑΝΟΝΙΣΤΙΚΗ
ΠΡΑΞΗ","organizationUid":"100010874",
"OrganizationName":{
"@type":"legalName",
"organizationLabel":"ΥΠΟΥΡΓΕΙΟ ΕΣΩΤΕΡΙΚΩΝ ΚΑΙ ΔΙΟΙΚΗΤΙΚΗΣ
ΑΝΑΣΥΓΚΡΟΤΗΣΗΣ"}
}
```

Fig. 6.22 The JSON-LD notation in the JSON data retrieved from the "Diagveia Program [36]" page

```
{
  "@graph": [
    {
      "@id": "_:b0",
      "http://schema.org/OrganizationName": {
        "@id": "_:b1"
      },
      "http://schema.org/decisionTypeLabel": "ΚΑΝΟΝΙΣΤΙΚΗ ΠΡΑΞΗ",
      "http://schema.org/decisionTypeUid": "A.2",
      "http://schema.org/doc_id": {
        "@id": "_:b2"
      },
      "http://schema.org/documentUrl": "https://diavgeia.gov.gr/doc/ΩΠΟΙ465ΦΘΕ-Τ4Γ",
      "http://schema.org/organizationUid": "100010874",
      "http://schema.org/subject": "Προσωρινή ρύθμιση κυκλοφορίας στη Ν.Ε.Ο. Αντιρρίου - Ιωαννίνων"
    },
    {
      "@id": "_:b1",
      "@type": "http://schema.org/legalName",
      "http://schema.org/organizationLabel": "ΥΠΟΥΡΓΕΙΟ ΕΣΩΤΕΡΙΚΩΝ ΚΑΙ ΔΙΟΙΚΗΤΙΚΗΣ ΑΝΑΣΥΓΚΡΟΤΗΣΗΣ"
    },
    {
      "@id": "_:b2",
      "@type": "@id",
      "http://schema.org/ada": "ΩΠΟΙ465ΦΘΕ-Τ4Γ",
      "http://schema.org/issueDate": "27/02/2015 02:00:00",
      "http://schema.org/protocolNumber": "2501/10/ 76-α",
      "http://schema.org/submissionTimestamp": "27/02/2015 13:40:58"
    }
  ]
}
```

Fig. 6.23 The graph of open data with semantic annotation in JSON-LD

In particular, with the notation "`@type`": "`@id`" in "`doc_id`" we give the sense of the document's identity, i.e. pairs of simple JSON information:

```
"ada":"ΩΠΟΙ465ΦΘΕ-4Γ",
"protocolNumber":"2501/10/ 76-α",
"issueDate":"27/02/201502:00:00",
"submissionTimestamp":"27/02/2015 13:40:58",
```

were tagged semantically and obtained the information that the original JSON data had.

At this point, we note that the notation with JSON LD was done manually in the original file retrieved from diavgeia.gov.gr.

6.4.3 *Producing RDF Triples out of JSON Data*

Both the size and differentiation of data published as open data require a more systematic and efficient processing. This can be achieved when the data is transformed in the same format, but without losing the information they carry. We also wish the semantics performance in the various data since it is not provided with the process of the data opening. The most popular and interesting data format that serves the above is the RDF format. This conversion poses a particular process in a heterogeneous data integration system, while until now the bibliographic search showed the need for further specialisation, especially in the case of JSON data. As mentioned JSON is the most popular web data format as it is one of the forms of open data of public administration in Greece. A key advantage of JSON data is the ease in parsing by machines using code in different programming languages. This gave us the possibility of direct semantic performance to this data without the use of JSON-LD. For this purpose, on the one hand we developed a relevant algorithm and on the other hand we used the Google Refine, which we presented in our relevant previous work [37].

6.4.3.1 Conversion Algorithm from JSON Data in RDF Triples

Regarding the conversion of JSON data in RDF triples, we propose the algorithm shown in Fig. 6.24.

Initially we retrieve the data from its source and then we temporarily store it in a variable. Since JSON data are similar to the key-value pairs of Python dictionaries, this variable is the corresponding type. Each key-value pair gives respectively the predicates and objects of RDF triples according to the general shape shown in Fig. 6.25.

```
1. JSON-Data retrieval from source
2. Parse JSON-Data
3. store data in data_variable
   #e.g. Python dictionary variable
4. for member in data_variable :
   a. subject<-member
   b. subjectURI<-datasetURI+"/member_value/"+member
   c. while key_value_exist :
      i. property(i)<-member(key(i))
     ii. propertyURI(i)<- vocabularyNameSpaceURI+
         member(key(i))
    iii. object(i)<-member(value(i))
     iv. objectURI<-subjectURI+member(key(i))_value+"/"+
         member(value(i))
```

Fig. 6.24 Conversion algorithm from JSON data in RDF triples

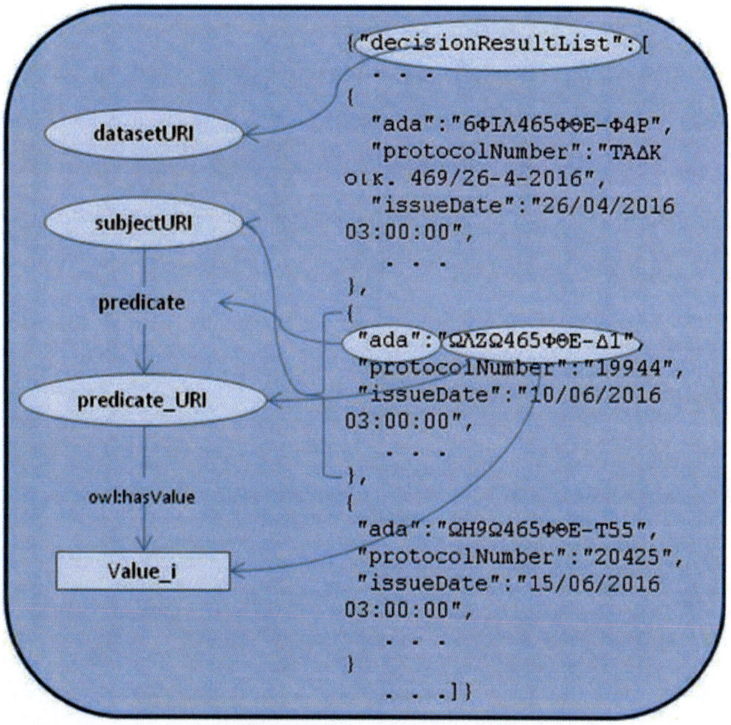

Fig. 6.25 The general shape for converting JSON data in RDF triples

As part of this work we codified this algorithm, using the Python core libraries: JSON, RDFlib as shown in Fig. 6.26. The results of running code are shown in Fig. 6.27.

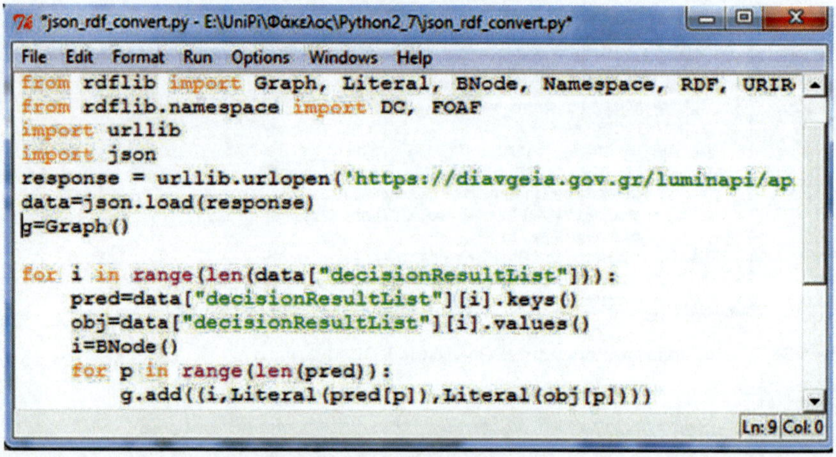

Fig. 6.26 Python code for convert JSON data in RDF triples

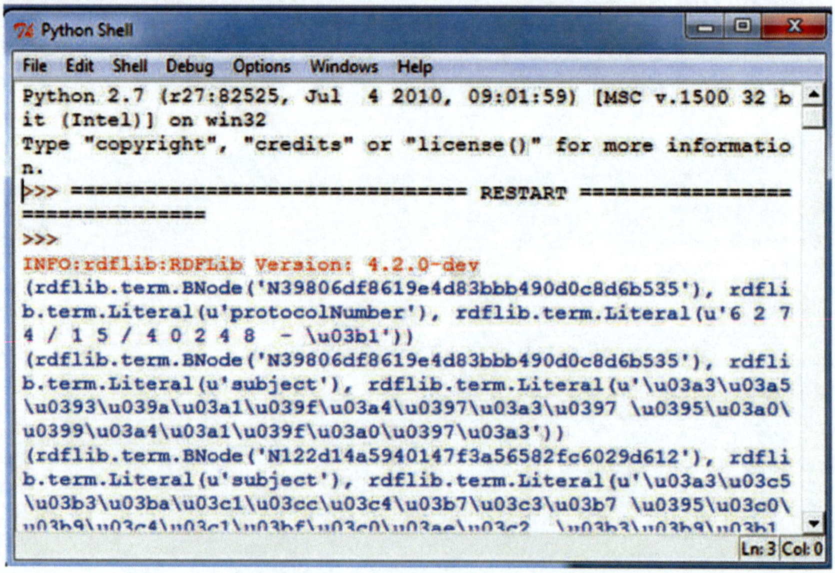

Fig. 6.27 Results from the code we propose in IDLE of Python

At this point, we note that in this code, we create the RDF triples and store them in the corresponding graph, but without making any connection with associated diagrams or namespaces.

6.4.3.2 Production of RDF Triples from JSON Data with Google Refine

In order to compare the algorithm we used in terms of effectiveness, we used the Google Refine. This tool we had used in our work in [37], so we had the relevant knowledge for its operation. In the Google Refine environment the procedure was done manually via the relevant API.

Step 1.

Initially we entered from the website of "diavgeia project" [36], a file on the metadata of administrative action posted in .JSON form.

Step 2.

Then, we made a necessary clean-up of data, to monitor any recurrence of contents.

Step 3.

Then, using the RDF extension, the declaration of prefixes was done in a specially designed tab managed by the user (Fig. 6.28). Here we note the statement by the prefix "pad" on the introduction of the vocabulary of the ontology of e-Government that we developed in our previous work [36]. The prefix refers to the internal address of our local computer. At the same time we declare additional and other popular vocabularies like DublicCore (dc/dcterms) and DataCube (qb).

Step 4.

The statement in the application of subjects, predicates, objects was made with the logic we described in the above algorithm. Figure 6.29 shows the state after the statements.

Step 5.

As a final step in this processing, we export using the corresponding option of the tool, the data in RDF/XML format. Part of the relevant output is shown in Fig. 6.30.

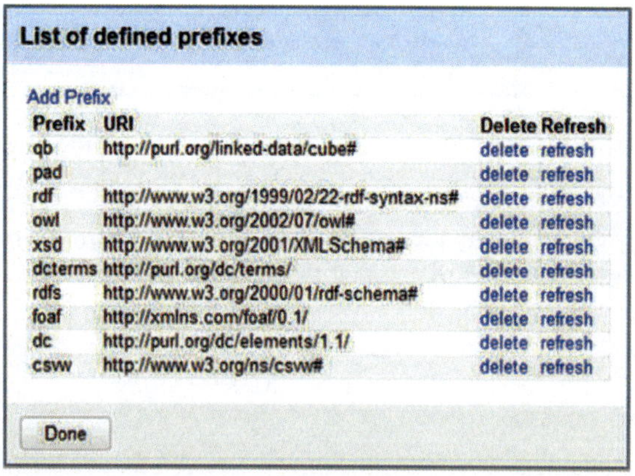

Fig. 6.28 Defining the necessary prefixes in Google Refine

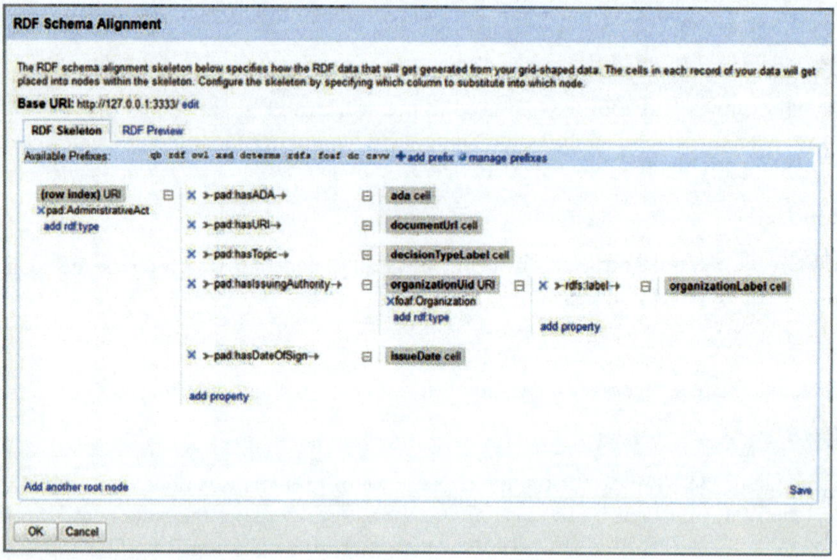

Fig. 6.29 The RDF triples definition in RDF extension

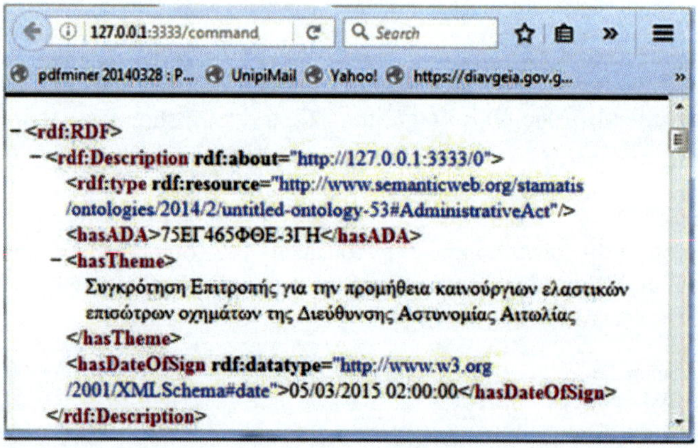

Fig. 6.30 Part from the output of the Google Refine RDF extension

As shown by the proceedings, the results through the application, are confirming the correctness of the algorithm. In addition, with the help of the application it is possible to publish the results of the transformation, on the web.

6.4.3.3 Producing RDF Triples out of JSON-LD

For completeness of this work, we searched Internet works or applications developed about data management for the production of RDF triples. As a result of this search, we present an application on the RDF serialisation data from various formats. In particular, it is about the application called "RDF translator". This service is provided by the E-Business and Web Science Research Group at Universität der Bundeswehr München [44]. This application can receive data either directly via the relevant API or through the introduction of relevant URLs of data. Among the input data JSON-LD data is included, but not JSON without semantic markup.

In order to control and utilise the semantic JSON data which we transformed into JSON-LD, we entered the relevant data related to the API of the application, and following the simple instructions, the application produced the corresponding RDF/XML code (Figs. 6.31 and 6.32).

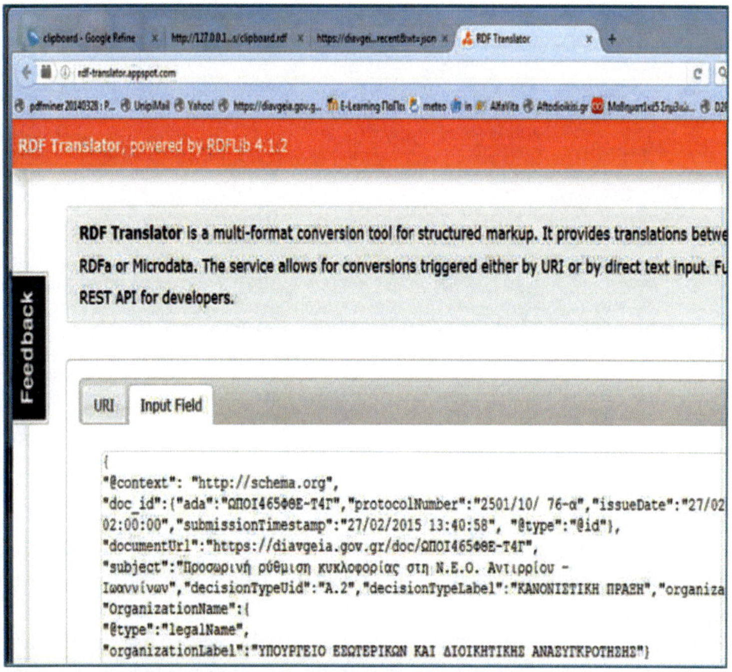

Fig. 6.31 The input JSON-LD data in RDF translator

Fig. 6.32 The RDF triples in the feedback of RDF translator

6.4.4 Conclusion—Future Work

Governments worldwide, within the Open government, have already published a huge number of raw data thus supporting open data. This data does not have semantic markup, or is supported by common vocabularies. These problems can be addressed with the technologies of the Semantic Web and their conversion into linked open data. In this work, we proposed the production of RDF triples from JSON data. The JSON data are the most common type of open data while RDF triples are the infrastructure to publish on the web linked open data. The linked open data we believe is the infrastructure for the development of new intelligent applications that use both citizens and the public administration itself for the management of public data and generating information with relevant knowledge.

From the study which preceded we can highlight the following conclusions:

- The JSON data due to their structure, is to be easily labelled with programming tools semantically and then converted to RDF triples.
- The basic tool used, serves to produce RDF triples by file types, such as JSON and .XLS files, but with a semi-automatic way. It requires the user to create a corresponding RDF schema with the help of which carries out the production of RDF triples. The statements of predicates and the subject and object as resources derive very easily from the functional environment of RDF extension.
- Highlighting data with JSON-LD produces linked data in an easier way, but devalues the huge crowd of already published data which do not bear semantic features.

6.5 Publication of Linked Data: The Case of the Open Ontology for Open Government

In order to publish the data related to the functioning of the public administration and in particular the ontologies that we developed and presented in the previous chapters, as linked data we used tools such as Protégé, Apache Jena Fuseki Server, Virtuoso OpenLink, RDF4J WorkBench and Pubby Server. Here we present the methodology, process and results from the publication of this data as well as a critical overview of these tools.

6.5.1 Create SparqL EndPoint and Publish Linked Open Data Using Fuseki and Pubby Server

Step 1: Generate linked data through ontology

The primary step for publishing the linked data is to produce an RDF or corresponding technology file (in the format: turtle, N3, etc.) so that the data has the desired format of triples according to the logic we have analysed in the previous chapters. In this work and as described above, through Protege we developed specific owl ontologies related to e-Government as well as other areas of public administration such as official statistics and the internal control system. Following the development of the ontologies, the production of the corresponding RDF files took place through the special tool provided by Protégé. In particular, at the stage of developing the ontology and producing its RDF file, we note the following points.

- As a general rule, when creating an ontology, the Ontology IRI must be declared, which is used as a prefix for naming and reporting all the entities (classes, properties, individuals) that are created. This statement may take the default Protégé value or an alternative according to the user's will.
- In the case presented in this work and since our purpose is to publish our ontologies as linked open data so that they can answer semantic queries through Fuseki and

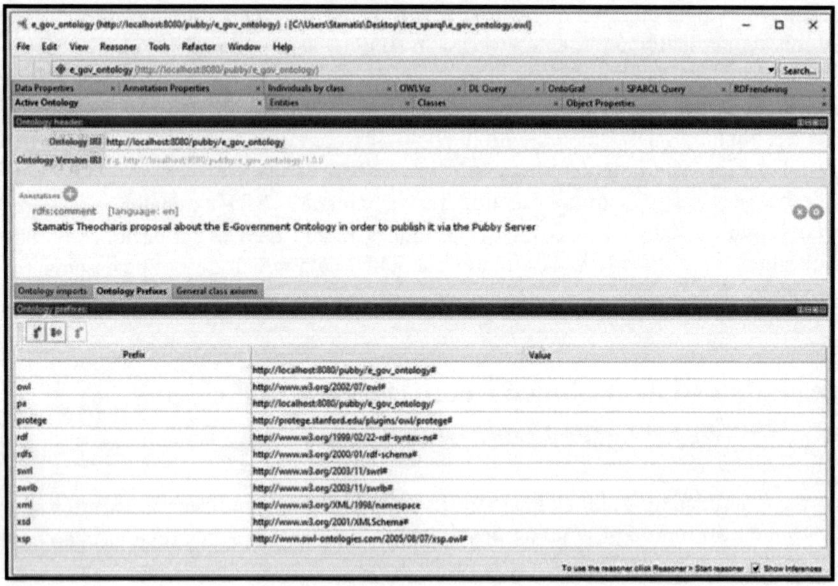

Fig. 6.33 The starting page of Protégé and the ontology IRI and prefixes declaration

can be accessed through the Pubby Server, the appropriate Ontology IRI statement in the relevant field provided by Protégé is important. This statement as well as the relevant prefixes are shown in Fig. 6.33 which is also the home page when designing the ontology in Protégé. Note that in this statement, the address to which the server "listens" (<...`localhost:8080/pubby`>) is used as a prefix, through which the data is accessible from the Internet, as well as the name of the relevant dataset in which we will "upload" the ontology in RDF file format.

- The creation of the RDF file could be done "manually" through any text editor or even from applications that produce relevant files as exports/conversions from primary data in various formats such as .xml, .xls, .csv, JSON, etc. Such programs are Open Refine, OpenLink Virtuoso, D2R through corresponding extensions provided and presented in previous paragraphs. The choice of the appropriate tool depends on the case of the primary data that we have at our disposal each time. In the case we present, we consider the choice of Protégé to be appropriate since our data came from the ontology we developed "from scratch". Through this program we were given the opportunity not only to define the entities that we considered as key and the properties that connect them but also specific instances, thus creating a knowledge base in RDF/XML format.
- A snapshot of the relevant RDF file produced by Protégé concerning "e-Government ontology" is shown in Fig. 6.34. In this file we can remark the use of the Ontology IRI as a part of the prefixes in the ontology concepts and properties.

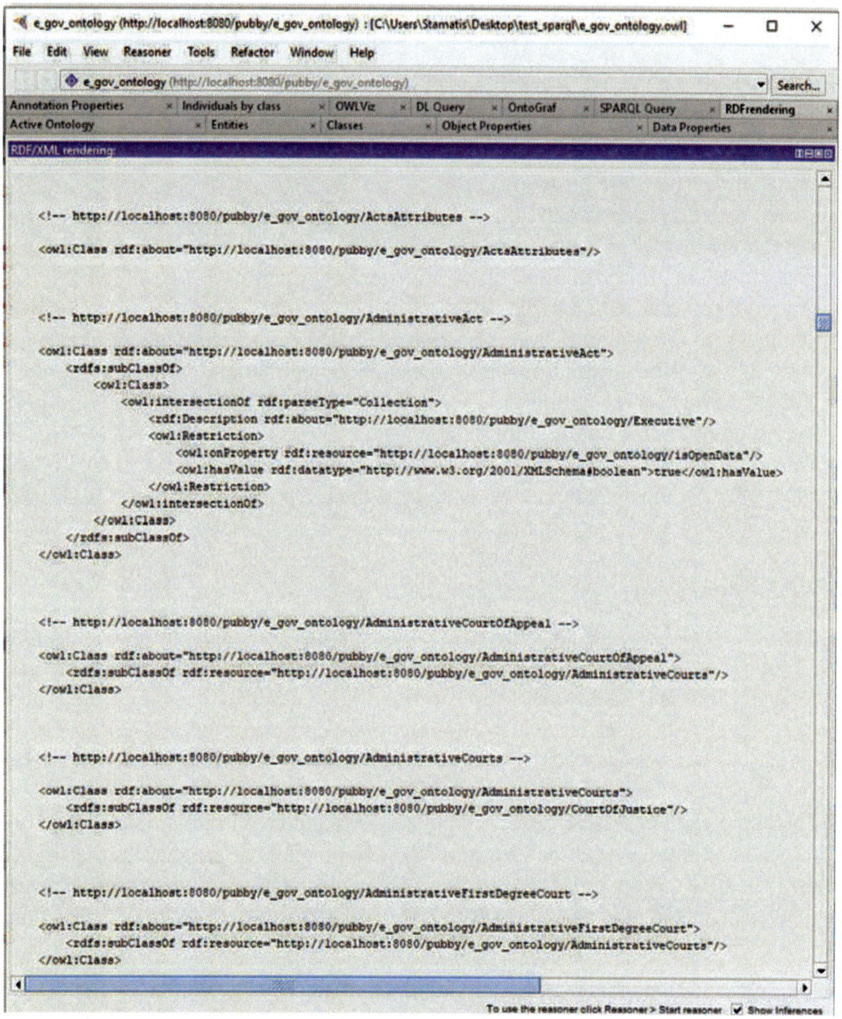

Fig. 6.34 Part of the RDF rendering file from the "e-Government ontology" in Protege

Step 2: Upload data in SparqL Server—SparqL Endpoint

In order to make our linked data open in a system of semantic queries at the level of the Semantic Web, we chose to upload our data from our local computer to Apache Jena Fuseki (Fuseki) which is a SparqL Server. In other words, through this server it is possible to develop a SparqL Endpoint in order to put the data in semantic queries (queries) either on a local computer level or from remote machines. For this choice we relied mainly on the following key features of Fuseki:

- Fuseki can run as an operating system service, as a Java web application (WAR file), and as a standalone server as well.
- It provides security (using Apache Shiro) and has a user interface for server monitoring and administration.
- Fuseki provides the SPARQL 1.1 protocols for query and update as well as the SPARQL Graph Store protocol.
- Fuseki is tightly integrated with TDB to provide a robust, transactional persistent storage layer, and incorporates Jena text query. It can be used to provide the protocol engine for other RDF query and storage systems.
- From our experience and based on the results in the interaction of various SparqL Servers with the Pubby Server, we consider Fuseki as the simplest system with limited capabilities, which, however, are fully functional and fully interoperable with the Pubby server. In addition, for its full operation, no special configurations were required in the configuration file.
- The ability to operate as a standalone server is considered to give more flexibility when opening the data and at the same time the ability to use different machines for publishing and querying data.

Step 2.1. Fuseki Start

In our case we chose the development of Fuseki as a standalone server on our local computer, with commitment/opening of the local port:3030 for its interaction via the Internet. The start of the server is done either by executing the batch file provided by the relevant distribution or by the command `<fuseki-server [--mem | --loc=DIR] [[--update] /NAME]>`, where /NAME is the dataset publishing name at this server in URI space.

The home page of Fuseki on which on the one hand its correct operation and on the other hand the datasets that we have already created can be seen in the figure below (Fig. 6.35). From the Fuseki homepage, the user can enter SparqL queries for each dataset created, add additional data and display the corresponding information via the appropriately programmed buttons.

Step 2.1. Create SPARQL EndPoint and publish linked data to Fuseki

As in the case of CKAN, in Fuseki the data belong to a specific dataset as its elements. For this reason, in order to upload our data, we follow the following steps.

Step 2.1.1. Create a new dataset

In case there is no dataset available or we want to create a new one, this is easily done through the menu of the link "manage datasets" and then from the tab "add new dataset". In our case, we created the dataset "e_gov" (Fig. 6.36). The relevant process requires the selection of the type of storage of the dataset (in memory/persistent) as well as the name through which it is called from its environment.

Step 2.1.2. Upload data

Once the corresponding dataset is created, the upload can be done in two ways.

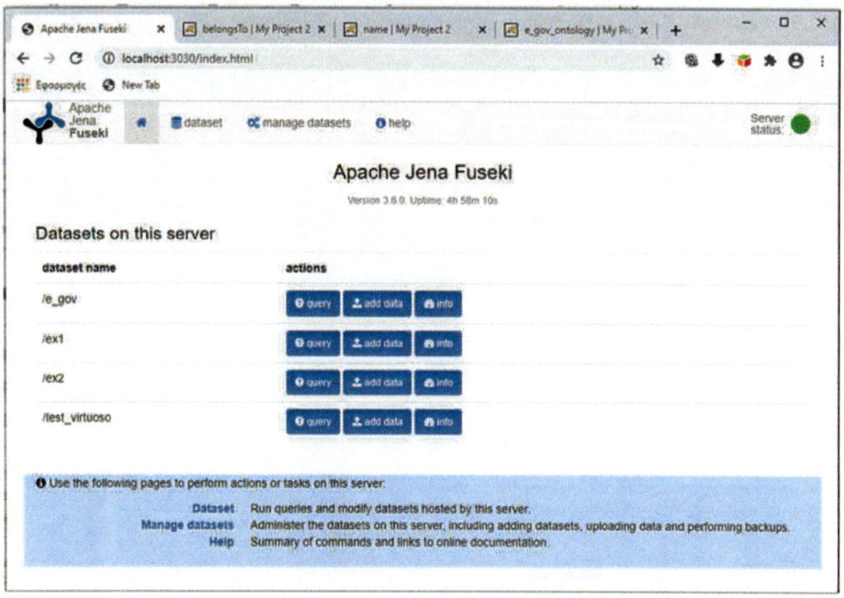

Fig. 6.35 Apache Jena Fuseki starting page

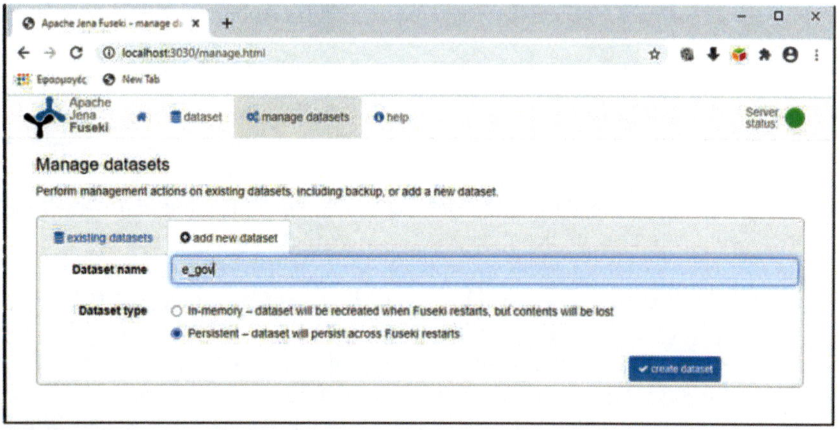

Fig. 6.36 The "add new dataset" page of Fuseki—creating a new dataset

A. In case the data is in the local computer (in the form of files that support the format of triples), in the machine that operates Fuseki, this is done by selecting the link "manage dataset" and then selecting the button "upload data" corresponding to the desired dataset. Alternatively, from the Fuseki home page we can select the "Add Data" button that corresponds to the desired dataset. In this process, we

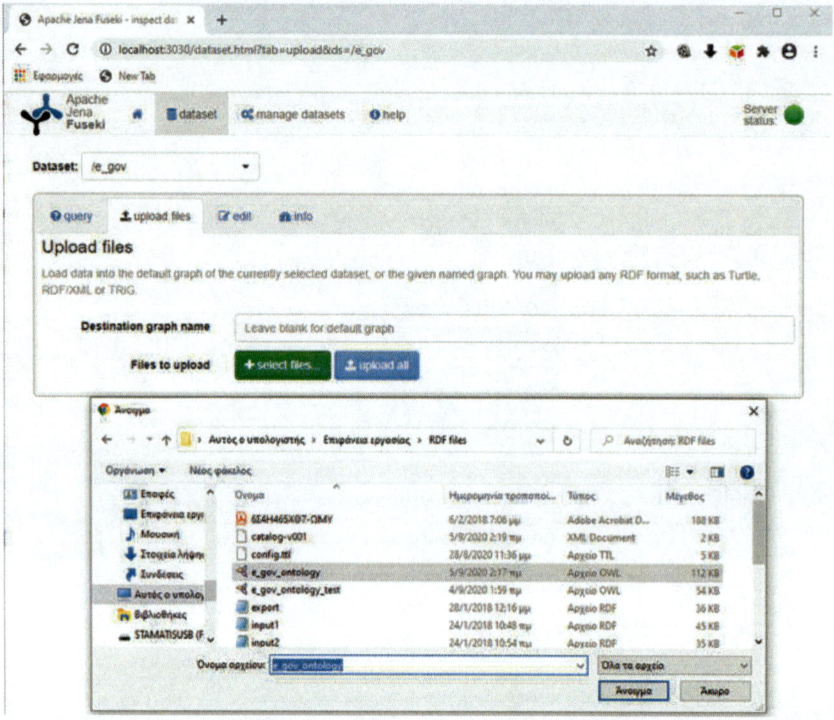

Fig. 6.37 Uploading data from local machine into the "e-gov" dataset in fuseki

uploaded the .owl file corresponding to the "e-Government ontology" we created with Protégé (Fig. 6.37).

The process is completed after selecting the "upload all" link and if the data meets the standards of the linked data. After that, a corresponding message for the success of the process is displayed which includes the number of triples identified-retrieved from the specific file. In the case of our ontology, 1375 triples were identified and retrieved (Fig. 6.38).

B. In case the data is in a remote location, we follow the procedure by submitting a SparqL query type "update". For example, from the fuseki home page we select the button "query" that corresponds to the desired dataset and in the page that opens we submit the query "`LOAD <http: // my_dataset / mydata>`" by properly configuring SPARQL ENDPOINT such as the address `<http://localhost:3030/e_gov/update>`.

We note the following points.

- In a specific dataset we can upload one or more files of different types, such as RDF, Turtle, N3, OWL, etc.

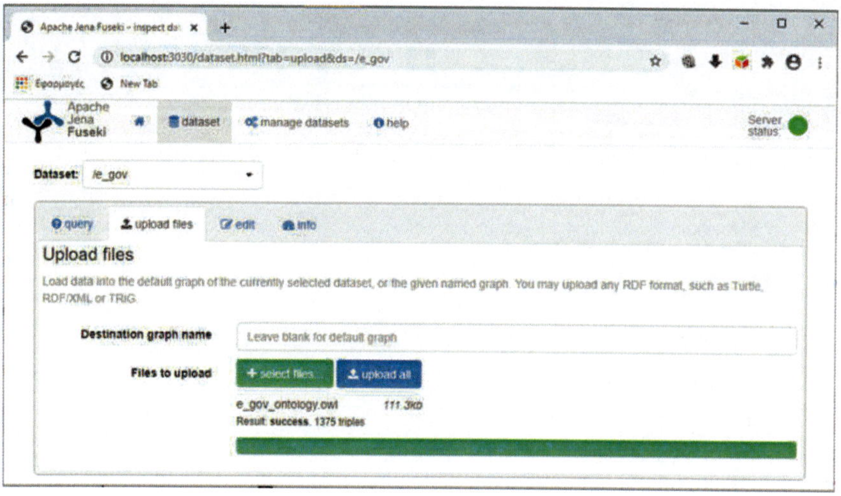

Fig. 6.38 Uploading e-Government ontology in Apache Jenna Fuseki server

- It is possible for the data we upload to belong either to a specific desired graph or to the default. In case no specific graph is declared, as much data as we upload, it is added to the default graph.
- Tests performed on remote datasets show that data retrieval is not always possible, which is influenced by the system that hosts the specific dataset on the remote computer. In general, Fuseki seems to interact satisfactorily with SparqL endpoints served by Virtuoso OpenLink.
- The possibility of queries on the data either those we have uploaded or others on remote machines is given through a corresponding page provided through the browser and the "query" button based on the SPARQL 1.1 protocol. The return to the most general query we can ask in the dataset "e_gov" in which we have already uploaded the relevant file of step 1 in order to return all possible triplets of this file is shown in Fig. 6.39. In the case of the e_gov dataset 1375 triplets are retrieved and returned.
- In any case of query submission or other action on the datasets, relevant POST or GET requests are submitted to the server, as shown in the Fig. 6.40.

Step 3. Linked data interface to SPARQL endpoints

An interface tool for accessing and presenting semantic information extracted from a SparqL EndPoint to datasets stored on it is Pubby. As we mentioned before, Pubby provides a simple HTML view of the stored data and manages content negotiation between the different representations. Users can navigate between the HTML views of the linked resources as linked data via the web interface. These views show all the semantic information of the resources, including their attributes and relationships

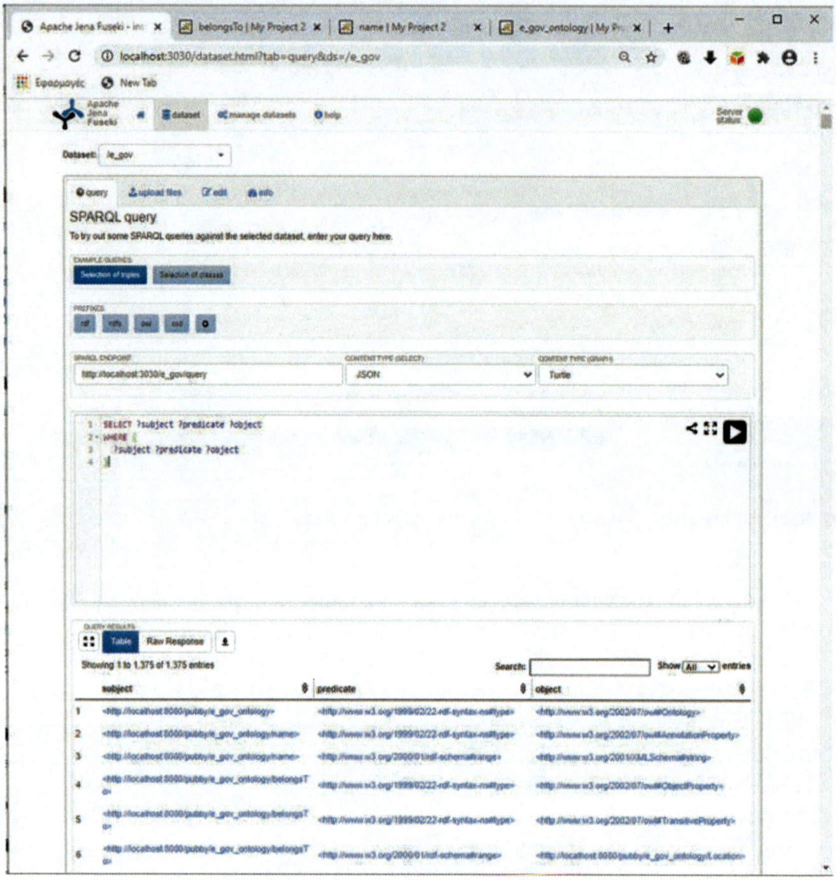

Fig. 6.39 Overall SparqL query in the "e-Government ontology" dataset

with other resources. In other words, Pubby makes it easy to turn a SPARQL endpoint into a linked data server.

In its basic function, every Pubby instance that is hosted on a local server is connected to a specific SparqL Endpoint (local or remote) that is declared in the configuration file. This link redirects the URLs received by Pubby and returns the corresponding SparqL EndPoint URLs. For example, when Pubby receives a `<http://mySparqLEndPoint/mydataset/data/file1>` URL for the `<file1>` resource, it redirects and returns its description to a corresponding URL of the format `<http://localhost:8080/mydataset/page/data/file1>` which is the address that appears in the html browser we use. This is because as we have said the resources in an RDF file are mapped to specific URIs and through them are recognized and return the information given to them.

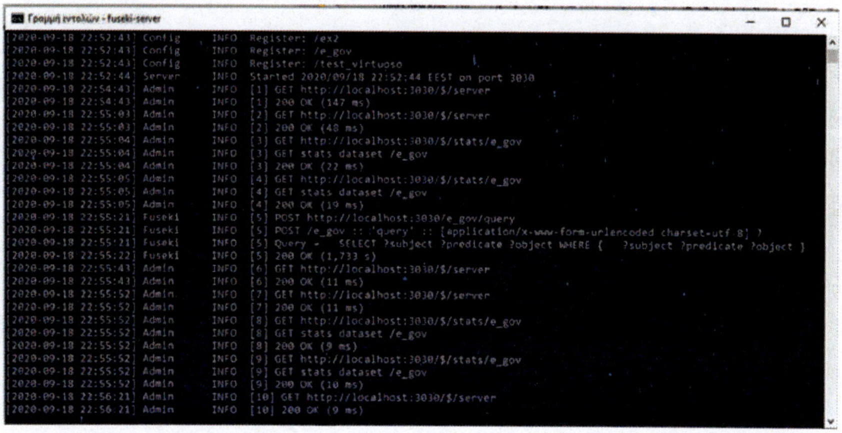

Fig. 6.40 POST and GET requests to Fuseki server

In order to use Pubby, it is necessary to install and service it through a local server such as Apache Tomcat or Jetty, depending on the operating system used by the local computer.

1. In this case, we installed Pubby on a Windows system and used Tomcat as the server. We note the special importance of the correct configuration of the Pubby configuration file (config.ttl or config.n3 depending on the pubby version) so that the dereferencing of the ontology resources works properly.
2. For the needs of our research, we first installed the necessary tools on the same computer: Fuseki, Tomcat and Pubby.
3. We chose the standalone mode of the Fuseki SparqL EndPoint to listen to the local port: 3030 and the Pubby service through the Apache Tomcat listening to the local port: 8080. For this function the Pubby configuration file is configured as shown in Appendix A.2.
4. Then we started the tools through the terminal (Fig. 6.41) and through the browser we connected to Tomcat and through the administration page (Fig. 6.42) we selected the Pubby service.
5. Based on the configuration we performed Pubby returned the expected homepage with information about the main resource of our ontology. As shown in the Fig. 6.43, during the background operation of the pubby, a "`get request`" is sent to fuseki server at the URL `<http://localhost:3030/e_gov>` (since we installed fuseki in our local machine). This request includes a DESCRIBE query related to the resource `<http://localhost:8080/pubby/e_gov_ontology>`. Fuseki's response is redirected to pubby's URL `<http://localhost:8080/pubby/page/e_gov_ontology>`.

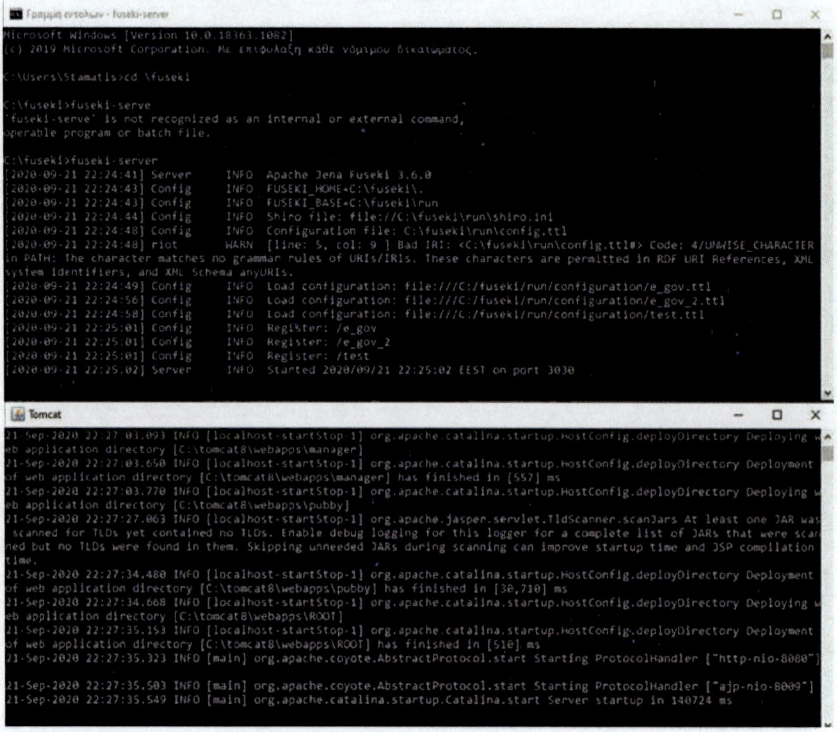

Fig. 6.41 Apache Jena Fuseki and Apache Tomcat startup screens

In order to check the dereferencing for the other entities of our ontology, through the browser we sent relevant requests to fuseki and received corresponding answers as expected, shown in the following images (Figs. 6.44 and 6.45).

Step 4: Evaluate Fuseki SparqL EndPoint

In order to evaluate the functionality of SparqL EndPoint that we developed with fuseki we followed two different methods.

A. The first method involves accessing SparqL EndPoint via Pubby from as an external-remote computer. Although the function of SparqL EndPoint seems to work properly when the tools communicate with each other on the local computer, it is very useful to have its image as seen from an external machine. For this purpose we installed Apache Tomcat and Pubby on another computer on a corresponding Windows system and properly configured the pubby configuration file. The essential difference of the configuration concerns the declaration of the URL of SparqL EndPoint, where in this case the declared value was the absolute address of the machine that hosted it. The relevant file is included in Appendix A.3.

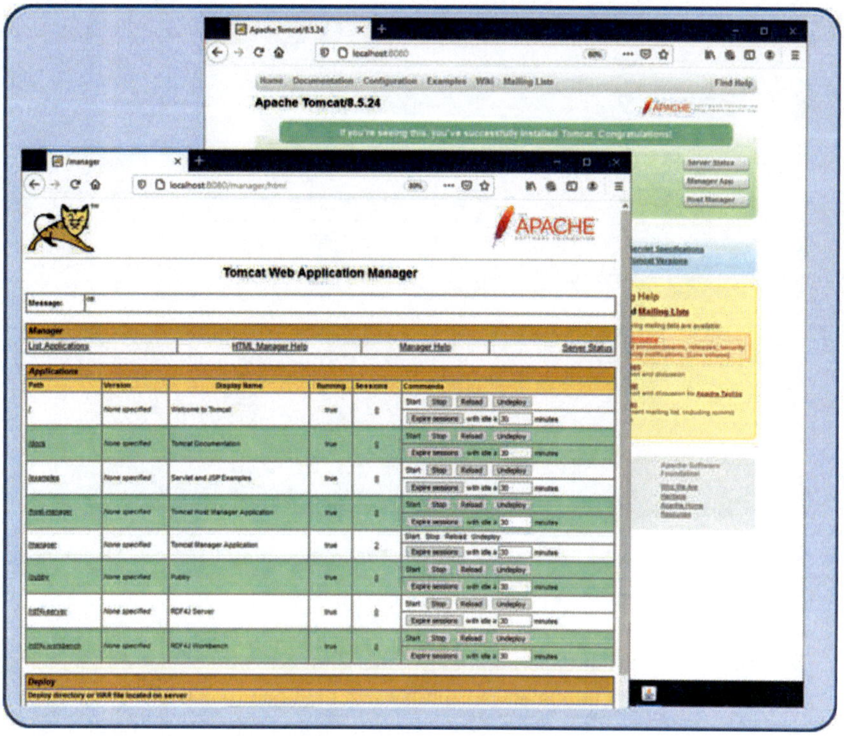

Fig. 6.42 Tomcat starting (in the background) and manager pages in our local machine

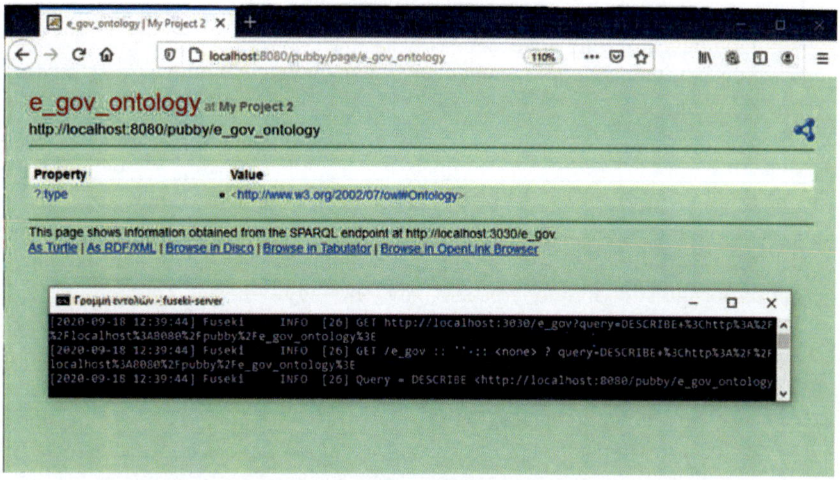

Fig. 6.43 Pubby starting page and the relevant “describe” query in Fuseki server

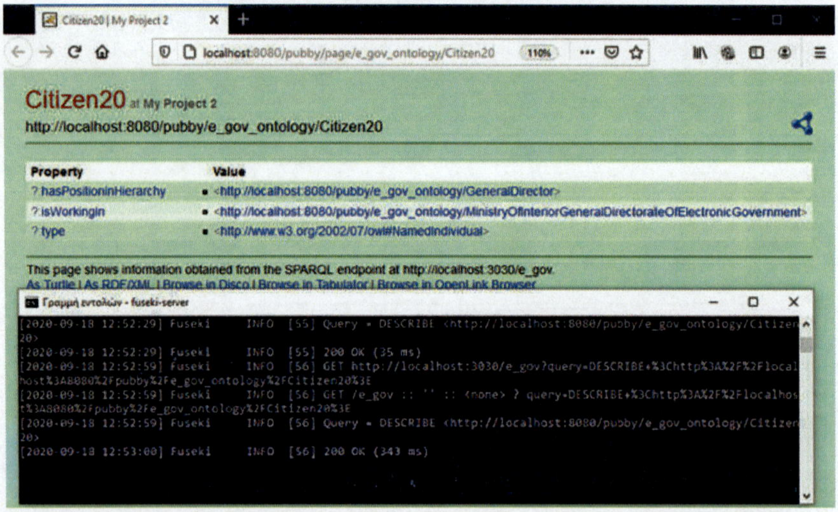

Fig. 6.44 A "describe query" to fuseki via a relevant get request from pubby about the resource "Citizen20" of the e-Government ontology

In order to first test the operation of SparqL EndPoint as well as dereferencing through Pubby, we connected to the remote machine that hosted Fuseki and then submitted specific queries. The answers were as expected while during the relevant procedure no dysfunction occurred. Then on the local machine we first connected to the Apache Tomcat Application Manager and through it to Pubby. Pubby homepage is shown in Fig. 6.46.

From the Fuseki terminal we can easily see the correct operation first of dereferencing and then the operation of SparqL EndPoint. We notice that SparqL EndPoint running at the URL `<http://5.54.141.78:3030/e_gov_2>` (Fuseki server) receives a "GET" request regarding the submission of a "DESCRIBE" query for the resource with URI `<http://5.54.141.78:8080/pubby/e_gov_ontology2>`.

Executing the query at the remote point returns the state shown in the Fig. 6.46 (in the part of command line) by redirecting to the URL <http://localhost:8080/pubby/page/e_gov_ontology2> hosted by Pubby.

Then we browsed through the browser, submitted additional requests for other resources of our ontology and got corresponding answers. Indicative results from the relevant navigation in our ontology through the remote SparqL EndPoint are shown in the following images (Figs. 6.47, 6.48, 6.49, 6.50 and 6.51) from where we see its correct operation.

B. The second method involves the use of tools such as Vapour Validator [45]. As mentioned above Vapour provides an objective and automated way to verify the proper functioning of a SparqL EndPoint or a point where linked data has been posted. In other words, it can serve specific data formats such as RDF/XML,

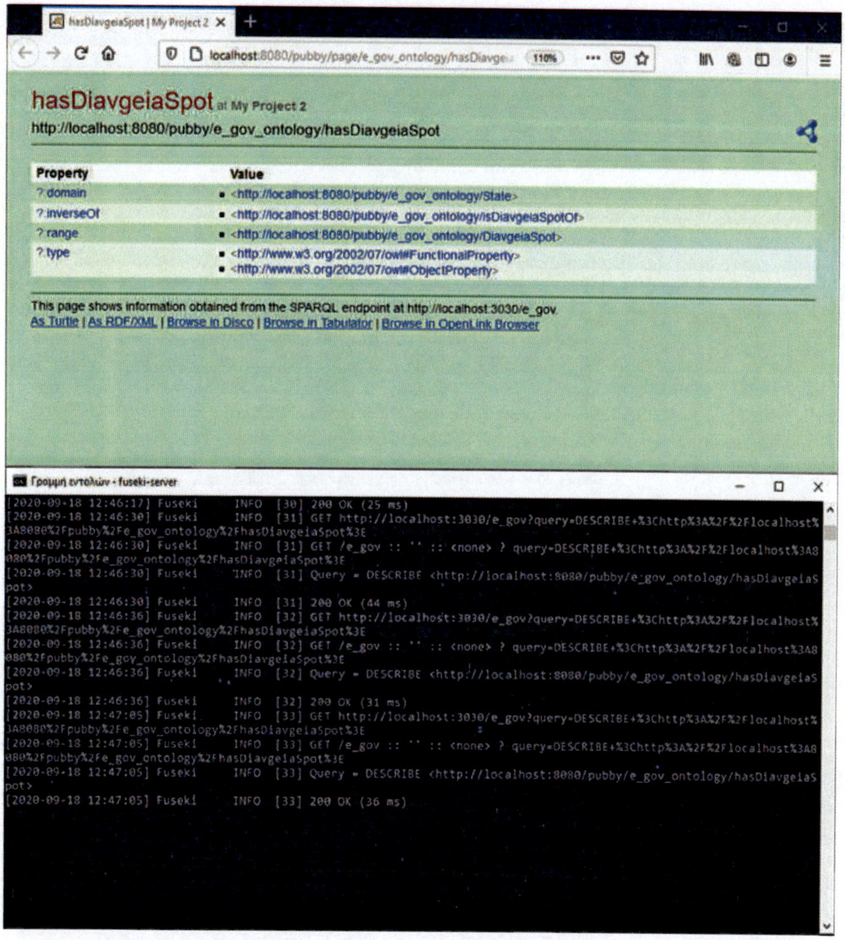

Fig. 6.45 A "describe query" to fuseki via a relevant get request from pubby about the property "hasDiavgeiaSpot" of the e-Government ontology

JSON, etc. On the homepage of the tool we gave in order to check, the URL through which we have access to the SparqL EndPoint that we developed, with a search parameter data of the RDF/XML format, i.e. the format that we followed in the development of our ontology. The answer through this tool is positive as shown in the relevant summary report (Fig. 6.52).

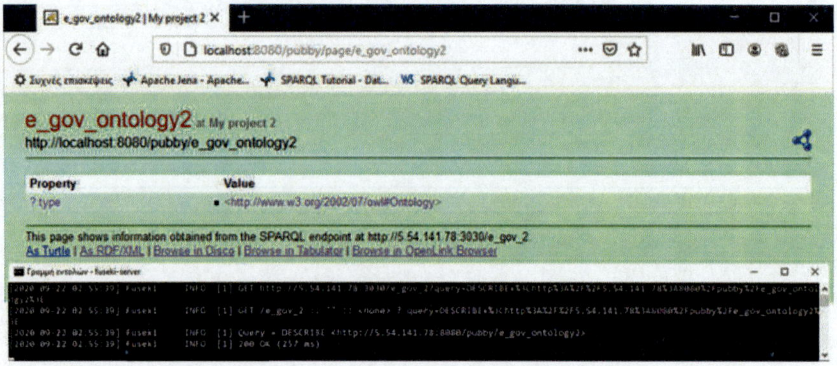

Fig. 6.46 Pubby starting page—the case of the remote SparqL EndPoint

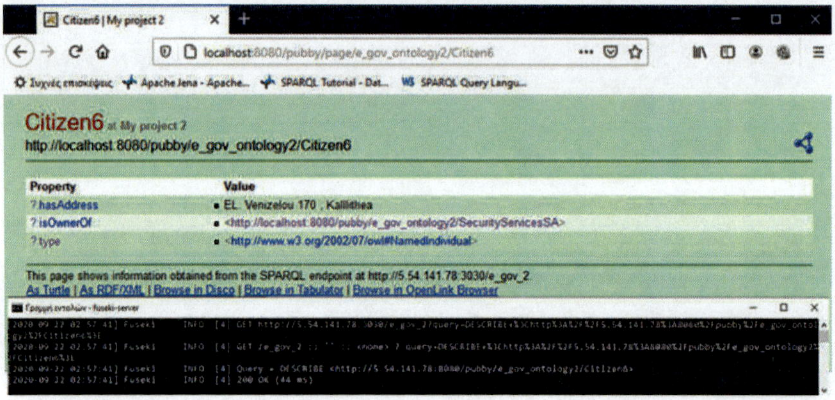

Fig. 6.47 The resource "Citizen6" description via the Pubby server

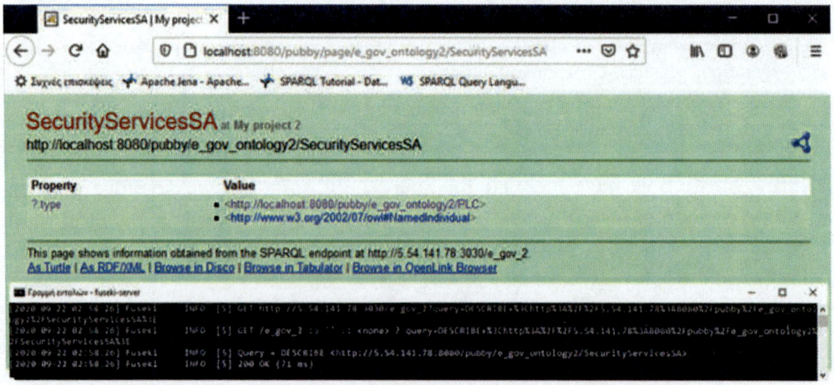

Fig. 6.48 The resource "SecurityServicesSA" description via the Pubby server

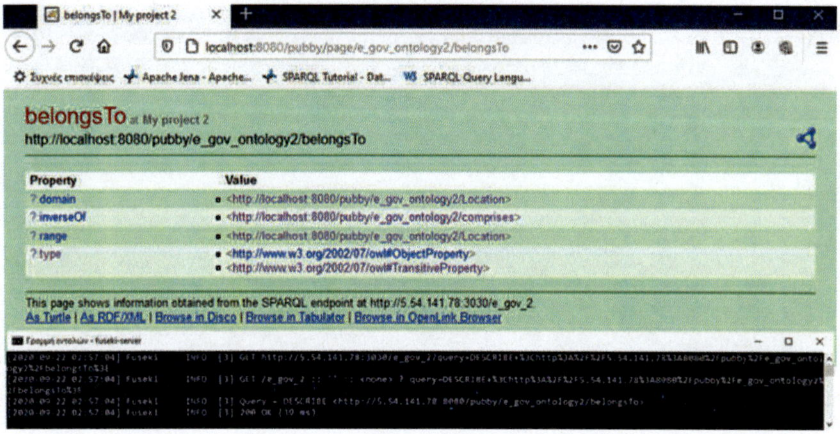

Fig. 6.49 The resourse—property "belongsTo" description via the Pubby server

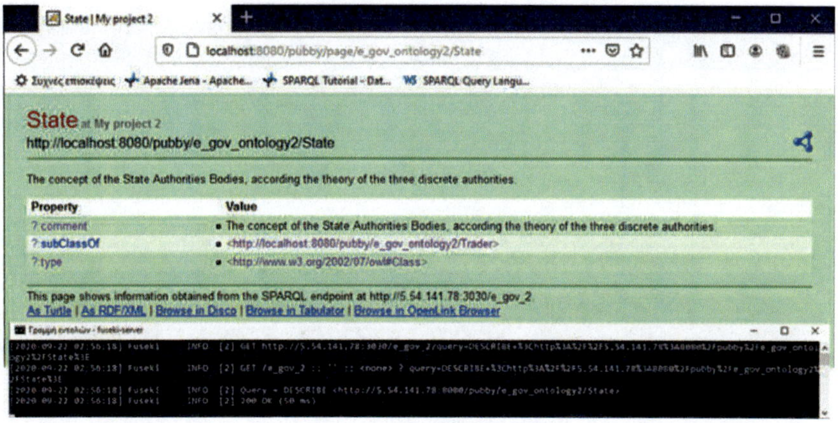

Fig. 6.50 The resource "State" description via the Pubby server

Vapour's response also includes a report from the dereferencing control process that was put to the test, which includes a relevant graphical representation of the tests performed (Fig. 6.53). The answer in the case of our ontology also confirms the results we obtained from the visual check when accessing the SparqL Endpoint described above.

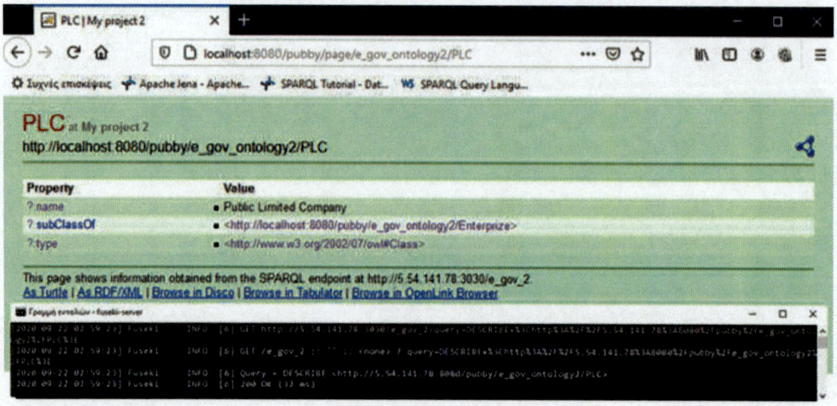

Fig. 6.51 The resources "PLC" description via the Pubby server

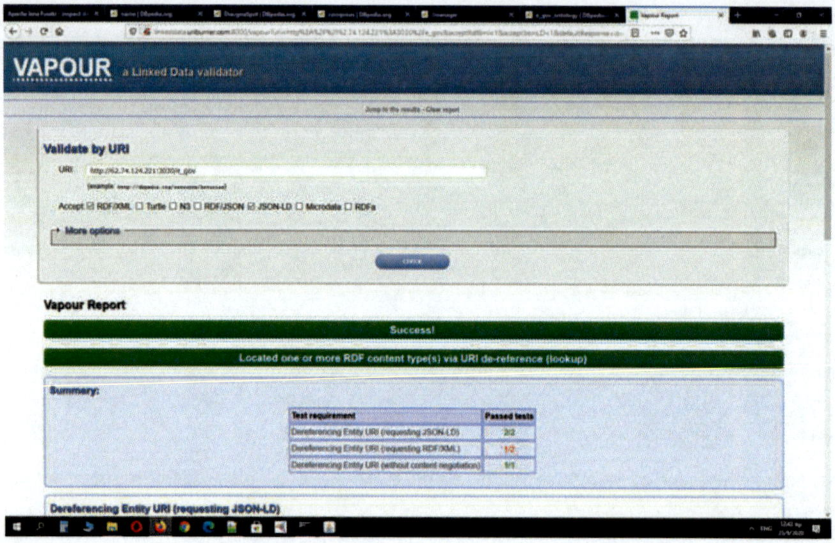

Fig. 6.52 The report screen (1/2) from VAPOUR Validator about the e-Government ontology in the relevant dataset we published in Fuseki

6.5.2 Ontology Publish as Linked Open Data

6.5.2.1 Upload Data to Virtuoso ODS Briefcase

In order to complete the publication of the "e-Government ontology" that we developed and presented above, we chose as an alternative way of publishing the OpenLink Virtuoso and in particular the ODS Briefcase package offered in version 8.3.

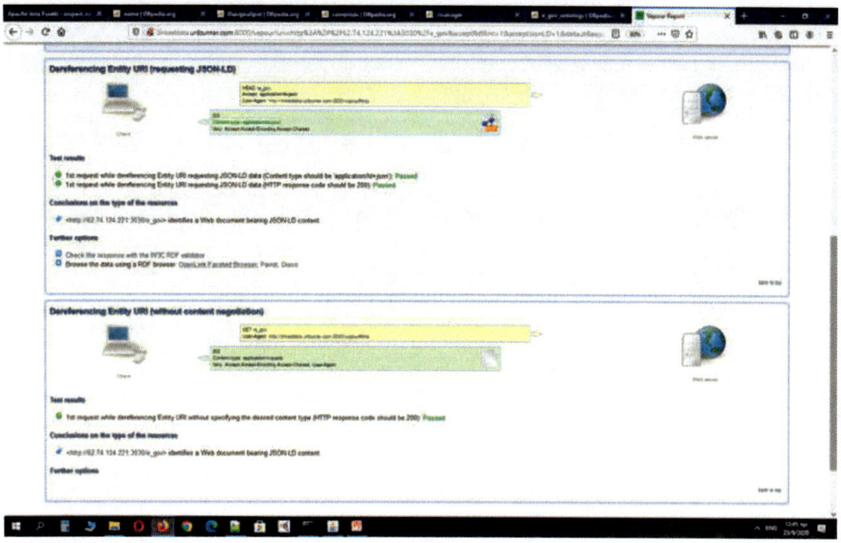

Fig. 6.53 The report screen (2/2) from VAPOUR Validator about the e-Government ontology in the relevant dataset we published in Fuseki

In particular, after downloading the necessary files for the installation of Virtuoso 8.3 and the relevant licence from its official website [46] we proceeded to install the necessary packages in order to be able to manage linked data. To install and manage the packages offered through Virtuoso, it is necessary to create and connect through special accounts that correspond to administrators or other types of users.

Among the packages we installed as administrators in Virtuoso is the ODS Briefcase. We used the possibilities offered by this package in order to upload the relevant .owl file containing the code and the statements for the "e-Government ontology" that we have already developed with Protégé (Figs. 6.54 and 6.55). Upload settings include the relevant licences for the owner and other users of this file.

After completing this process, the status of the respective public folder indicates the uploaded file as well as its basic characteristics including its type (RDF data) as well as its content type (application/RDF + xml) (Fig. 6.56). In order for this data to be practically accessible via the Web, it is essential that the network infrastructure is properly configured. In our case, the virtuoso was hosted on a local home computer so the necessary settings for port-forwarding were made through the corresponding page of the router provided by the ISP from where it was retrieved and a URL through which access to virtuoso is achieved (via the local port:8890). For the proper operation of port-forwarding, settings in the firewall of the local machine are also necessary.

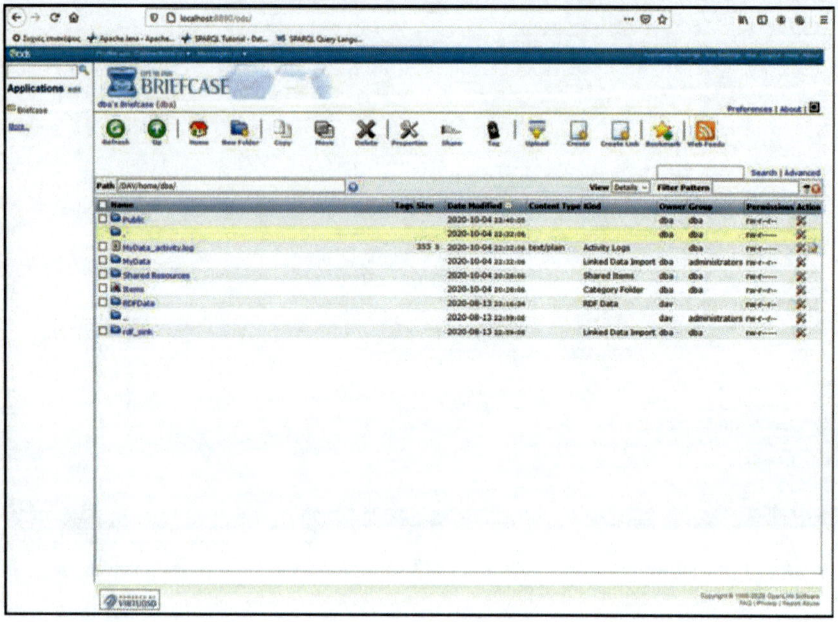

Fig. 6.54 The ODS Briefcase home page

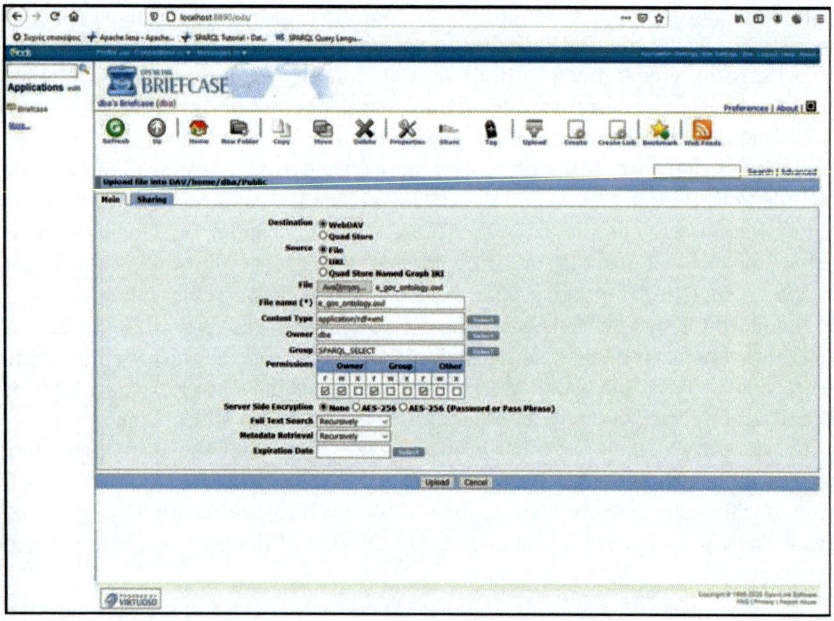

Fig. 6.55 Uploading the “e-Government ontology” file in ODS Briefcase

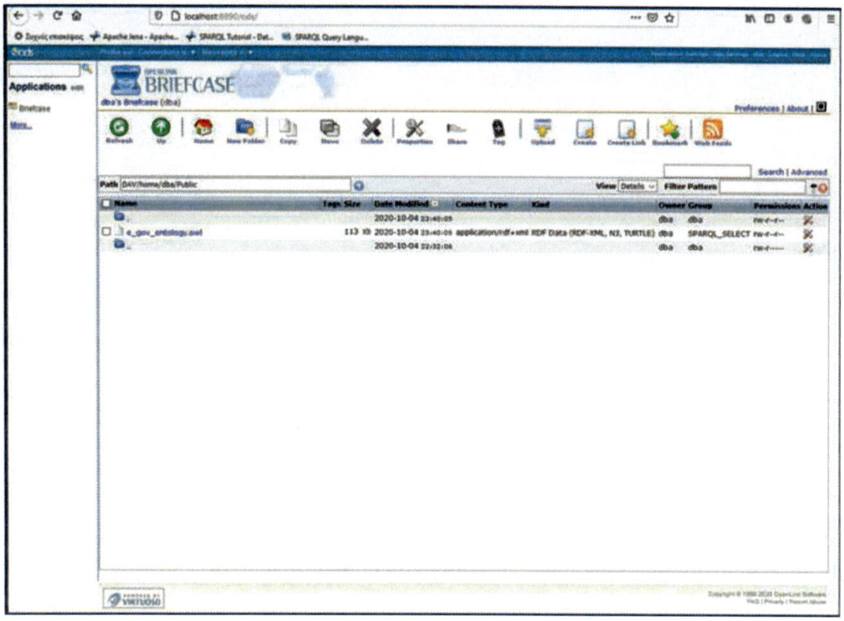

Fig. 6.56 The status of the "public" folder of the ODS Briefcase in Virtuoso 8.3

6.5.2.2 RDF Data Retrieval from Virtuoso

In order to check the accessibility and availability of the public file that we uploaded through Virtuoso ODS Briefcase we used two different tools: the Fuseki server and the RDF4J Workbench which we installed and operated on a different machine and a different network from Virtuoso. Note that Fuseki works as a standalone server and listens on the local port: 3030, while RDF4J is served as a war application through the Apache Tomcat admin page that listens on the local port: 8080.

Regarding fuseki, first we activate the server (by calling fuseki-server.bat) then we create a new dataset in order to store the data to be retrieved from Virtuoso and put test SparqL queries on them. In order to retrieve the data from virtuoso we run on the corresponding page of fuseki an update query which is sent to the address that virtuoso listens to as shown in the figure below (Fig. 6.57). As shown in the relevant message, the communication with virtuoso and the data recovery was completed successfully.

Then we run the general query shown in the Fig. 6.58 in order to have all the triples retrieved. Through the answer we have from fuseki we confirm the number of triples as well as their content which is as expected.

As for the RDF4J workbench, we first activate it through the management page of the Apache Tomcat that already works and listens to the local port: 8080 and then we create a new dataset, in the form of "In Memory Store Repository" (Fig. 6.59).

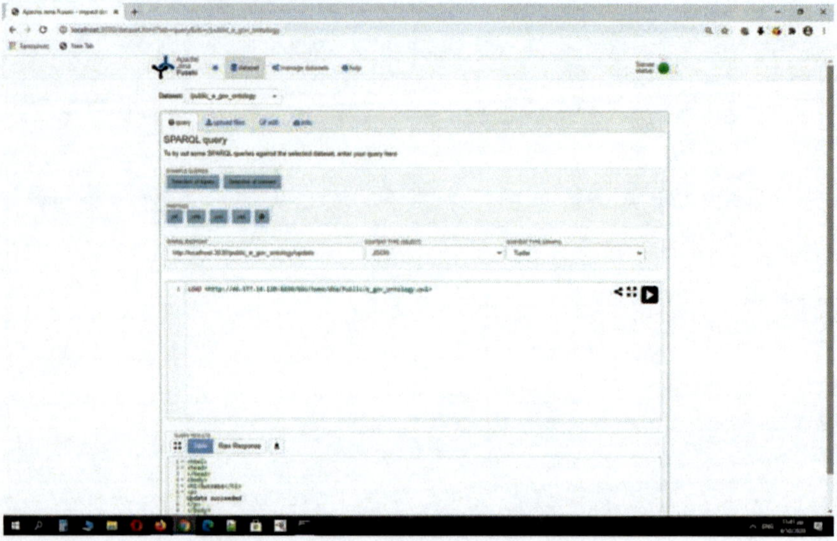

Fig. 6.57 RDF data retrieval from ODS Briefcase repository in Fuseki server

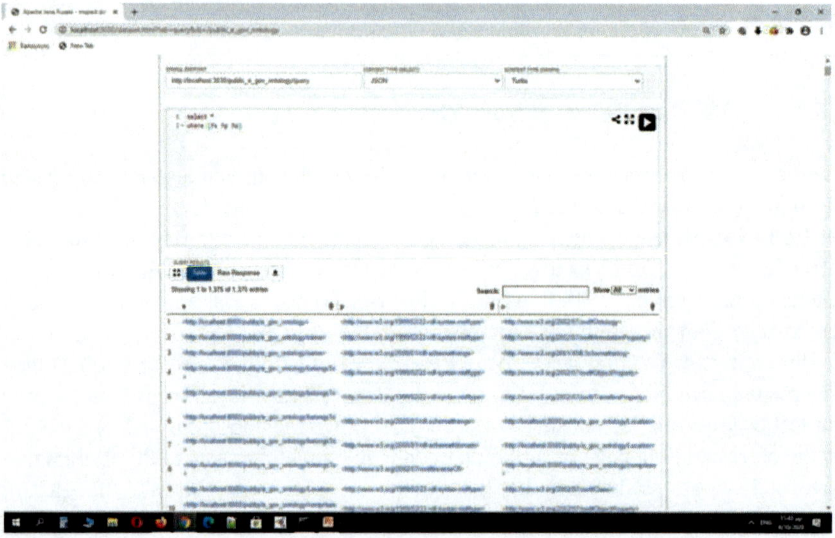

Fig. 6.58 Query results in Fuseki server

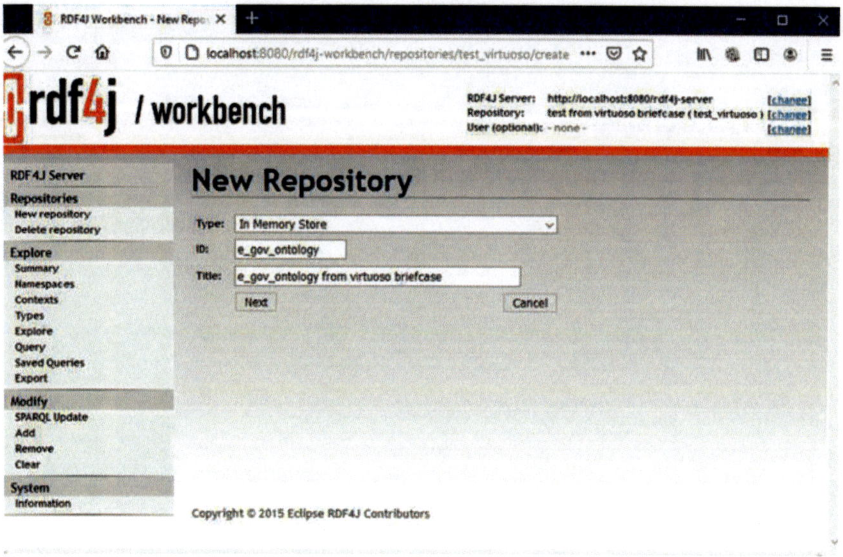

Fig. 6.59 New repository creation in RDF4J Workbench

The first step of creating the repository is completed successfully as shown in the Fig. 6.60. However, the data has not been retrieved yet, so at this step the repository has no content.

In the repository we created we add the data we have uploaded to virtuoso through the "add" option of the repository configuration menu. The key point in the process is the correct declaration of the base URI and RDF data URL. Here is the actual address where the virtuoso is hosted and listened to on our local machine (Fig. 6.61).

As shown in the figure below (Fig. 6.62), all the triples that were retrieved by RDF4J, are that have been recovered in the corresponding procedure described above for the fuseki server.

In order to present the triples included in the relevant repository we perform the general SparqL query (Fig. 6.63). The relevant answer confirms the correct retrieval both numerically and in terms of the content of the triples of the repository (Fig. 6.64).

6.5.2.3 Ontology Retrieval

In order to check for the correct retrieval of our ontology from the corresponding file we published through Virtuoso, we used Protégé. Through this tool we have the ability to open .owl files either from the local machine or from a specific URI via the web. From the Protégé homepage and the "File" menu we choose to open an existing ontology with the option "Open from URL..." where we state the address from which the relevant .owl file will be retrieved. As shown in the Fig. 6.65 we state

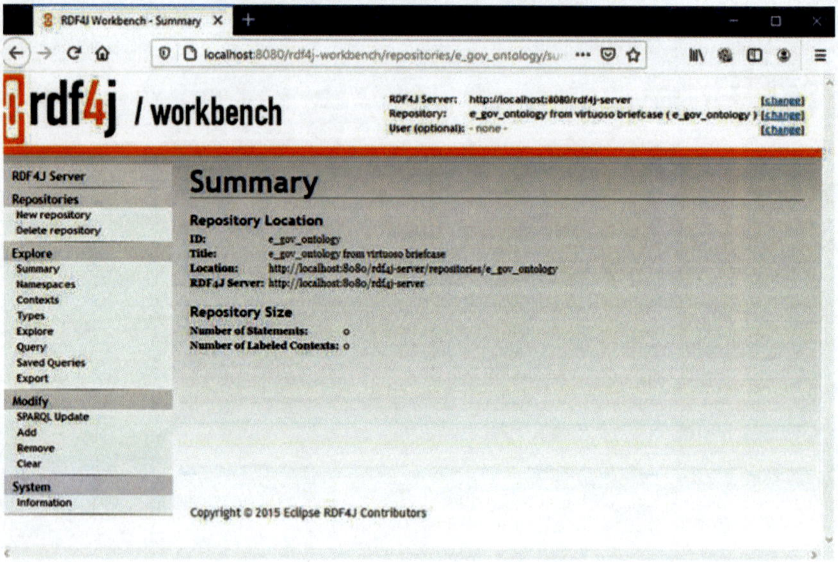

Fig. 6.60 Summary about the e_gov_ontology repository

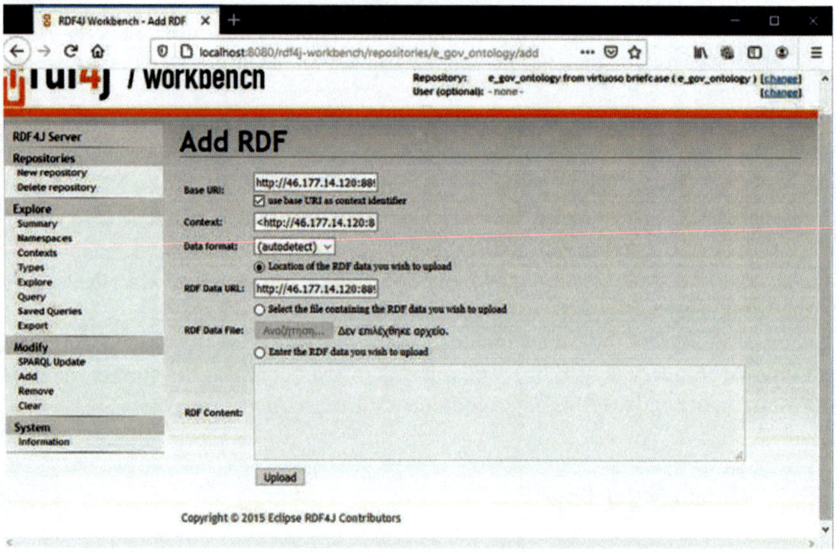

Fig. 6.61 Fill in the necessary elements for data retrieval from Virtuoso

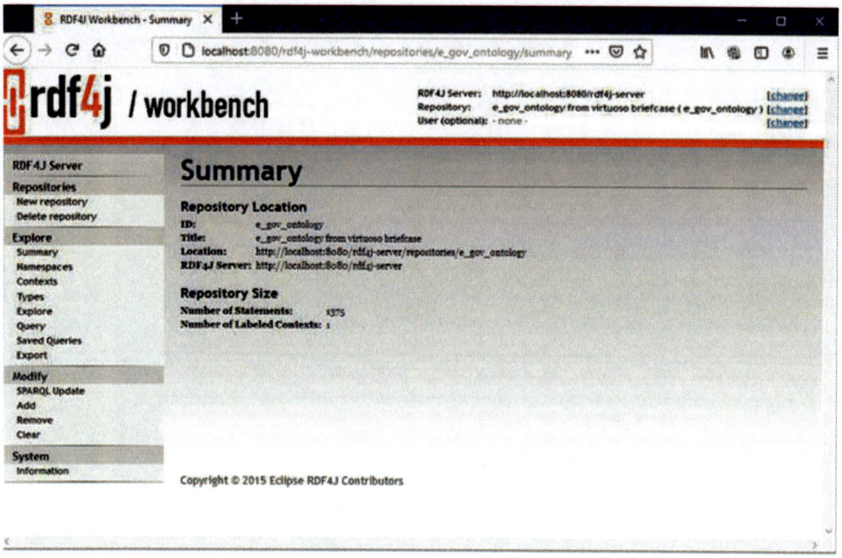

Fig. 6.62 The summary page of the repository created in RDF4J

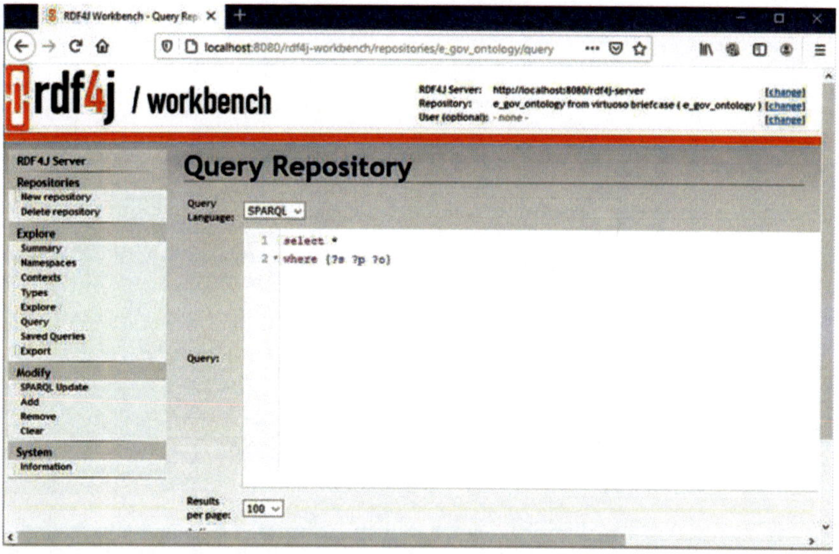

Fig. 6.63 General query in RDF4J repository

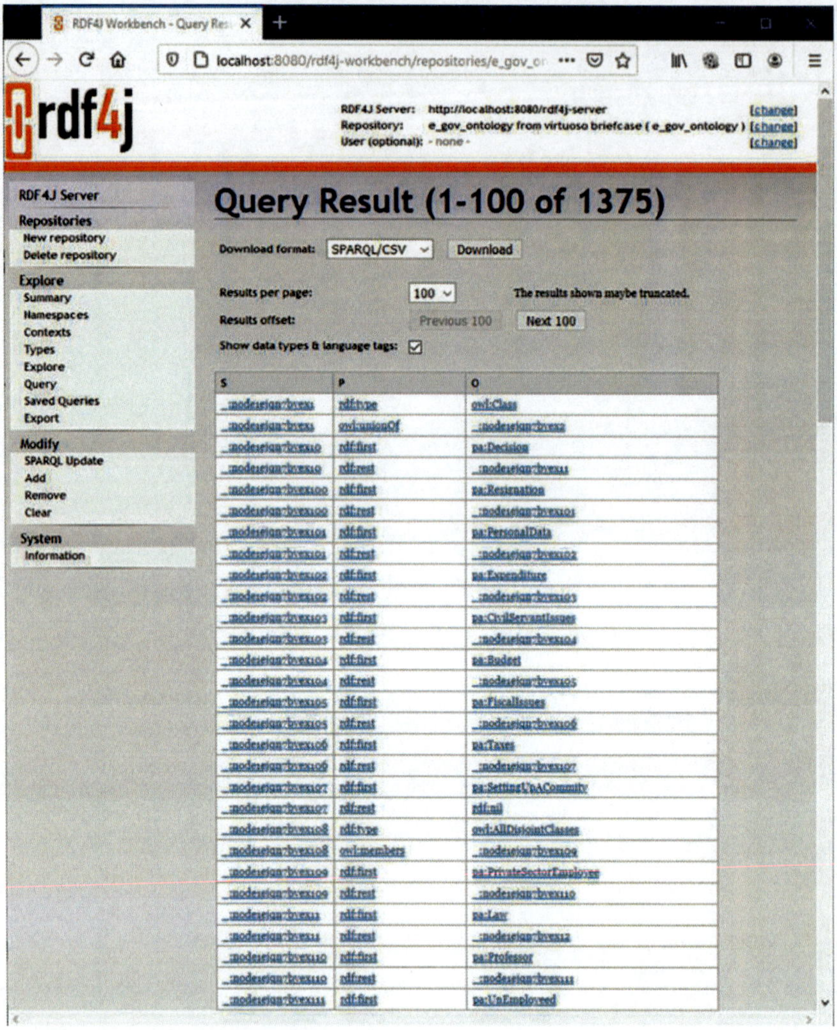

Fig. 6.64 Query results from the e_gov_ontology repository in RDF4J

the address that Virtuoso listens to and in which we have published the file with the code of our ontology.

In order for the ontology to be accessible via the Web as linked data using pubby, we declare the appropriate IRI ontology with the relative configuration we can do in Protégé (Fig. 6.66).

The result of the retrieval (Fig. 6.67), relates to all the classes, snapshots, properties and rules of e-Government ontology that we had coded and published through the corresponding .owl file in Virtuoso's ODS Briefcase.

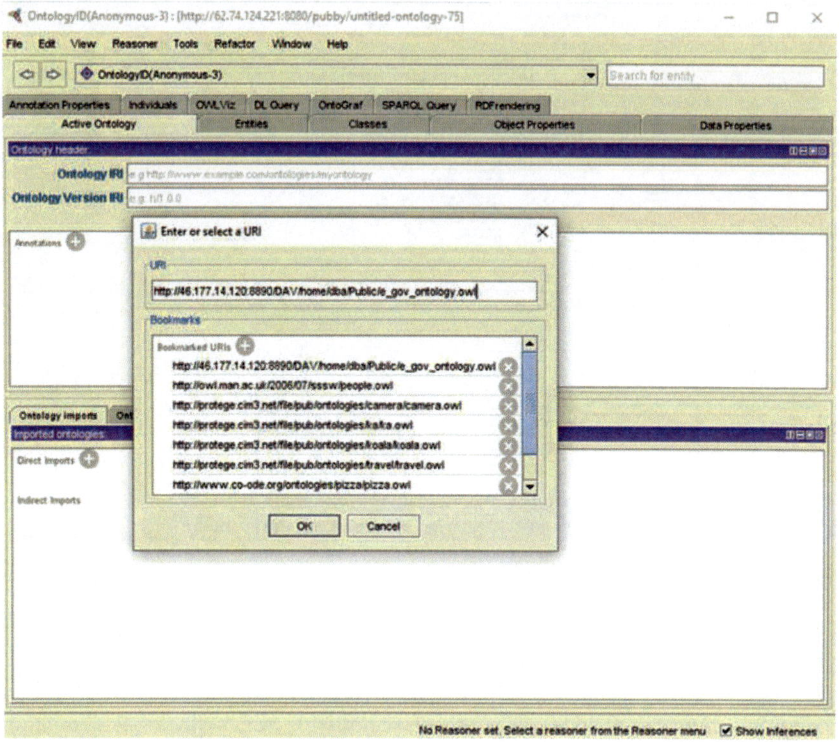

Fig. 6.65 Ontology retrieval in Protégé from specific URL

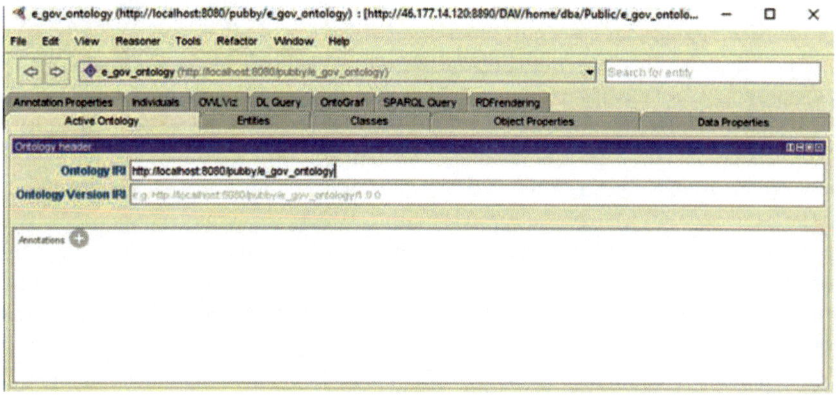

Fig. 6.66 Ontology IRI declaration in Protégé

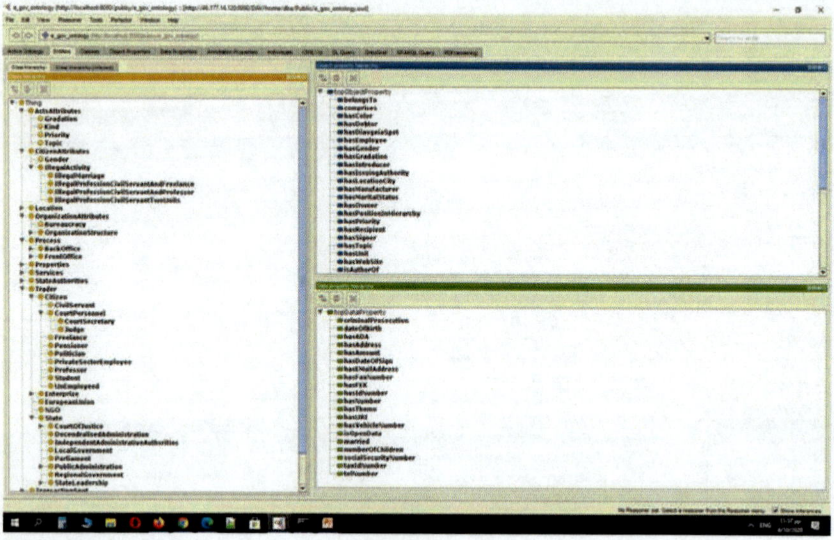

Fig. 6.67 General view of the e-Government ontology retrieved from the ODS Briefcase

6.5.2.4 Ontology and RDF Data Validation

As above, we used the Vapour Validator tool in order to verify the correct operation of our ontology as published and can be retrieved from Virtuoso's ODS Briefcase. On the homepage of the tool we state the URL which is to be evaluated (Fig. 6.68).

The response of the tool is positive for the RDF/XML content that was tested and for the dereferencing as shown in the image below (Fig. 6.69).

An additional check—evaluation of the correctness of the triples contained in our ontology can be performed using the tool called "RDF validator" provided by the respective website of the W3C [47]. This tool provides the ability to validate RDF files uploaded by the user from their local machine or RDF files that have been published and are located at a specific URL. In our case on the homepage of this tool, we declare the URL we have reserved for our RDF data via virtuoso (Fig. 6.70). The response of the tool, which seems to be positive, includes in addition to the relevant message and all the triples retrieved from the specific URL. This result is fully in line with the relevant results from the respective process via fuseki and RDF4J presented above.

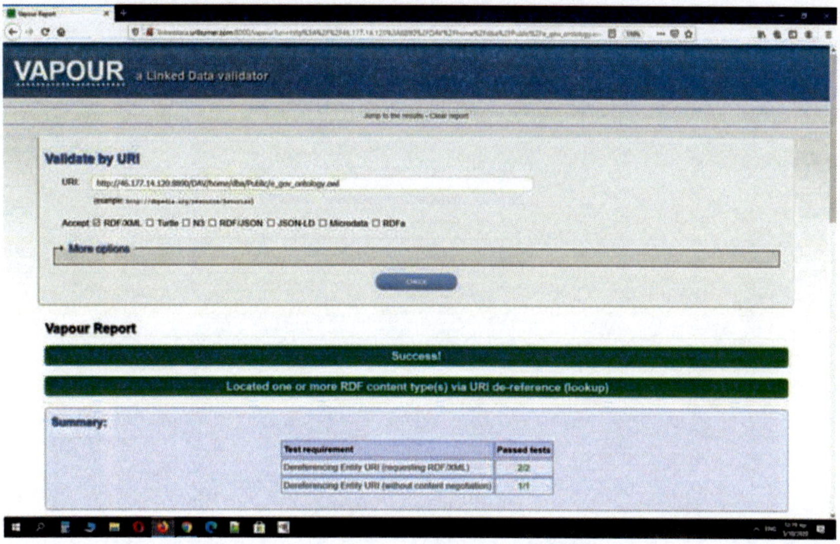

Fig. 6.68 Results from Vapour Validator about the e-Government ontology file in virtuoso (1/2)

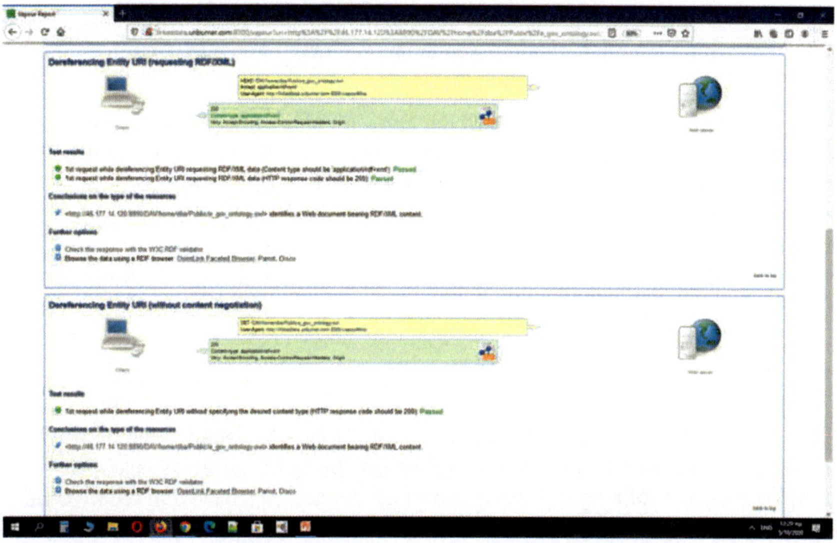

Fig. 6.69 Results from Vapour Validator about the e-Government ontology file in virtuoso (2/2)

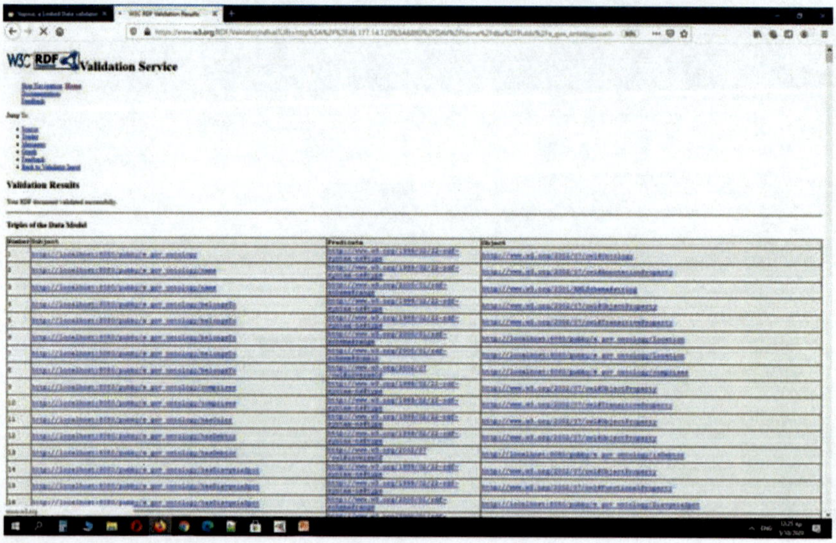

Fig. 6.70 RDF Validator results about the RDF data in Virtuoso

6.6 Publication of Open Ontology in the LOD Cloud

In order to publish our ontologies in the form of linked open data, achieving the highest score according to the five-point rating scale "5 stars" published by Tim Berners Lee [2], specific terms and conditions are required. Among other things, the data is required to be accessible to the general public, to be easy to search and retrieve in specific formats and at the same time to link and refer to other also published data. For this reason, we have chosen to publish our ontologies through the most popular point-of-search network for linked open data, the "linked open data Cloud" (LOD Cloud) [48]. We also published our ontologies in another popular repository of linked open vocabularies and ontologies, Linked Open Vocabularies [49]. The publication process at these points has specific operational requirements, which are analysed below. We also present and comment on the results from the publication of this data.

In particular, as functional requirements for the publication of linked data we mean the operation of a SparqL Endpoint so that queries can be submitted to retrieve strong triples on a case by case basis as well as a repository, such as a triple store from which it is possible to retrieve all the data or in our case the open ontologies. In this case, we used Open Link Virtuoso as the basic support infrastructure for both requirements, according to the procedure described below. In particular, the case of publishing ontologies in full format as .owl files through the Virtuoso Briefcase has been presented in a previous section.

6.6.1 *Create a SparqL Endpoint Through OpenLink Virtuoso*

Step 1: Upload ontology in the form of linked data

We first connect via the appropriate role (account) and permissions on the Virtuoso Conductor through the virtuoso homepage. Then, using the linked data service and in particular through the Quad Store Upload tab, we properly configure the relevant page in order to retrieve the corresponding data set from the local computer and store it in the special triple store repository provided by virtuoso (Fig. 6.71).

Step 2: Create a graph

Virtuoso allows us to define a specific graph for the data we upload as linked, through the definition of a specific IRI, which will be used to draw the corresponding triples from. This IRI is accessible both locally and remotely via the server we have developed, as long as the interface of the router used for the Internet is properly configured (Fig. 6.72).

We confirm the creation of the graph through the corresponding tab of virtuoso which displays all the stored graphs of our system (Fig. 6.73).

Step 3: Retrieve triples via Virtuoso SparqL Endpoint

OpenLink virtuoso provides the user with a special tool for semantic query on stored graphs which works as a SparqL Endpoint which is accessible both locally and remotely. From the home page of virtuoso it is possible to call the corresponding

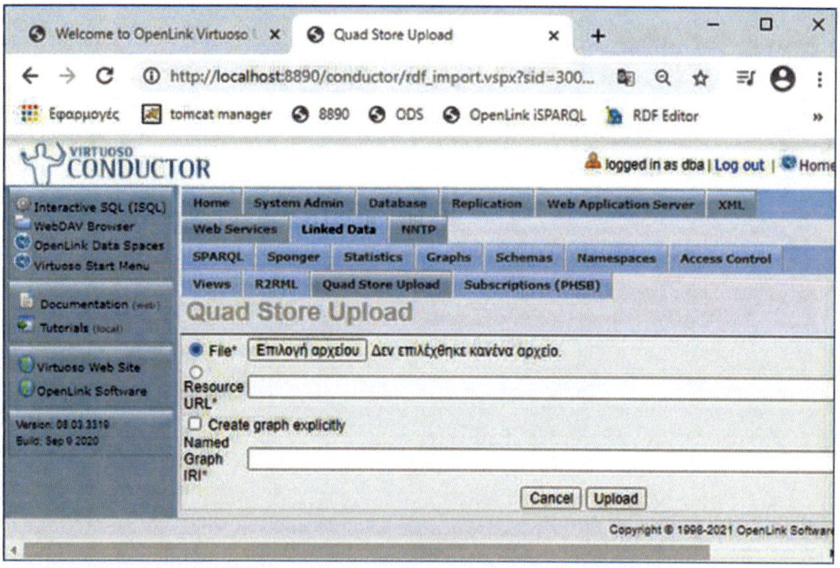

Fig. 6.71 Ontology upload via the Quad Store Upload Service in Virtuoso

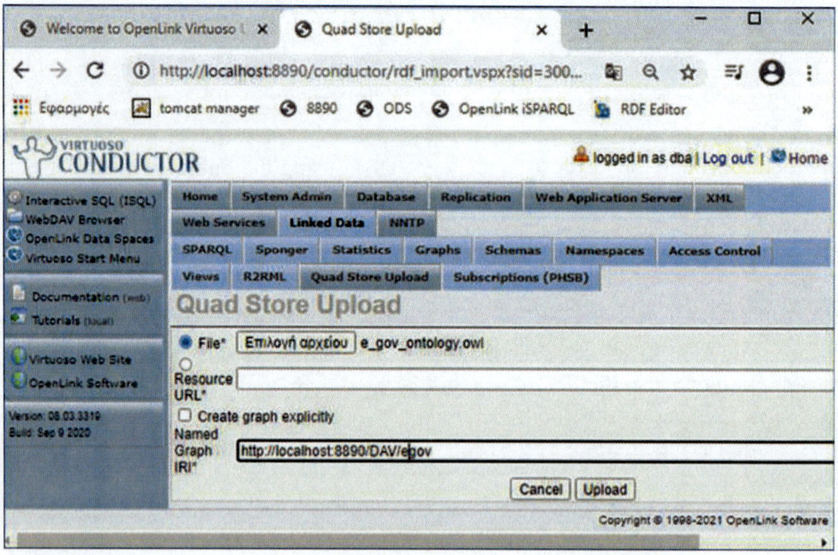

Fig. 6.72 The graph declaration via the Quad Store Upload Service in Virtuoso

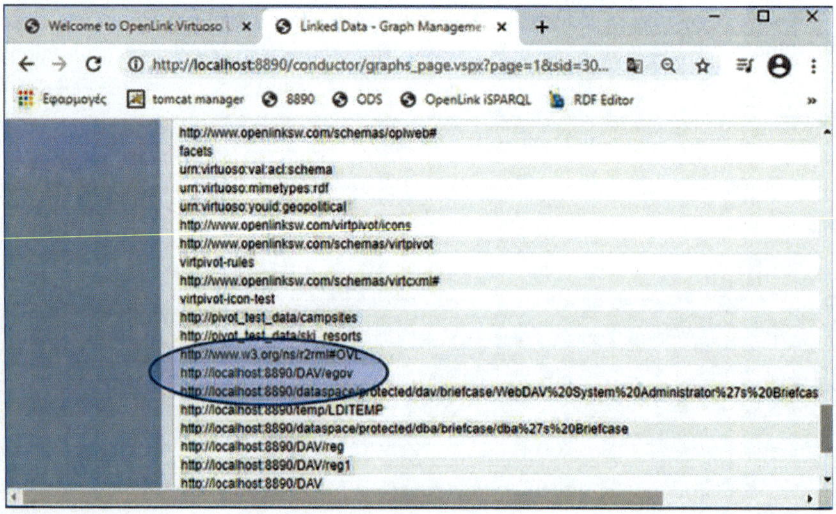

Fig. 6.73 Verification of the creation of the new graph

service of “OpenLink Virtuoso SparqL Query Editor” or alternatively we can have direct access via the address <http://localhost:host/SparqL>. The page we have already configured shows by default the relevant graph and a basic question about it. On this page it is possible to ask any other question based on the SparqL protocol or/and to state another graph that we are interested in and that is stored in our system. We also note the possibility of configuring the sponging method and the shape of the input and output triplets. The result of the relevant process by submitting a basic query to our ontology graph returns all the triples and is shown in the image below (Figs. 6.74 and 6.75).

Access to this service is possible both from the local machine and from a remote point after the router interface has been properly configured.

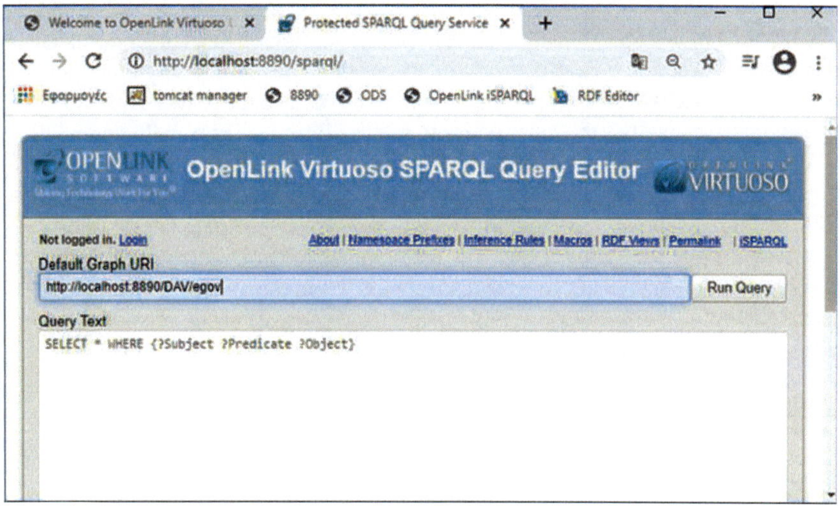

Fig. 6.74 Our pre-configured SparqL Endpoint Home Page via the virtuoso

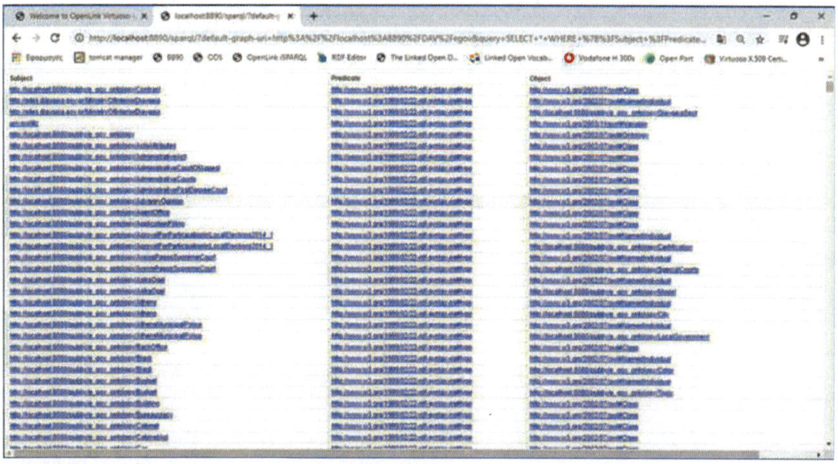

Fig. 6.75 Part of the results via our SparqL Endpoint

Step 4: Browsing through iSPARQL

As we saw in a previous section, it is important to enable the regular user to browse our linked data without necessarily knowing the Semantic Web protocols and without submitting specific semantic queries. One such tool is provided by virtuoso which has the characteristic title "OpenLink iSPARQL", which has been developed as a package—service of virtuoso. Access is achieved either indirectly via a specific link from the SparqL Query Editor tool provided via virtuoso or directly via the <http://localhost:port/iSparqL/> address. The respective website first provides the ability to submit specific questions and then the ability to navigate with the method of links which hide from the ordinary user the submission of special queries such as "select" and "describe". Results from the use of this tool are shown in the pictures below. In our case we have properly configured the home page to load by default the relevant graph as well as the basic query to return all triples (Figs. 6.76 and 6.77).

Access to this service is possible both from the local machine and from a remote point after the router interface has been properly configured.

As mentioned above, through this tool it is possible to navigate the returned data in order to provide the ordinary user with the semantic information that is "hidden" in the triplets of the graph. Relevant results from the use of this feature are shown in the pictures below. We also note that this function is analogous to the service provided by the Pubby Server that we presented in the previous section (Fig. 6.78).

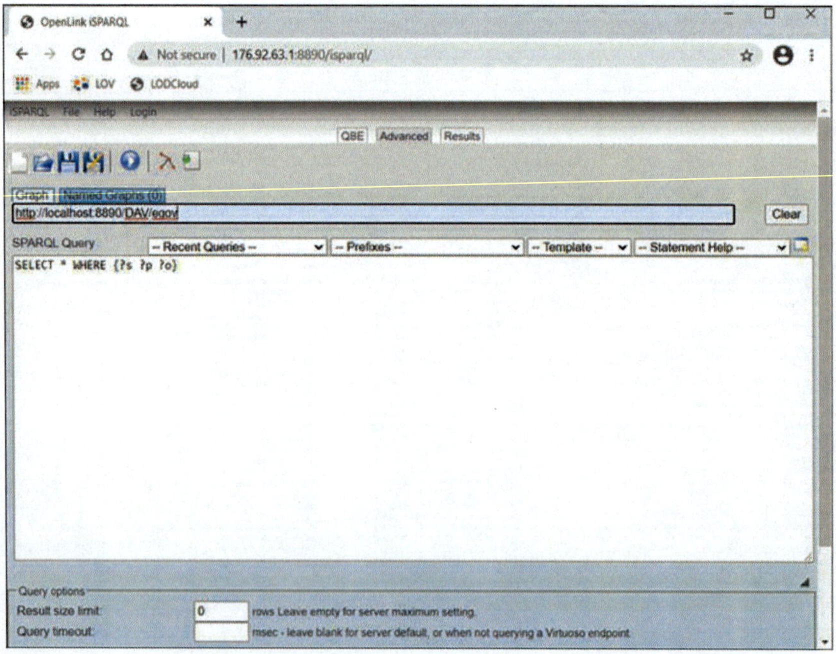

Fig. 6.76 The iSPARQL pre-configured home page in order to query our ontology (remote access)

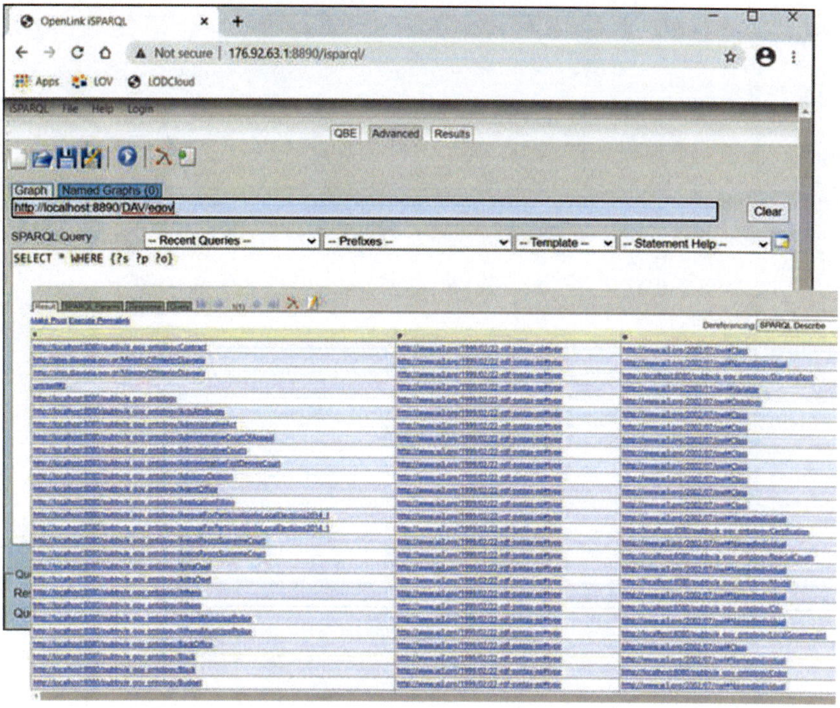

Fig. 6.77 The iSPARQL results from our e-Government ontology (remote access)

6.6.2 Publication of the e-Government Ontology in the LOD Cloud

As we mentioned in the previous section, the point at which the data is published is important, since through it the data becomes known and accessible to the general public. One such point is the linked open data cloud, the size of which is increasing geometrically [48]. The LOD cloud dataset currently contains 1269 datasets with 16,201 links (as of May 2020).

In order to publish our data on this site, the operation of a SparqL Endpoint, a repository is required in order to enable the download as well as the relevant data website. The corresponding entry in order to complete the publication of the e-Government ontology is shown in the image below. We note that the required infrastructure has already been developed and is operating as we have already mentioned previously (Figs. 6.79, 6.80 and 6.81).

The results from the remote access to the LOD Cloud are shown in the following images, from which the expected correct publication of our ontology emerges (Fig. 6.82).

Fig. 6.78 Browsing results through the triples via the iSPARQL (describe method)

6.6.3 Publication of the Ontology in Linked Open Vocabularies

An alternative site for publishing ontologies and vocabularies in a specialised repository is Linked Open Vocabularies (LOV). The vocabulary collection is maintained by the LOV team of curators in charge of validating and inserting vocabularies in the LOV database and assigning them a detailed review (updated on a yearly basis). Before a vocabulary is inserted, LOV team contacts the authors to make sure the vocabulary is published following the best practices and meets quality requirements of the overall LOV ecosystem. When some metadata fails to be extracted automatically (such as creators of a vocabulary), curators try to add them manually by harvesting information from the documentation or from direct communication with

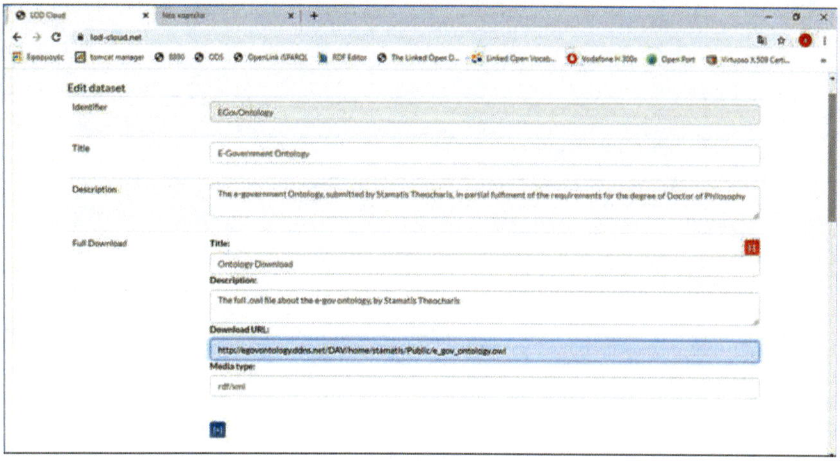

Fig. 6.79 Publishing our ontology in the LOD Cloud (1/2)

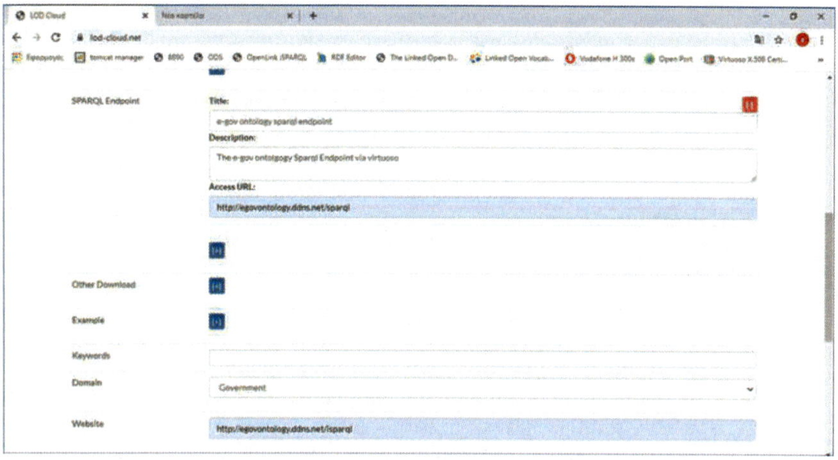

Fig. 6.80 Publishing our ontology in the LOD Cloud (2/2)

the publisher. Once included, an automatic script checks for vocabulary updates on a daily basis. The documentation assists any user in the task of understanding the semantics of each vocabulary term and therefore of any data using it [49].

In order to publish our ontology in the LOV repository, we visit the relevant home page [49] and through the option "suggest" we declare the URI that corresponds to our ontology. In our case we declare the URI that we have committed through the virtuoso briefcase (Fig. 6.83).

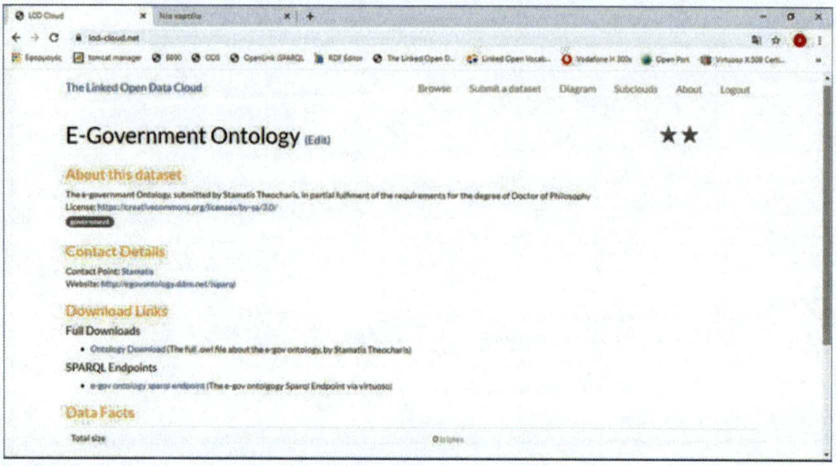

Fig. 6.81 The e-Government ontology published in the LOD Cloud

The result from the automatic test is positive as shown in the image below. The relevant answer returns specific information regarding metadata such as the URI, Namespace and the number of classes, properties, instances etc. (Fig. 6.84).

6.7 Conclusions

6.7.1 Pubby Operation

- A key advantage of Pubby is its basic use for the regular user. The user does not need to know the semantic protocols required to submit queries to the corresponding SparqL Endpoint. It is possible to simply use a regular browser to browse the data and to automatically display the linked entities, the relationships on which they are connected as well as any properties or limitations that they have.
- However, we note that Pubby does not offer the ability to submit dynamic and optional semantic queries but only the ability to present the semantic description of the resources visited by the user based on the triplets stored in the connected SparqL EndPoint.
- We consider the use of this tool to be essential for the regular user, but on the part of the developer at the level of the Semantic Web offers specific features that are limited to the simple display of any semantic information of each dataset and the submission of a specific type of queries which it is not possible to adjust dynamically.
- The communication between Pubby, Tomcat and Fuseki tools in our case did not show any malfunction. In particular, the operation of the Fuseki server did not

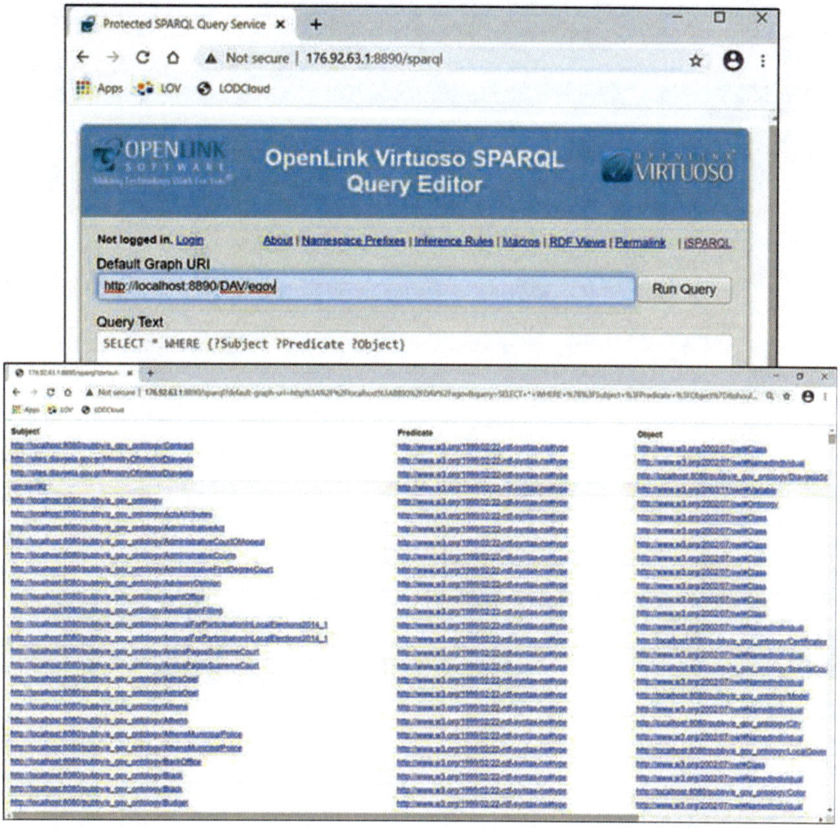

Fig. 6.82 Results from the the relevant links of our e-Government ontology in LOD Cloud

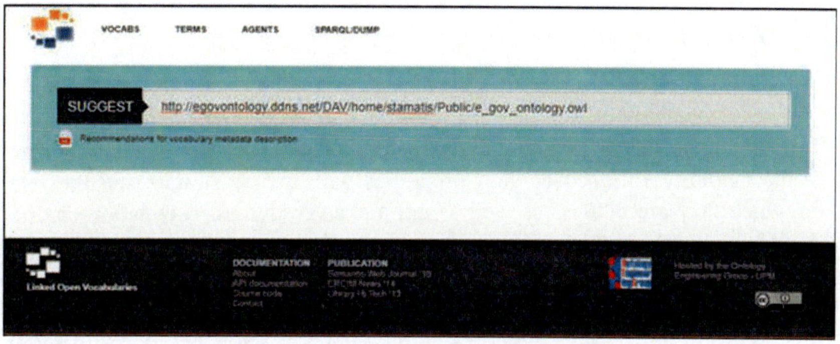

Fig. 6.83 Uploading our e-Government ontology in Linked Open Vocabularies

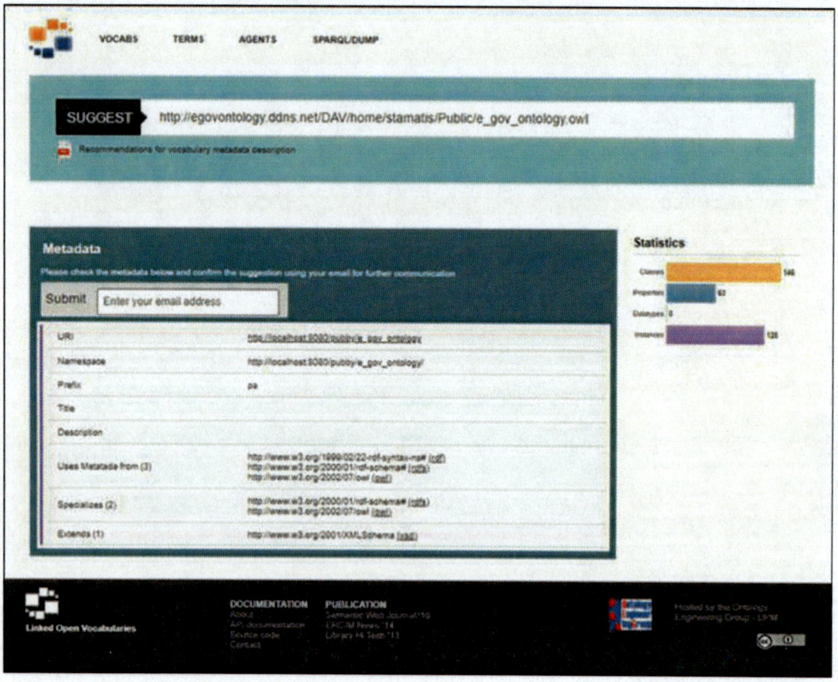

Fig. 6.84 Validation and results of our ontology via the LOV

require any special configuration while it was continuous and without problems both during the access from our local system and during the remote access.

6.7.2 *Benefits of Publishing the Ontology for Open Government as Open Data*

- Through Pubby, the ordinary user is given the opportunity to browse the resources of the ontology, to discover their properties with the constraints and the sets on which they are defined as well as any linked resources. Navigation can be guided through Pubby in addition to ontology classes and subclasses and individuals and links between them. For example, as we saw above, starting with the hypothetical resource—individual "Citizen6" we discover in addition to its type (i.e. it is an individual), its address and the property "`isOwnerOf`" with value "`SecurityServiceSA`", i.e. a specific company. The tour of the company (which is an affiliated resource) leads us to the appearance of its properties and so on. Also, starting from the "`State`" resource, one can discover the structure of the state, public bodies and the administrative hierarchy through the linked resources and the properties they have.

- For users who may have stronger credentials than the regular user, such as authorised public officials, through Pubby it is possible to browse further properties related to non-public (visible to all) data and properties.
- Through the now published ontology their concepts and properties are made accessible through the usual search engines.
- With the use of the published concepts—terms of the ontology, the semantic interoperability of the systems is achieved since the data stored in the knowledge base of the ontology is shared as public.
- A key advantage that emerged from the use of virtuoso to publish our ontology in the LOD Cloud and LOV, is that it provides a specific tool, iSPARQL, for browsing specific data from the server side. This results in ease of use by the client, who unlike the pubby server is not obliged to install any special software on his computer. Also, access to this service does not require the connection to the server, but the assignment of specific rights when installing the corresponding graph so that the browsing is either completely free or controlled on the basis of special credentials.
- Following the publication of our ontology in LOD Cloud and LOV, the respective data acquire all the characteristics required in order to be rated with the maximum possible grade, i.e. that of 5 stars.

References

1. The World Wide Web Consortium—W3C—Official site: www.w3c.org
2. Berners-Lee, Linked Data (2006). https://www.w3.org/DesignIssues/LinkedData.html
3. Openlink Software, Linked Data & ODBC Similarities (2010). https://virtuoso.openlinksw.com/presentations/Creating_Deploying_Exploiting_Linked_Data2/Creating_Deploying_Exploiting_Linked_Data2_TimBL_v6.html#(11)
4. Lewis, R. (2007). Volantis Systems Ltd., Dereferencing HTTP URIs-Draft Tag Finding 31 August 2007, W3C. https://www.w3.org/2001/tag/doc/httpRange-14/2007-08-31/httpRange-14.html
5. Bizer, C., Cyganiak, R., & Heath, T. (2007). *How to publish linked data on the Web?* http://wbsg.informatik.uni-mannheim.de/bizer/HowtoPublishLinkedData.htm
6. Openlink Software, Linked Data & ODBC Similarities (2010). https://virtuoso.openlinksw.com/presentations/Creating_Deploying_Exploiting_Linked_Data2/Creating_Deploying_Exploiting_Linked_Data2_TimBL_v6.html#(32)
7. Bauer, F., & Kaltenböck, M. A quick start guide for decision makers. *Linked open data: The Essentials*. https://www.reeep.org/LOD-the-Essentials.pdf
8. W3C working group, Linked Data Cookbook (2013). https://www.w3.org/2011/gld/wiki/Linked_Data_Cookbook
9. Heath, T. (2009). *An introduction to Linked Data*. https://web.archive.org/web/20181031024638/http://tomheath.com/slides/2009-02-austin-linkeddata-tutorial.pdf
10. The linked open data cloud site: https://lod-cloud.net/
11. The datahub.org central site: https://www.thedatahub.org/
12. The Apache-Jena project central site: https://jena.apache.org
13. The Eclipse/RDF4J project central site: https://RDF4j.org/
14. The Pubby project central site: http://wifo5-03.informatik.uni-mannheim.de/pubby/

15. The tomcat apache project central site: https://tomcat.apache.org/
16. Bizer, C., & Cyganiak, R. (2015). *D2R server: Accessing databases with SPARQL and as Linked Data*. http://d2rq.org/d2r-server
17. Cyganiak, R., et al. (2012). The D2RQ mapping language. *D2RQ documentation*. http://d2rq.org/d2rq-language#examples
18. OpenLink Virtuoso central site: https://virtuoso.openlinksw.com/
19. OpenLink Software, OpenLink Documentation. http://docs.openlinksw.com/virtuoso/whatisnewto2x/
20. OpenLink Software, OpenLink Documentation. http://vos.openlinksw.com/owiki/wiki/VOS/VOSIntro
21. Solid IT, The DB-Engines Ranking. https://db-engines.com
22. Ministry of Interior, Hellinic Republic. (2013). *Circular of Ministry of Interior for the national population register*. www.ypes.gr
23. Pasha, M., & Sattar, A. (2012). Building domain ontologies from relational database using mapping rules. *IJIES, 5(1)*, 20–27.
24. Berardi, R., et al. (2013). StdTrip+K: Design rationale in the RDB-to-RDF process. In H. Decker, et al. (Eds.), *DEXA 2013*. Part I. LNCS (Vol. 8055, pp. 303–310).
25. Būmans, G., & Čerāns, K. (2011). Advanced RDB-to-RDF/OWL mapping facilities in RDB2OWL. In J. Grabis & M. Kirikova (Eds.), *BIR 2011, LNBIP 90* (pp. 142–157).
26. Būmans, G., & Cerans, K. (2010). RDB2OWL: A practical approach for transforming RDB data into RDF/OWL. In *6th International Conference on Semantic Systems*. https://doi.org/10.1145/1839707.1839739
27. Salas, P. E., et al. (2011). RDB2RDF plugin: Relational databases to RDF plugin for eclipse. In *Proceedings of TOPI'11*.
28. Hert, M., et al. (2011). A comparison of RDB-to-RDF mapping languages. In *Proceedings of I-SEMANTICS 2011, 7th International Conference on Semantic Systems*.
29. Sahoo, S. S., et al. (2009). A survey of current approaches for mapping of relational databases to RDF. www.w3.org/2005/Incubator/rdb2RDF/RDB2RDF_SurveyReport_01082009.pdf
30. Arenas, M., et al. (2012). *A direct mapping of relational data to RDF, W3C*. http://www.w3.org/TR/rdb-direct-mapping/
31. W3C, Implementations (2012). http://www.w3.org/2001/sw/rdb2RDF/wiki/Implementations
32. Vergoulis, A., et al. (2014). *Data Governance in the era of the Web of Data: Generate, manage, preserve, share and protect resources in the Web of Data*. Technical Report, Athena R. Greece.
33. Openlink Software, Linked Data & ODBC Similarities (2010). https://virtuoso.openlinksw.com/presentations/Creating_Deploying_Exploiting_Linked_Data2/Creating_Deploying_Exploiting_Linked_Data2_TimBL_v6.html#(30)
34. The OGP project Web Page in http://www.opengovpartnership.org/countries
35. Berners-Lee, T. (2006). *Linked data—Design issues*. http://www.w3.org/DesignIssues/LinkedData.html
36. The "Diavgeia Project" web page in http://www.diavgeia.gov.gr
37. Theocharis, S., & Tsihrintzis, G. (2014). Ontology development to support the Open Public Data—The Greek case. In *Proceedings of IISA 2014*. IEEE. https://doi.org/10.1109/IISA.2014.6878820
38. Theocharis, S., & Tsihrintzis, G. (2015). Open data triplification—The case of the Greek Open Public Data. In *Proceedings of IISA 2015*. IEEE. https://doi.org/10.1109/IISA.2015.7388063
39. Jeremy Tandy, Met Office, Ivan Herman, W3C, Gregg Kellogg, Kellogg Associates, Generating RDF from Tabular Data on the Web. http://w3c.github.io/csvw/csv2RDF/
40. Tomaszuk, D. (2010). Institute of Computer Science, University of Bialystok, Poland, Flat triples approach to RDF graphs in JSON. https://www.w3.org/2009/12/RDF-ws/papers/ws02
41. Stolz, A., Rodriguez-Castro, B., & Hepp, M. (2013). RDF translator: A RESTful multi-format data converter for the Semantic Web. https://www.researchgate.net/publication/259335163_RDF_Translator_A_RESTful_Multi-Format_Data_Converter_for_the_Semantic_Web
42. Kellogg, G. JSON-LD: JSON for Linked Data. http://www.slideshare.net/gkellogg1/JSON-for-linked-data

43. The JSON data web page: http://JSON.org
44. RDF translator website: http://RDF-translator.appspot.com/
45. Vapour Validator website: www.linkeddata.uriburner.com:8000/vapour
46. OpenLink Virtuoso web site: https://shop.openlinksw.com/license_generator/virtuoso-download/
47. W3C RDF validator web site: www.w3c.org/RDF/validator
48. The Linked open data cloud web site: http://lod-cloud.net
49. The Linked Open Vocabularies web site: https://lov.linkeddata.es/dataset/lov/

Chapter 7
Education and e-Government—The Case of a Moodle Based Platform for the Education and Evaluation of Civil Servants

Abstract In this chapter, issues of training and evaluation of public servants are examined, as well as the case of the use of learning management systems in the context of the development of e-Government. In particular, issues related to the prevailing situation in the education of public servants in the case of the Greek public administration are presented, and the case of e-learning of employees in the context of e-Government. Finally, the development of a wiki as a means of education and knowledge management in public organisations is proposed and a specific wiki implementation through open software is presented.

Keywords e-Learning · Wiki for e-Government · Learning management systems

7.1 Introduction

The introduction of e-Government in the majority of transactions with the administration has significant benefits for citizens and businesses in recent years. The procedures and tools used by the administration are changing and transforming in order to support the project of e-Government. The technological solutions implemented with the help of information and communication technologies require special familiarisation and skills by officials that work for the administration. So for the further successful implementation and introduction of e-Government human factors and its adaptation to the new conditions become significant. A key tool in this direction is continuous and lifelong education of civil servants.

The training of officials is one of the most important issues to be faced by the administration of an organisation. We could use either classical education methods in classroom organised departments or innovative learning methods with the help of new technologies that do not require the physical presence of the learners and the instructor in a particular place. Particularly interesting for us is the investigation of distance education and e-learning capabilities of civil servants. And this is because the classical method of physical presence in the area of education, requires special organising at the premises and at times training as well as support to provide resources both financial and human.

S. Theocharis and G. A. Tsihrintzis, *Semantic Knowledge Modelling via Open Linked Ontologies*, Artificial Intelligence-Enhanced Software and Systems Engineering 4,
https://doi.org/10.1007/978-3-031-20585-9_7

The e-training can contribute significantly to reducing the resources required as in increasing the participation and interest of the trainees employees. In the following paragraphs we analyse the advantages of e-learning and present the Moodle as a means for making the e-training program with an emphasis on innovation and personalised education.

7.2 Current Situation in the Education of Civil Servants in Greece

Under current law (Civil Servant Code as applicable), training of civil servants aims to provide the necessary expertise for the efficient and effective performance of their duties. Starting from the moment of their entry into the public service and continued throughout their term of office at this. There are seven (7) basic educational and training categories concerning civil servants:

1. Pre-introductory training
2. Introductory training
3. Service training
4. Specialisation
5. Promotional training
6. Retraining
7. Postgraduate Education.

Public educational providers responsible for the formulation and implementation of education policies:

- National School of Public Administration.
- Training Institute.
- The Directorate of Education in the Ministry of Administrative Reform.
- Units of Education in the Ministries.
- Ministries' Schools.

The training of employees in the above bodies is most often in the form of seminars for a few days or weeks, demanding the physical presence of trainees and trainers. In the process of conducting seminars, the education directorates, the providers who perform the seminars and the trainees servants are all implemented. In this process there are some dysfunctions among which we mention the following:

- When an organisation must train its employees on a particular issue, the seminars must be planned in such a way, so as not to cause any dysfunctions in the organisation itself by the simultaneous absence of employees. On the other hand all employees must be trained in a short time on this issue so that there will be no employees with different levels of knowledge.
- The planning of seminars is not always apropos to the needs of each organisation or trainee. For example, when changing a basic law that covers all of services

and training needs are enormous, some employees will be trained much later than others.

- Due to the large number of employees, few manage to eventually be trained according to their actual needs and the time period necessary.
- Seminar participants belong to various departments so that there are different expectations and needs to be covered.
- All employees who participate in a seminar do not have the same skills or the same knowledge. So education outcomes are not as expected for all participants.

7.3 e-Training in e-Government

Public administration in recent years faces administrative reorganisation efforts in the e-Government. The change in operational practices and organisational structures are always intertwined with the familiarisation of employees with the changes planned. The current methodology for the training of employees is insufficient and needs supplementing with solutions coming from the field of e-training. Without doubt the administration would benefit from such a synthetic approach, both in terms of time savings, and in relation to saving money.

The e-Government uses technology to provide services tailored to the needs of citizens following new procedures so the e-training uses information systems to support and transfer knowledge, increasing the benefits and effectiveness of traditional models of learning [1, 2].

The benefits of e-Government from implementing e-training system are numerous. Relevant examples include:

- Reduces the obstacles which employees often face in various areas of administration, in an effort to gain on an equal footing high quality education and support services that integrate e-Government technology.
- Assists in developing tools and applications and strengthens the strategy of e-Government and changing work culture.
- Diffuses knowledge and makes it continuous and timely.
- Supports transmission of skills and knowledge through giving access to specialists and the possibility of cooperation and exchange of views (e.g. through community-forum) among employees.

Distance learning is the key factor for the successful implementation of e-Governance as it changes the way employees gain access to knowledge and skills, reduces barriers between them for access to high quality information and helps organisations to respond quickly in an environment of change [3]. This is accomplished by making learning more accessible (in time, spatially and immediately whenever requested), continuous (formal or informal and lifelong), relative (not occasional but when necessary, exciting and personalised), collaborative (with access to specialists and in cooperation with all employees having the opportunity of interaction) [4, 5].

So we are able to achieve results faster, increase productivity, change work culture, get comparative advantage and measurable results.

The e-training helps to transform the administrative functions of public administration and affects the training of citizens, the optimisation of service delivery and the development of human resources.

Major challenges for the future are the implementation of reward systems learning progression, the acquisition or development of remarkable electronic content distance learning, the selection of the best tools and the introduction of reliable infrastructure upgrading, the targeted investment for measurable results and the dealing with matters of administrative organisation and culture of services.

The modern trend leads them today to follow the strategy of blended learning, combining traditional education programs with mentoring, coaching collaborative tools and systems of knowledge management.

7.4 The Wiki as a Means of Education and Knowledge Management in Public Institutions

With the concept Wiki usually is meant as a website that allows users to add, remove, or edit content, very quickly and easily, without having made compulsory registration. Thus, it facilitates cooperation of many individuals for writing a project. What actually makes a wiki system is to simplify the process of creating HTML pages and record each individual change that occurs over time, so that at any time a page can be returned to any of the previous statements.

A. *Advantages*:

- Anyone can participate in the drafting-editing of content.
- Easy to learn and use.
- Saves time in obtaining and renewal of content.
- Possibility of cooperation on the same project of people located in different parts of the globe.
- The software keeps data for any amendment made restoring a previous edition an article is a simple process.
- Opens access to online editions for users with limited technical knowledge.
- The wiki has no predefined structure—thus is a versatile tool that can be used for wide range of applications.
- There is a wide variety of open source software wiki, so the cost of the right of use should not be an obstacle to the establishment of a wiki from a foundation.

B. *Disadvantages*:

The characteristics that under some conditions are considered advantages under different conditions may be drawbacks.

- Anyone can modify the content and this can be excessive discretion for some applications, for example when it comes to confidential documents. However there is the possibility of regulating the access users have to the content.
- The wiki is vulnerable to spam and vandalism if not make the necessary adjustments and protection handling.
- The flexibility of the structure of a wiki can mean that the information is unorganised. When the size of a wiki gets bigger, the community plans and manages its structure collectively.

C. *Benefits arising from the use of wiki as a learning tool*:

- Links-Connections
  - Develop greater linkages between the new and the older knowledge allowing learners to create structures for information and ideas.
  - Trainees use the functions of synthesis and evaluation constantly and therefore when working on a wiki.
- Creativity
  - Development of creative skills, especially the skills of processing and fluency. Development of creative flexibility by accepting the amendments of others.
  - Encouragement of 'elicitation' ideas (a type of creative process and analytical thinking: if applies X what applies then for Y).
  - Introduction and strengthening of the idea that a creative work is never "complete".
- Commitment
  - Increase the commitment of all learners.
  - Trainees no longer hold the passive role of 'consumer' that receives the information presented (thus lose interest and ignore it) as factors of the wiki answer, they make changes and improvements.
  - Projects that are in a very advanced stage it is not necessary to complete.
- Interpersonal Relations
  - Develop interpersonal and communication skills, particularly achieve consensus and compromise, in an environment where the result motivates interpersonal problem solving.
  - Develop the essential skills of teamwork.
- Writing
  - Improving the capacity of revision and review.
  - Increased flexibility in finding alternative ways to express the same meaning.
  - Adapting to an environment of a more and more authentic audience.

D. Ways of using in public administration

- Informal learning (informal learning)—The wiki helps employees to search and share the knowledge that has been developed within the organisation, thus enhancing organisational performance.
- Broaden communication among employees, both in the training process and during the performance of their duties.
- Average feedback to applicable procedures. Employees can benefit from direct communication that provides a wiki.
- Scheduling management actions.
- Channel of communication and consulting objectives of management.
- Means of communication and consultation with citizens and businesses.

7.5 Suggested Wiki Implementation Through Moodle

The Moodle (Modular Object Oriented Developmental Learning Environment) [6] is a free educational content management software. Mainly used for the purpose of Asynchronous e-Learning. However it can complement conventional education in various ways. Through the graphical interface of Moodle, which does not require specialised knowledge to build the course and its follow-up, the teacher can present the subject in a way that is interesting with the introduction of educational material in various forms, assign tasks to students, communicating with them through asynchronous tools or modern communication and evaluation of the performance of trainees [6]. Main features are communication and collaboration and tracing of the activity of the learner on the platform [7, 8].

In this part we examine the opportunities given by the Moodle to create wiki in a course. So typically we created a page in Moodle for training civil servants and for the course Civil Servants Code we created a wiki as an activity to develop relevant issues (Figs. 7.1, 7.2 and 7.3). From our experience we note the following points:

- The administrator and the instructor of the course have the ability to register each trainee in the wiki.
- There are several options to access the wiki. Depending on the setting and whether or not groups, one student may have access only to its own wiki or to those of other students or groups.
- All registered users have the ability to process changes to the original publication of the wiki depending on the settings of the administrator or trainer. They also have the ability to access files on hanging.
- There is the possibility to monitor the successive changes or additions to publications (history) as well as the individuals who have made them.
- There is a possibility to restore the page to a previous state in case where there have been undesired changes in a page.
- It is possible to search topics by some keywords, a tool extremely useful if the content of the wiki is bulky.

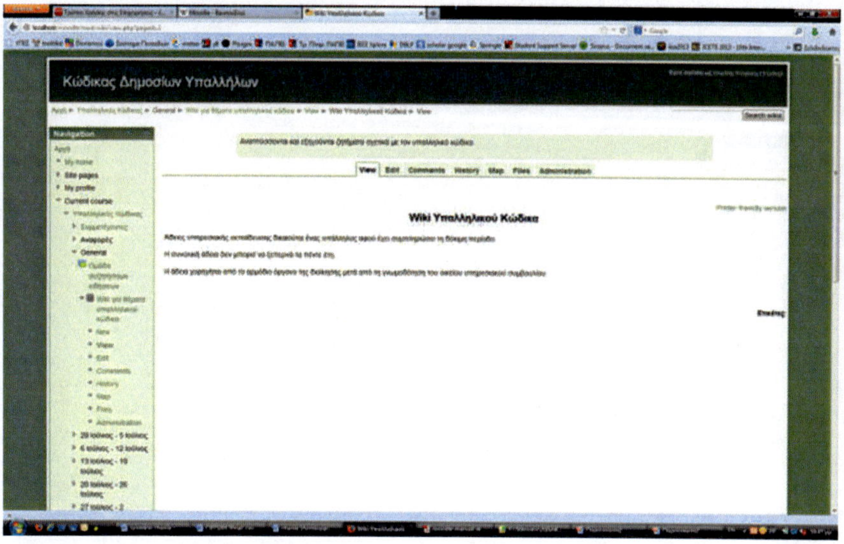

Fig. 7.1 Moodle page for training civil servants

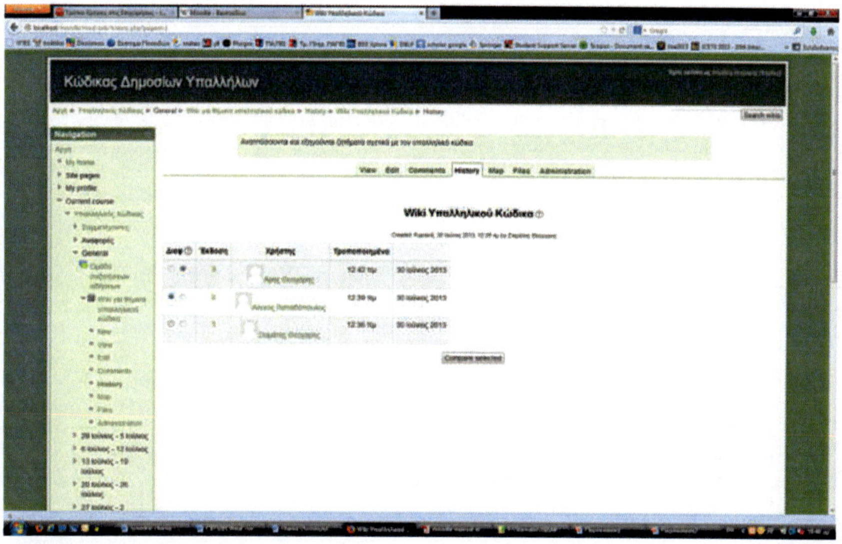

Fig. 7.2 Wiki page created within Moodle

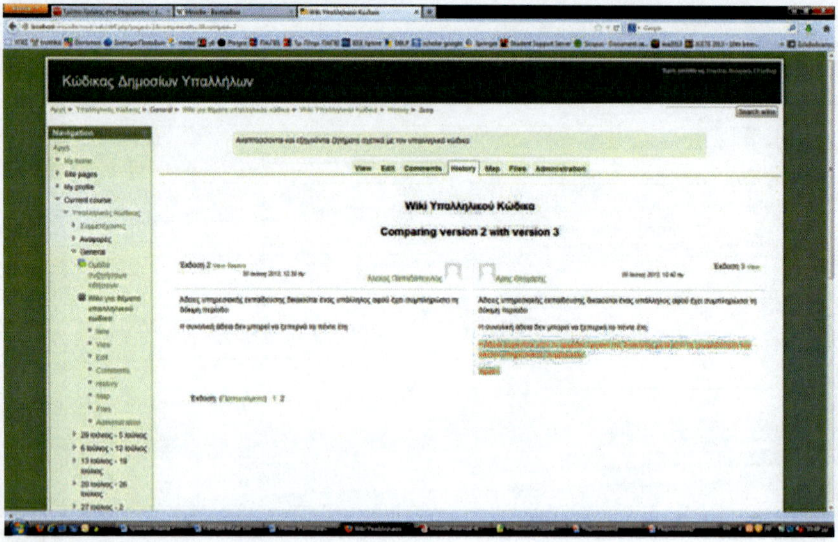

Fig. 7.3 Comparing versions of Wiki pages created within Moodle

References

1. Indzhov, H., Totkov, G., & Doneva, R. (2011). e-Learning in a Moodle-based adaptive and accumulative system. In *Proceedings of International Conference on Computer Systems and Technologies—CompSysTech'11.*
2. Palkovots, S., Lorente, J. M., & Karagiannis, D. (2002). e-Learning and e-Government—An approach for process-based training and learning in the public administration. https://citeseerx.ist.psu.edu/viewdoc/download?doi=10.1.1.492.7326&rep=rep1&type=pdf
3. MOD.SA, E-Learning & E-Government in Public Administration, *Newsletter.* http://newsletter.elearning.mou.gr/NewsItemFullPage.aspx?newsitem-id=199&newsletter-id=55
4. Novo-Corti, I., Varela-Candamio, L., & Ramil-Dvaz, M. (2013). E-learning and face to face mixed methodology: Evaluating effectiveness of e-learning and perceived satisfaction for a microeconomic course using the Moodle platform. *Computers in Human Behavior, 29*, 410–415.
5. Braim, S. (2004). Role of e-Learning in e-Government implementation and adoption. In *Proceedings of International Workshop on Improving e-Learning Policies and Programs*, Asian Development Bank Institute (ADBI). Lecture Series (pp. 30–31). http://www.adbi.org/files/2004.08.09.cpp.linking.elearning.egovernment.korea.pdf
6. Moodle platform web site. https://moodle.org/
7. Despotović-Zrakić, M., Marković, A., Bogdanović, Z., Barać, D., & Krčo, S. Providing adaptivity in Moodle LMS courses. *Educational Technology & Society, 15*(1), 326–338.
8. Escobar-Rodriguez, T., & Monge-Lozano, P. (2012). The acceptance of Moodle technology by business administration students. *Computers & Education 58,* 1085–1093.

Chapter 8
Conclusions—Future Work

Abstract General conclusions from those presented in the previous chapters are presented in this chapter. In particular, the conclusions regarding the applications of open data and linked open data in the fields of e-Government and Open Government, benefits and opportunities in these fields are presented. Conclusions are also drawn on the process and perspectives of publishing open ontologies in the form of linked open data.

Keywords e-Government · Linked open data · Open ontologies · Semantic tools

The transition from e-Government towards Open Government and Digital State is connected to the production and publishing of Open Government Data and their linkage to the benefit of learning from them. The huge number, the diversity and the complexity of terms used to all public sector bodies, has already created a confusion when searching for data and trying to interconnect them. The Semantic Web and the technologies that support it are expected to give solutions in this area. The information environment that can support this endeavour is associated with the ontologies and the particular languages RDF/RDFs and OWL. We consider that the first step is the appropriate modelling and representation of data of interest with the help of relative ontologies and the use of common vocabularies and terminologies with the rest of the Internet users.

Linked open data and their applications in the e-Government and Open Government sectors were the focus of our study. Their development requires the use of a host of new technologies, methodologies and standards while further promoting and engaging the scientific community and the public sector. In their main positive characteristics, the following are recorded as the conclusion of this work.

- Linked data lies at the heart of what Semantic Web is all about: large scale integration of, and reasoning on, data on the Web. Almost all applications listed in, say collection of Semantic Web case studies and Use Cases are essentially based on the accessibility of, and integration of linked data at various levels of complexities.

S. Theocharis and G. A. Tsihrintzis, *Semantic Knowledge Modelling via Open Linked Ontologies*, Artificial Intelligence-Enhanced Software and Systems Engineering 4,
https://doi.org/10.1007/978-3-031-20585-9_8

- By providing bridges between areas of interest and knowledge, the linked data Web can reveal relationships where information might otherwise be seen as unrelated. For instance, one might explore which industry and government leaders had previously been classmates or were somehow connected by family.
- Particular tools such as semantic browsers enable a user to browse from one "thing" to another just as easily as they would click from one page to another with standard hyperlinked document browsing, but with greater understanding and comprehension from the first click.
- The importance of great linked datasets is not only that it includes a great amount of specific data, but also that it incorporates links to other datasets on the Web. By providing those extra links (in terms of RDF triples) applications may exploit the extra (and possibly more precise) knowledge from other datasets when developing an application; by virtue of integrating facts from several datasets, the application may provide a much better user experience.

The following are the main benefits from the application of the solutions of the Semantic Web.

- Provide a new experience when browsing websites as human–computer interaction is now based on the intelligence acquired by the computer. This is achieved by using the technologies and methods of the linked data to support the automatic retrieval and presentation of relevant information on a case-by-case basis.
- With the help of semantic markup of websites and the use of ontologies the search for information will become more effective as it will focus on pages that are semantically absolutely related to the concept or entity of interest. Along with more intelligent access, data mining is served more efficiently and effectively, especially by dynamic websites.
- The overall vision concerns the transition from the existing Internet of static government pages to a network of dynamic public service providers (Web-services) that automatically discover the information or services sought and negotiate for goods or services.
- Areas that are expected to benefit from the evolution of the current web using the technologies of the Semantic Web include Knowledge Management as well as all the areas it refers to such as business administration and public administration through the evolution of e-Government, the e-commerce through the development of new agents for the extraction of useful information, e-learning as an evolution of the now widespread distance learning as well as the field of education in general in various fields.

Among the benefits and opportunities in the specific areas of e-Government and Open Government we conclude as follows.

- Regarding the operation of e-Government, opportunities through Semantic Web solutions are linked to the fifth level of its development, i.e. the provision of personalised services, according to the needs of each user. At the same time, the provision of personalised services and pre-completed forms and applications will make online services more attractive to the general public.

- Further substantial support for Open Government and the further expansion of open data production combined with the adoption of linked data principles will help transform e-Government into a truly "smart" digital state.

Regarding the back-office of the public sector, the search for information from the involved bodies and the sharing of data based on common terminology and semantics will improve both the internal delivery times of the data and the overall performance of the bodies, since when all public systems "speak the same language", sharing problems are easier to solve.

- In terms of serving the citizens, the search for the provided services will be significantly facilitated both electronically through the government portals (development of semantic portals) and with the physical presence at the service point (within the operation of the one-stop shops). Searching for the right service from the relevant body can become more efficient and effective.
- In the field of knowledge modelling in the public sector, ontology management tools have to offer special benefits for the creation and management of the knowledge base, regarding public data and the procedures followed, issues that we consider particularly critical.
- Particular benefits may arise in the field of management of the various registers maintained by public sector services in the areas of trade, codification of legislation, procedures, etc. using ontologies and knowledge base management tools.
- The tools and technologies of the Semantic Web can assist in addressing the problems recorded by the opening of public data through the "CLEAR" program in Greece, as highlighted above.

As the main object of this Book we highlighted the production and publication of the Ontology for e-Government in the form of linked open data using corresponding tools of the Semantic Web. Also, two more ontologies were developed and presented, one in the field of Official Statistics of the Greek Statistical System and the other in the field of Internal Audit. Through the process we followed, in order to develop and publish the aforementioned ontology, we followed the general guidelines for opening data in conjunction with the goals of the 5 stars outlined by the W3C. Among the main conclusions during the development of ontologies and their publication as linked open data, we record the following:

- Regarding the benefits of this process, we record the following:
  - Dissemination of knowledge to both citizens and executives of public organisations regarding the vocabulary used without the need to use special software.
  - Providing the possibility of discovering information regarding the provided services and the competent bodies which would be more difficult to discover with the classical technologies.
  - Opportunities for educational use mainly for new employees but also for younger citizens. At the same time, the citizen's awareness of government

  functions is increasing in order to achieve greater credibility for the procedures followed.
  - Achieving semantic interoperability from the use of ontology.
  - The ontologies developed can be a guide for the development of other sub-ontologies in the field of public administration.
  - Forming a new perspective for the development of new innovative applications and services at the level of the Semantic Web.

- Linked data is the most appropriate solution for publishing open government data on the Web, as their rules are in line with the rules of open government data.
- The opening up of data and especially government data can lead to the production of new information and knowledge and consequently added value as well as increased entrepreneurship from it.
- Language interoperability issues arise that need to be addressed in the future as linked data comes from different sources and countries.
- The main advantage of linked data is that it is scalable. It is easy to add additional linked data to existing ones, even when the terms and definitions used change over time. The central idea is that regardless of the data format the RDF model should be used as a link. When data is published in RDF format, it becomes available for processing, not just for public access.
- The use of URIs and RDF, which are globally accessible formats, involves the standardisation of the identification and representation of government resources. In this way, barriers to the connection of open government data on a global scale will be reduced.
- RDF connects better than any other model because:
  - Uses URIs, thus allowing the connection of objects and concepts.
  - Allows separate systems that are independently designed to be then merged.
  - Allows interoperability to be added where it is beneficial.
  - Allows all data to be expressed as a mixture of vocabularies.

Future work

Our future plans for further research and development of semantic applications in the field of Knowledge Modeling and linked open data include the following.

- Further enrichment of the developed ontologies, their publication in a public place as well as an effort for the production and reuse of linked open ontologies.
- Given that until nowadays, the available software does not adequately deal with vocabularies in individual languages such as Greek, in the future we will focus on the production and publication of linked open data and open ontologies with support in the Greek language.
- Study and development of a process for the creation of consistent and coherent links between ontologies (ontology alignment) in order to achieve their mutual agreement in the context of reuse and connection of open ontologies.
- Development of a process for the automatic production of ontologies and the supply of a relevant knowledge database from open government data.

- Development of educational software using built-in semantic tools such as linked ontologies and linked data. The purpose of this work is to provide personalised distance education with emphasis on adult education—retraining of executives of companies and organisations.

Appendix

A.1 Customization of the Configuration CKAN File (*production.ini*)

```
#
# CKAN - Pylons configuration
#
# These are some of the configuration options available for your CKAN
# instance. Check the documentation in 'doc/configuration.rst' or at
the
# following URL for a description of what they do and the full list
of
# available options:
#
# http://docs.ckan.org/en/latest/maintaining/configuration.html
#
# The %(here)s variable will be replaced with the parent directory of
this file
#

[DEFAULT]

# WARNING: *THIS SETTING MUST BE SET TO FALSE ON A PRODUCTION
ENVIRONMENT*
debug = false

[server:main]
use = egg:Paste#http
host = 0.0.0.0
port = 5000
```


S. Theocharis and G. A. Tsihrintzis, *Semantic Knowledge Modelling via Open Linked Ontologies*, Artificial Intelligence-Enhanced Software and Systems Engineering 4,
https://doi.org/10.1007/978-3-031-20585-9

```
[app:main]
use = egg:ckan
full_stack = true
cache_dir = /tmp/%(ckan.site_id)s/
beaker.session.key = ckan

# This is the secret token that the beaker library uses to hash the
cookie sent
# to the client. `paster make-config` generates a unique value for
this each
# time it generates a config file.
beaker.session.secret = CsK2Rdshlhg0Ic8bdEscv7+OY

# `paster make-config` generates a unique value for this each time it
generates
# a config file.
app_instance_uuid = 671e998b-8524-4810-a7a5-3c2afe1b54b1

# repoze.who config
who.config_file = %(here)s/who.ini
who.log_level = warning
who.log_file = %(cache_dir)s/who_log.ini
# Session timeout (user logged out after period of inactivity, in
seconds).
# Inactive by default, so the session doesn't expire.
# who.timeout = 86400

## Database Settings
sqlalchemy.url = postgresql://ckan_default:123@localhost/ckan_default

ckan.datastore.write_url =
postgresql://ckan_default:123@localhost/datastore_default
ckan.datastore.read_url =
postgresql://datastore_default:123@localhost/datastore_default


# PostgreSQL' full-text search parameters
ckan.datastore.default_fts_lang = english
ckan.datastore.default_fts_index_method = gist


## Site Settings

ckan.site_url =http://127.0.0.1
#ckan.use_pylons_response_cleanup_middleware = true

## Authorization Settings

ckan.auth.anon_create_dataset = false
ckan.auth.create_unowned_dataset = false
ckan.auth.create_dataset_if_not_in_organization = false
ckan.auth.user_create_groups = false
ckan.auth.user_create_organizations = false
ckan.auth.user_delete_groups = true
ckan.auth.user_delete_organizations = true
ckan.auth.create_user_via_api = false
ckan.auth.create_user_via_web = true
ckan.auth.roles_that_cascade_to_sub_groups = admin
```

```
## Search Settings

ckan.site_id = default
solr_url = http://127.0.0.1:8983/solr


## Redis Settings

# URL to your Redis instance, including the database to be used.
#ckan.redis.url = redis://localhost:6379/0


## CORS Settings

# If cors.origin_allow_all is true, all origins are allowed.
# If false, the cors.origin_whitelist is used.
# ckan.cors.origin_allow_all = true
# cors.origin_whitelist is a space separated list of allowed domains.
# ckan.cors.origin_whitelist = http://example1.com
http://example2.com


## Plugins Settings

# Note: Add ``datastore`` to enable the CKAN DataStore
#       Add ``datapusher`` to enable DataPusher
#           Add ``resource_proxy`` to enable resorce proxying and get
around the
#           same origin policy

ckan.plugins = datapusher datastore stats text_view image_view
recline_view

# Define which views should be created by default
# (plugins must be loaded in ckan.plugins)
ckan.views.default_views = image_view text_view recline_view

# Customize which text formats the text_view plugin will show
#ckan.preview.JSON_formats = JSON
#ckan.preview.xml_formats = xml RDF RDF+xml owl+xml atom rss
#ckan.preview.text_formats = text plain text/plain

# Customize which image formats the image_view plugin will show
#ckan.preview.image_formats = png jpeg jpg gif

## Front-End Settings

# Uncomment following configuration to enable using of Bootstrap 2
#ckan.base_public_folder = public-bs2
#ckan.base_templates_folder = templates-bs2

ckan.site_title = CKAN
ckan.site_logo = /base/images/ckan-logo.png
ckan.site_description =
ckan.favicon = /base/images/ckan.ico
ckan.gravatar_default = identicon
ckan.preview.direct = png jpg gif
ckan.preview.loadable = html htm RDF+xml owl+xml xml n3 n-triples
turtle plain atom csv tsv rss txt JSON
ckan.display_timezone = server
```

```
# package_hide_extras = for_search_index_only
#package_edit_return_url = http://another.frontend/dataset/<NAME>
#package_new_return_url = http://another.frontend/dataset/<NAME>
#ckan.recaptcha.publickey =
#ckan.recaptcha.privatekey =
#licenses_group_url =
http://licenses.opendefinition.org/licenses/groups/ckan.JSON
# ckan.template_footer_end =


## Internationalisation Settings
ckan.locale_default = en
ckan.locale_order = en pt_BR ja it cs_CZ ca es fr el sv sr sr@latin
no sk fi ru de pl nl bg ko_KR hu sa sl lv
ckan.locales_offered =
ckan.locales_filtered_out = en_GB

## Feeds Settings

ckan.feeds.authority_name =
ckan.feeds.date =
ckan.feeds.author_name =
ckan.feeds.author_link =

## Storage Settings

ckan.storage_path = /var/lib/ckan/default
#ckan.max_resource_size = 10
#ckan.max_image_size = 2


## Datapusher settings

# Make sure you have set up the DataStore

#ckan.datapusher.formats = csv xls xlsx tsv application/csv
application/vnd.ms-excel application/vnd.openxmlformats-
officedocument.spreadsheetml.sheet
ckan.datapusher.url = http://0.0.0.0:8800/
#ckan.datapusher.assume_task_stale_after = 3600

# Resource Proxy settings
# Preview size limit, default: 1MB
#ckan.resource_proxy.max_file_size = 1048576
# Size of chunks to read/write.
#ckan.resource_proxy.chunk_size = 4096

## Activity Streams Settings

#ckan.activity_streams_enabled = true
#ckan.activity_list_limit = 31
#ckan.activity_streams_email_notifications = true
#ckan.email_notifications_since = 2 days
ckan.hide_activity_from_users = %(ckan.site_id)s
```

```
## Email settings

#email_to = errors@example.com
#error_email_from = ckan-errors@example.com
#smtp.server = localhost
#smtp.starttls = False
#smtp.user = username@example.com
#smtp.password = your_password
#smtp.mail_from =


## Logging configuration
[loggers]
keys = root, ckan, ckanext

[handlers]
keys = console

[formatters]
keys = generic

[logger_root]
level = WARNING
handlers = console

[logger_ckan]
level = INFO
handlers = console
qualname = ckan
propagate = 0

[logger_ckanext]
level = DEBUG
handlers = console
qualname = ckanext
propagate = 0

[handler_console]
class = StreamHandler
args = (sys.stderr,)
level = NOTSET
formatter = generic

[formatter_generic]
format = %(asctime)s %(levelname)-5.5s [%(name)s] %(message)s
```

A.2 Configuration File of Pubby Server (Access in a SparqL Endpoint in the Same Machine)

```
# Pubby Example Configuration
#
# This configuration connects to the DBpedia SPARQL endpoint and
# re-publishes on your local machine, with dereferenceable
# localhost URIs.
#
# This assumes you already have a servlet container running
# on your machine at http://localhost:8080/ .
#
# Install Pubby as the root webapp of your servlet container,
# and make sure the config-file parameter in Pubby's web.xml
# points to this configuration file.
#
# Then browse to http://localhost:8080/ .

# Prefix declarations to be used in RDF output
@prefix conf: <http://richard.cyganiak.de/2007/pubby/config.RDF#> .
@prefix meta: <http://example.org/metadata#> .
@prefix RDF: <http://www.w3.org/1999/02/22-RDF-syntax-ns#> .
@prefix RDFs: <http://www.w3.org/2000/01/RDF-schema#> .
@prefix xsd: <http://www.w3.org/2001/XMLSchema#> .
@prefix owl: <http://www.w3.org/2002/07/owl#> .
@prefix dc: <http://purl.org/dc/elements/1.1/> .
@prefix dcterms: <http://purl.org/dc/terms/> .
@prefix foaf: <http://xmlns.com/foaf/0.1/> .
@prefix skos: <http://www.w3.org/2004/02/skos/core#> .
@prefix geo: <http://www.w3.org/2003/01/geo/wgs84_pos#> .
@prefix dbpedia: <http://localhost:8080/resource/> .
@prefix p: <http://localhost:8080/property/> .
@prefix yago: <http://localhost:8080/class/yago/> .
@prefix units: <http://dbpedia.org/units/> .
@prefix geonames: <http://www.geonames.org/ontology#> .
@prefix prv:      <http://purl.org/net/provenance/ns#> .
@prefix prvTypes: <http://purl.org/net/provenance/types#> .
@prefix doap:     <http://usefulinc.com/ns/doap#> .
@prefix void:     <http://RDFs.org/ns/void#> .
@prefix ir:
<http://www.ontologydesignpatterns.org/cp/owl/informationrealization.
owl#> .
```

```
# Server configuration section
<> a conf:Configuration;
# Project name for display in page titles
conf:projectName "DBpedia.org";
# Homepage with description of the project for the link in the
page header
conf:projectHomepage <http://dbpedia.org>;
# The Pubby root, where the webapp is running inside the servlet
container.
conf:webBase <http://localhost:8080/pubby/>;
# URL of an RDF file whose prefix mapping is to be used by the
# server; defaults to <>, which is *this* file.
conf:usePrefixesFrom <>;
# If labels and descriptions are available in multiple languages,
# prefer this one.
conf:defaultLanguage "en";
# When the homepage of the server is accessed, this resource will

# be shown.
conf:indexResource <http://localhost:8080/pubby/e_gov_ontology>;
# Will be appended to the conf:webBase to form the public
# resource URIs; if not present, defaults to ""
#conf:webResourcePrefix "page/";

# Dataset configuration section (for DBpedia resources)
#
# URIs in the SPARQL endpoint: http://dbpedia.org/resource/*
# URIs on the Web:             http://localhost:8080/resource/*
conf:dataset [
# SPARQL endpoint URL of the dataset
conf:SparqLEndpoint <http://localhost:3030/e_gov>;
# Default graph name to query (not necessary for most
endpoints)
#conf:SparqLDefaultGraph <http://dbpedia.org>;
# Common URI prefix of all resource URIs in the SPARQL
dataset
conf:datasetBase <http://localhost:8080/pubby/>;
# Fixes an issue with the server running behind an Apache
proxy;
# can be disabled otherwise
conf:fixUnescapedCharacters "(),'!$&*+;=@";
# DBpedia uses URIs, not IRIs
conf:supportsIRIs "true";


];
.
```

A.3 Configuration File of Pubby Server (Access in a Remote SparqL Endpoint)

```
# Pubby Example Configuration
#
# This configuration connects to the Stamatis Theocharis's "e-
Government ontology" SPARQL endpoint
# via Fuseki and re-publishes on your local machine, with
dereferenceable
# localhost URIs.
#
# This assumes you already have a servlet container running
# on your machine at http://localhost:8080/ .
#
# Install Pubby as the root webapp of your servlet container,
# and make sure the config-file parameter in Pubby's web.xml
# points to this configuration file.
#
# Then browse to http://localhost:8080/ .

# Prefix declarations to be used in RDF output
@prefix conf: <http://richard.cyganiak.de/2007/pubby/config.RDF#> .
@prefix meta: <http://example.org/metadata#> .
@prefix RDF: <http://www.w3.org/1999/02/22-RDF-syntax-ns#> .
@prefix RDFs: <http://www.w3.org/2000/01/RDF-schema#> .
@prefix xsd: <http://www.w3.org/2001/XMLSchema#> .
@prefix owl: <http://www.w3.org/2002/07/owl#> .
@prefix dc: <http://purl.org/dc/elements/1.1/> .
@prefix dcterms: <http://purl.org/dc/terms/> .
@prefix foaf: <http://xmlns.com/foaf/0.1/> .
@prefix skos: <http://www.w3.org/2004/02/skos/core#> .
@prefix geo: <http://www.w3.org/2003/01/geo/wgs84_pos#> .
@prefix dbpedia: <http://localhost:8080/resource/> .
@prefix p: <http://localhost:8080/property/> .
@prefix yago: <http://localhost:8080/class/yago/> .
@prefix units: <http://dbpedia.org/units/> .
@prefix geonames: <http://www.geonames.org/ontology#> .
@prefix prv:      <http://purl.org/net/provenance/ns#> .
@prefix prvTypes: <http://purl.org/net/provenance/types#> .
@prefix doap:     <http://usefulinc.com/ns/doap#> .
@prefix void:     <http://RDFs.org/ns/void#> .
@prefix ir:
<http://www.ontologydesignpatterns.org/cp/owl/informationrealization.
owl#> .

# Server configuration section
<> a conf:Configuration;
# Project name for display in page titles
conf:projectName "My project 2";
# Homepage with description of the project for the link in the
page header
```

```
conf:projectHomepage <http://example.org>;
# The Pubby root, where the webapp is running inside the servlet
container.
conf:webBase <http://localhost:8080/pubby/>;
# URL of an RDF file whose prefix mapping is to be used by the
# server; defaults to <>, which is *this* file.
conf:usePrefixesFrom <>;
# If labels and descriptions are available in multiple languages,
# prefer this one.
conf:defaultLanguage "en";
# When the homepage of the server is accessed, this resource will
# be shown.
conf:indexResource
<http://5.54.141.78:8080/pubby/e_gov_ontology2>;
#conf:webResourcePrefix "e_gov_ontology2/";

# Dataset configuration section #1 (for DBpedia resources)
#
# URIs in the SPARQL endpoint: http://dbpedia.org/resource/*
# URIs on the Web:             http://localhost:8080/resource/*
conf:dataset [
# SPARQL endpoint URL of the dataset
conf:SparqLEndpoint <http://5.54.141.78:3030/e_gov_2>;
# Default graph name to query (not necessary for most
endpoints)
# conf:SparqLDefaultGraph <http://dbpedia.org>;
# Common URI prefix of all resource URIs in the SPARQL
dataset
conf:datasetBase <http://5.54.141.78:8080/pubby/>;
# Will be appended to the conf:webBase to form the public
# resource URIs; if not present, defaults to ""
# conf:webResourcePrefix "resource/";
# Fixes an issue with the server running behind an Apache
proxy;
# can be ignored otherwise
conf:fixUnescapedCharacters "(),'!$&*+;=@";
conf:addSameAsStatements "true";
# include metadata
# conf:metadataTemplate "metadata.ttl";

# configure your metadata here
# Use properties with the meta: prefix where the property
name
# corresponds to the placeholder URIs in metadata.ttl that
begin
# with about:metadata:metadata:
# Examples for such properties are:
#       meta:pubbyUser <URI of the data publisher who uses this
Pubby>;
#       meta:pubbyOperator <URI of the service provider who operates
this Pubby>;
#       meta:endpointUser <URI of the data publisher who uses the
SPARQL endpoint queried by this Pubby>;
#       meta:endpointOperator <URI of the service provider who
operates the SPARQL endpoint>;
#       meta:endpointDataset <URI of the linked dataset that is
exposed via the SPARQL endpoint>;
];
```

A.4 Summary of Key Open Government Licence Sites

- The UK Open Government Licence was created to enable any public sector information holder to make their information available for use and reuse under its terms.
- The Open Database Licence (ODbL) is an open licence for databases and data that includes explicit attribution and share-alike requirements.
- Public Domain Dedication and Licence (PDDL) is a document intended to allow you to freely share, modify, and use a particular data for any purpose and without any restrictions.
- Open Data Commons Attribution Licence is a database specific licence requiring attribution for databases.
- The Creative Commons Licences are several copyright licences that allow the distribution of copyrighted works.
- The Open Licence compatible with the following licences:
  - Open government Licence (OGL) of the UK
  - Creative Commons Attribution 2.0 (CC-BY 2.0)
  - Open Data Commons Attribution (ODC-BY).